Turbulent Combustion

The combustion of fossil fuels remains a key technology for the foreseeable future. It is therefore important that we understand the mechanisms of combustion and, in particular, the role of turbulence within this process. Combustion always takes place within a turbulent flow field for two reasons: Turbulence increases the mixing process and enhances combustion, but at the same time combustion releases heat, which generates flow instability through gas expansion and buoyancy, thus enhancing the transition to turbulence.

The four chapters of this book present a thorough introduction to the field of turbulent combustion. After an overview of modeling approaches, the three remaining chapters consider the three distinct cases of premixed, nonpremixed, and partially premixed combustion, respectively.

This book will be of value to researchers and students of engineering and applied mathematics by demonstrating the current theories of turbulent combustion within a unified presentation of the field.

Norbert Peters has been a professor in mechanics at RWTH Aachen in Germany since 1976. Currently he is deputy editor of *Combustion and Flame* and editor of *Flow, Turbulence and Combustion*. In 1990 he received the Leibniz-Prize of the Deutsche Forschungsgemeinschaft, and in 1994 he received an honorary doctorate from the Université Líbre de Bruxelles.

CAMBRIDGE MONOGRAPHS ON MECHANICS

FOUNDING EDITOR

G. K. Batchelor

GENERAL EDITORS

S. Davis

Walter P. Murphy Professor
Applied Mathematics and Mechanical Engineering
Northwestern University

L. B. Freund

Henry Ledyard Goddard University Professor
Division of Engineering
Brown University

S. Leibovich

Sibley School of Mechanical & Aerospace Engineering
Cornell University

V. Tvergaard

Department of Solid Mechanics
The Technical University of Denmark

TURBULENT COMBUSTION

NORBERT PETERS

Institut für Technische Mechanik
Rheinisch-Westfälische Technische
Hochschule Aachen, Germany

PUBLISHED BY THE PRESS SYNDICATE OF THE UNIVERSITY OF CAMBRIDGE
The Pitt Building, Trumpington Street, Cambridge, United Kingdom

CAMBRIDGE UNIVERSITY PRESS
The Edinburgh Building, Cambridge CB2 2RU, UK   http://www.cup.cam.ac.uk
40 West 20th Street, New York, NY 10011-4211, USA   http://www.cup.org
10 Stamford Road, Oakleigh, Melbourne 3166, Australia
Ruiz de Alarcón 13, 28014 Madrid, Spain

© Cambridge University Press 2000

This book is in copyright. Subject to statutory exception
and to the provisions of relevant collective licensing agreements,
no reproduction of any part may take place without
the written permission of Cambridge University Press.

First published 2000

Printed in the United Kingdom at the University Press, Cambridge

Typeface Times Roman 10/13 pt.   *System* LATEX 2_ε [TB]

A catalog record for this book is available from the British Library.

Library of Congress Cataloging in Publication Data
Peters, Norbert.
Turbulent combustion / N. Peters.
p. cm. – (Cambridge monographs on mechanics)
Includes bibliographical references.
ISBN 0-521-66082-3
1. Combustion engineering. 2. Turbulence. I. Title. II. Series.
TJ254.5. .P48 2000
621.402′3 – dc21   99-089451

ISBN 0 521 66082 3   hardback

To Cordula

Contents

Preface		*page* xi
1	**Turbulent combustion: The state of the art**	1
	1.1 What is specific about turbulence with combustion?	1
	1.2 Statistical description of turbulent flows	5
	1.3 Navier–Stokes equations and turbulence models	10
	1.4 Two-point velocity correlations and turbulent scales	13
	1.5 Balance equations for reactive scalars	18
	1.6 Chemical reaction rates and multistep asymptotics	22
	1.7 Moment methods for reactive scalars	29
	1.8 Dissipation and scalar transport of nonreacting and linearly reacting scalars	30
	1.9 The eddy-break-up and the eddy dissipation models	33
	1.10 The pdf transport equation model	35
	1.11 The laminar flamelet concept	42
	1.12 The concept of conditional moment closure	53
	1.13 The linear eddy model	55
	1.14 Combustion models used in large eddy simulation	57
	1.15 Summary of turbulent combustion models	63
2	**Premixed turbulent combustion**	66
	2.1 Introduction	66
	2.2 Laminar and turbulent burning velocities	69
	2.3 Regimes in premixed turbulent combustion	78
	2.4 The Bray–Moss–Libby model and the Coherent Flame model	87

	2.5	The level set approach for the corrugated flamelets regime	91
	2.6	The level set approach for the thin reaction zones regime	104
	2.7	A common level set equation for both regimes	107
	2.8	Modeling premixed turbulent combustion based on the level set approach	109
	2.9	Equations for the mean and the variance of G	114
	2.10	The turbulent burning velocity	119
	2.11	A model equation for the flame surface area ratio	127
	2.12	Effects of gas expansion on the turbulent burning velocity	137
	2.13	Laminar flamelet equations for premixed combustion	146
	2.14	Flamelet equations in premixed turbulent combustion	152
	2.15	The presumed shape pdf approach	156
	2.16	Numerical calculations of one-dimensional and multidimensional premixed turbulent flames	157
	2.17	A numerical example using the presumed shape pdf approach	162
	2.18	Concluding remarks	168
3	**Nonpremixed turbulent combustion**		**170**
	3.1	Introduction	170
	3.2	The mixture fraction variable	172
	3.3	The Burke–Schumann and the equilibrium solutions	176
	3.4	Nonequilibrium flames	178
	3.5	Numerical and asymptotic solutions of counterflow diffusion flames	186
	3.6	Regimes in nonpremixed turbulent combustion	190
	3.7	Modeling nonpremixed turbulent combustion	194
	3.8	The presumed shape pdf approach	196
	3.9	Turbulent jet diffusion flames	198
	3.10	Experimental data from turbulent jet diffusion flames	203
	3.11	Laminar flamelet equations for nonpremixed combustion	207
	3.12	Flamelet equations in nonpremixed turbulent combustion	212

	3.13	Steady versus unsteady flamelet modeling	219
	3.14	Predictions of reactive scalar fields and pollutant formation in turbulent jet diffusion flames	222
	3.15	Combustion modeling of gas turbines, burners, and direct injection diesel engines	229
	3.16	Concluding remarks	235
4	**Partially premixed turbulent combustion**		**237**
	4.1	Introduction	237
	4.2	Lifted turbulent jet diffusion flames	238
	4.3	Triple flames as a key element of partially premixed combustion	245
	4.4	Modeling turbulent flame propagation in partially premixed systems	251
	4.5	Numerical simulation of lift-off heights in turbulent jet flames	255
	4.6	Scaling of the lift-off height	258
	4.7	Concluding remarks	261

Epilogue   263
Glossary   265
Bibliography   267
Author Index   295
Subject Index   302

Preface

Fossil fuels remain the main source of energy for domestic heating, power generation, and transportation. Other energy sources such as solar and wind energy or nuclear energy still account for less than 20% of total energy consumption. Therefore combustion of fossil fuels, being humanity's oldest technology, remains a key technology today and for the foreseeable future. It is well known that combustion not only generates heat, which can be converted into power, but also produces pollutants such as oxides of nitrogen (NO_x), soot, and unburnt hydrocarbons (HC). Ever more stringent regulations are forcing manufacturers of automotives and power plants to reduce pollutant emissions, for the sake of our environment. In addition, unavoidable emissions of CO_2 are believed to contribute to global warming. These emissions will be reduced by improving the efficiency of the combustion process, thereby increasing fuel economy.

In technical processes, combustion nearly always takes place within a turbulent rather than a laminar flow field. The reason for this is twofold: First, turbulence increases the mixing processes and thereby enhances combustion. Second, combustion releases heat and thereby generates flow instability by buoyancy and gas expansion, which then enhances the transition to turbulence.

This book addresses gaseous turbulent flows only. Although two-phase turbulent flows such as fuel sprays are also of much practical interest, they are omitted here, because their fundamentals are even less well understood than those of turbulent combustion. We also restrict ourselves to low Mach number flows, because high speed turbulent combustion is an area of its own, with practical applications in supersonic and hypersonic aviation only.

Technical processes in gaseous turbulent combustion can be subdivided in terms of mixing: premixed, nonpremixed, or partially premixed turbulent combustion. For example, combustion in homogeneous charge spark-ignition engines or in lean-burn gas turbines occurs under premixed conditions. In contrast, combustion in a diesel engine or in furnaces essentially takes place under nonpremixed or partially premixed conditions.

In the spark-ignition engine, fuel and oxidizer are mixed by turbulence for a sufficiently long period of time before the electrical spark ignites the mixture. The deposition of electrical energy from the spark ionizes the gas and heats it to several thousand degrees Kelvin. At temperatures above one thousand degrees Kelvin chemical reactions are initiated. These generate a flame kernel that grows at first by laminar, then by turbulent flame propagation. The turbulent burning velocity and its prediction from first principles will play a central role in the chapter on premixed turbulent combustion.

In stationary lean-burn gas turbines that have recently been developed for power generation, the fuel is prevaporized and premixed with air before entering into the combustion chamber. To homogenize the temperature field and thereby reduce emissions, in particular those of NO_x, turbulent mixing is very strong and dominates the combustion process. In the postflame region mixing with secondary air reduces the temperature to levels that are low enough not to destroy the turbine blades downstream of the exhaust. The reduction in temperature also freezes the NO_x formation in the postflame region.

In the rich-to-lean-burn gas turbines used in aircraft engines, liquid fuel is injected into the combustion chamber where it evaporates and burns in a turbulent diffusion flame, which is stabilized by swirl and recirculation. As in lean-burn gas turbines secondary air is mixed into the product stream of the flame to oxidize the remaining hydrocarbons, CO, and soot and to reduce the temperature at the exhaust.

In the diesel engine, several liquid fuel sprays are injected into hot compressed air; the fuel evaporates and mixes partially with the air before autoignition occurs. The auto-ignition process happens quite randomly and independently at several locations within the sprays. These ignition kernels will then initiate the burnout of the partially premixed fuel–air mixture. The final burnout occurs essentially under nonpremixed conditions.

In furnaces, finally, gaseous, liquid, or solid fuels are injected separately from the air into the combustion chamber. The air may be preheated or partially diluted by hot exhaust gases. Once the mixture is ignited, the flame propagates toward the nozzle until it stabilizes at a distance, called the lift-off height, downstream of the nozzle. Partial premixing takes place in the region between the nozzle and the lift-off height and determines the stabilization of the turbulent flame. Further downstream, combustion again occurs under nonpremixed conditions.

It is clear that in addition to premixed and nonpremixed combustion, partially premixed combustion plays, at least locally, an important role in practical applications. Another example for partially premixed combustion is the modern direct-injection spark-ignition engine where the charge is stratified such that the flame initiated at the spark propagates through a partially premixed inhomogeneous mixture.

A second criterion for the subdivision of turbulent combustion relates to the ratio of turbulent to chemical time scales. Because of chain-branching reactions, hydrocarbon oxidation occurs if the temperature is above a certain crossover temperature, but it ceases if the temperature falls below that temperature. The crossover temperature is defined as the temperature where the effects of chain-branching reactions just balance that of chain-breaking reactions. At ambient pressure this temperature lies between 1,300 K and 1,500 K for hydrocarbon flames but it increases with pressure. At temperatures lower than the crossover temperature, extinction occurs, which must in general be avoided in practical applications. Therefore, combustion engines typically operate safely apart from extinction at conditions where temperatures are high enough and chemical reactions occur rapidly. This is often referred to as fast chemistry.

Slow chemistry does not very often occur in practical processes. Although NO_x formation is a relatively slow process, it is nearly always linked to the fast chemistry of the main combustion reactions, since it depends on the presence of radicals. There are a few applications of slow oxidation chemistry just above the crossover temperature between chain-branching and chain-breaking reactions, for example low NO_x burners, which operate at temperatures where both oxidation and NO_x chemistry are relatively slow. This process has been called MILD combustion. An example is the MILD combustion mode, which will be discussed in Chapter 3.

The four chapters in this book are organized as follows:

In Chapter 1, after a general introduction, the more prominent current modeling approaches for turbulent flows with combustion will be presented. At the end of that chapter an overview of the existing models in terms of mixing and in terms of infinitely fast or finite rate chemistry will be presented.

In Chapter 2, which deals with premixed turbulent combustion, emphasis will be placed on a combustion model that uses the level set approach to determine the location of the premixed flame surface. Damköhler's regimes of large scale and small scale turbulence will be associated with the corrugated flamelets regime and the thin reaction zones regime, respectively. Finally, based on an equation for the flame surface area ratio, an expression will be derived for the turbulent burning velocity, valid in both regimes. Other models and experimental data will be discussed in the light of this formulation. Although the mathematics behind the level set approach is rather demanding, the author believes that it is necessary in order to capture the physics of a propagating premixed turbulent flame surface.

In Chapter 3, models for nonpremixed combustion will be presented. Emphasis is placed on models that are based on the mixture fraction as independent variable. The definition and the role of the mixture fraction will be discussed in detail. The derivation of the flamelet equations will be based on two-scale asymptotic arguments. Filtering of the mixture fraction field and that of the reactive scalars outside of the thin reaction zone will justify the use of flamelet equations in the entire available mixture fraction range. The flamelet model will be compared to experiments and other models used for nonpremixed turbulent combustion. Scaling laws for round turbulent jet flames including NO_x formation and the influence of buoyancy will be presented.

Chapter 4 is concerned with partially premixed combustion. The classical problem in partially premixed combustion is flame stabilization at the lift-off height in a turbulent jet diffusion flame, which will be discussed in detail. The key element of the instantaneous flame front is the triple flame structure. The level set formulation for premixed combustion and the mixture fraction formulation for nonpremixed combustion will be combined to obtain an expression for the turbulent burning velocity in partially premixed systems. This expression is validated with respect to its capability to predict lift-off heights of lifted turbulent diffusion flames.

Finally, in the epilogue the principles behind turbulent combustion modeling are emphasized.

The turbulence models used in the four chapters of this book rely on modeling procedures that were developed for nonreacting constant density flows. They are highly disputed even for those applications because they rely on empiricism and some kind of intuition supplemented by physical arguments. It will become clear that with combustion, empiricism and the number of necessary simplifications increase. This is reflected by the large variety of different combustion models that have been formulated and that are pursued and continuously improved by different groups in the combustion community. In some cases competing models are based on the same physical concepts, with different modeling strategies leading to different formulations. In other cases, the concepts are fundamentally different. For instance, Lagrangian models such as those used in the pdf transport equation model differ in their physical and mathematical formulation from classical flamelet models formulated in the Eulerian framework. As a consequence, if one needs to solve a combustion problem numerically, an a priori choice has to be made as to which model to use. This is regrettable, since no model is infallible and certainly no single approach will provide definite answers to the large variety of problems in turbulent combustion.

In view of these difficulties one may be tempted to give up trying to model turbulent combustion and revert to direct numerical simulations (DNS). However, as pointed out by Bray (1996), "DNS are an extremely valuable research tool ... from which much can be learned.... However, DNS cannot and will not meet the pressing need for improved predictive methods to aid ... the design of practical high Reynolds number combustion systems." The reason for this lies in the large range of scales that would have to be resolved if a full numerical simulation were to be performed. Since combustion requires molecular mixing on the smallest scale of turbulence, the Kolmogorov scale, this scale and the even smaller scales of the thin reaction layers would have to be resolved. The inertial subrange between the integral scale ℓ and the Kolmogorov scale η extends typically over two orders of magnitude in engineering applications with combustion. The size of the combustion chamber L is estimated as at least one order of magnitude larger than the integral scale. Therefore there are at least three orders of magnitude between L and the mesh size Δ needed to resolve processes occuring at the Kolmogorov scale by a direct numerical simulation. With constant mesh sizes and a three-dimensional (3D) mesh this would result in at least 10^9 grid points, which makes such computations prohibitive for many years to come.

However, quite elaborate numerical codes using adaptive gridding to account for complex geometries have been developed in recent years and are commercially available. These codes use RANS (Reynolds averaged Navier–Stokes equations) models and resolve the flow by using mesh sizes of the order of the integral length scale ℓ. Industrial users of these codes naturally demand that they are also applicable to flows with combustion. This sets a framework for turbulent combustion models that is quite restrictive and may require compromises.

A way to improve the predictive capability of these models, in particular with respect to the flow field, is to employ large eddy simulations (LES) for flows with combustion. LES models employ a larger number of grid points to resolve the energy containing scales within the inertial subrange. It will be argued at the end of Chapter 1 that some RANS models, in particular flamelet models combined with the presumed shape pdf approach, can easily be extended to LES.

I am indebted to many friends and colleagues in the combustion community for sharing with me over the past twenty years their stimulating ideas on turbulent combustion. Their papers are referenced in this book to the best of my knowledge. The book could also not have been written without the thesis work of many of my students and collaborators who have amply contributed to many of the sections in Chapters 2 to 4. That work has been funded generously by the Deutsche Forschungsgemeinschaft, which is gratefully acknowledged. Early versions of the manuscript have received critical reading by Bill Ashurst, Bernd Binninger, Philip de Goey, Johannes Janicka, Alain Kerstein, Rupert Klein, Moshe Matalon, Heinz Pitsch, Martin Oberlack, Luc Vervisch, and many others. This has considerably improved the text and is also gratefully acknowledged. Last but not least, I want to thank Beate Dieckhoff for her considerate and careful preparation of the manuscript.

1

Turbulent combustion: The state of the art

1.1 What Is Specific about Turbulence with Combustion?

In recent years, nothing seems to have inspired researchers in the combustion community so much as the unresolved problems in turbulent combustion. Turbulence in itself is far from being fully understood; it is probably the most significant unresolved problem in classical physics. Since the flow is turbulent in nearly all engineering applications, the urgent need to resolve engineering problems has led to preliminary solutions called turbulence models. These models use systematic mathematical derivations based on the Navier–Stokes equations up to a certain point, but then they introduce closure hypotheses that rely on dimensional arguments and require empirical input. This semiempirical nature of turbulence models puts them into the category of an art rather than a science.

For high Reynolds number flows the so-called eddy cascade hypothesis forms the basis for closure of turbulence models. Large eddies break up into smaller eddies, which in turn break up into even smaller ones, until the smallest eddies disappear due to viscous forces. This leads to scale invariance of energy transfer in the inertial subrange of turbulence. We will denote this as *inertial range invariance* in this book. It is the most important hypothesis for large Reynolds number turbulent flows and has been built into all classical turbulence models, which thereby satisfy the requirement of Reynolds number independence in the large Reynolds number limit. Viscous effects are of importance in the vicinity of solid walls only, a region of minor importance for combustion.

The apparent success of turbulence models in solving engineering problems has encouraged similar approaches for turbulent combustion, which consequently led to the formulation of turbulent combustion models. This is, however, where problems arise.

Combustion requires that fuel and oxidizer be mixed at the molecular level. How this takes place in turbulent combustion depends on the turbulent mixing process. The general view is that once a range of different size eddies has developed, strain and shear at the interface between the eddies enhance the mixing. During the eddy break-up process and the formation of smaller eddies, strain and shear will increase and thereby steepen the concentration gradients at the interface between reactants, which in turn enhances their molecular interdiffusion. Molecular mixing of fuel and oxidizer, as a prerequisite of combustion, therefore takes place at the interface between small eddies. Similar considerations apply, once a flame has developed, to the conduction of heat and the diffusion of radicals out of the reaction zone at the interface.

While this picture follows standard ideas about turbulent mixing, it is less clear how combustion modifies these processes. Chemical reactions consume the fuel and the oxidizer at the interface and will thereby steepen their gradients even further. To what extent this will modify the interfacial diffusion process still needs to be understood.

This could lead to the conclusion that the interaction between turbulence and combustion invalidates classical scaling laws known from nonreacting turbulent flows, such as the Reynolds number independence of free shear flows in the large Reynolds number limit. To complicate the picture further, one has to realize that combustion involves a large number of elementary chemical reactions that occur on different time scales. If all these scales would interact with all the time scales within the inertial range, no simple scaling laws could be found. Important empirical evidence, however, does not confirm such pessimism:

- The difference between the turbulent and the laminar burning velocity, normalized by the turbulence intensity, is independent of the Reynolds number. It is Damköhler number independent for large scale turbulence, but it becomes proportional to the square root of the Damköhler number for small scale turbulence (cf. Section 2.10).
- The flame length of a nonbuoyant turbulent jet diffusion flame, for instance, is Reynolds number and Damköhler number independent (cf. Section 3.9).
- The NO emission index of hydrogen–air diffusion flames is independent of the Reynolds number but proportional to the square root of the Damköhler number (cf. Section 3.14).
- The lift-off height in lifted jet diffusion flames is independent of the nozzle diameter and increases nearly linearly with the jet exit velocity (cf. Section 4.6).

1.1 What is specific about turbulence with combustion?

Power law Damköhler number scaling laws may be the exception rather than the rule, but they indicate that there are circumstances where only a few chemical and turbulent time scales are involved. As far as Reynolds number independence is concerned, it should be noted that the Reynolds number in many laboratory experiments is not large enough to approach the large Reynolds number limit. A remaining Reynolds number dependence of the turbulent mixing process would then show up in the combustion data. Apart from these experimental limitations (which become more serious owing to the increase of viscosity with temperature) it is not plausible that there would be a Reynolds number dependence introduced by combustion, because chemical reactions introduce additional time scales but no viscous effects. Even if chemical time scales interact with turbulent time scales in the inertial subrange of turbulence, these interactions cannot introduce the viscosity as a parameter for dimensional scaling, because it has disappeared as a parameter in that range. This does not preclude that ratios of molecular transport properties, Prandtl or Lewis numbers, for instance, would not appear in scaling laws in combustion. As we have restricted the content of this book to low speed combustion, the Mach number will not appear in the analysis.

There remains, however, the issue of to what extent we can expect an interaction between chemical and turbulent scales in the inertial subrange. Here, we must realize that combustion differs from isothermal mixing in chemically reacting flows by two specific features:

- heat release by combustion induces an increase of temperature, which in turn
- accelerates combustion chemistry. Because of the competition between chain branching and chain breaking reactions this process is very sensitive to temperature changes.

Heat release combined with temperature sensitive chemistry leads to typical combustion phenomena, such as ignition and extinction. This is illustrated in Figure 1.1 where the maximum temperature in a homogeneous flow combustor is plotted as a function of the Damköhler number, which here represents the ratio of the residence time to the chemical time. This is called the S-shaped curve in the combustion literature. The lower branch of this curve corresponds to a slowly reacting state of the combustor prior to ignition, where the short residence times prevent a thermal runaway. If the residence time is increased by lowering the flow velocity, for example, the Damköhler number increases until the ignition point I is reached. For values larger than Da_I thermal runaway

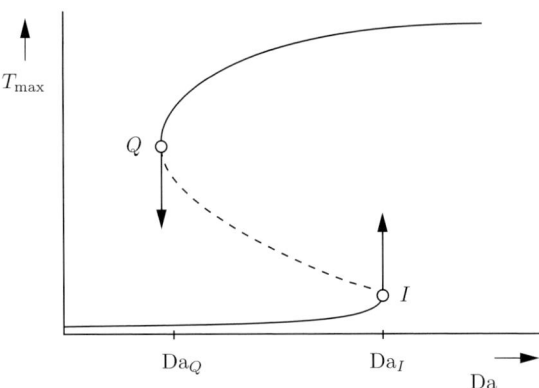

Figure 1.1. The S-shaped curve showing the maximum temperature in a well-stirred reactor as a function of the Damköhler number.

leads to a rapid unsteady transition to the upper close-to-equilibrium branch. If one starts on that branch and decreases the Damköhler number, thereby moving to the left in Figure 1.1, one reaches the point Q where extinction occurs. This is equivalent to a rapid transition to the lower branch. The middle branch between the point I and Q is unstable.

In the range of Damköhler numbers between Da_Q and Da_I, where two stable branches exist, any initial state with a temperature in the range between the lower and the upper branch is rapidly driven to either one of them. Owing to the temperature sensitivity of combustion reactions the two stable branches represent strong attractors. Therefore, only regions close to chemical equilibrium or close to the nonreacting state are frequently accessed. In an analytic study of stochastic Damköhler number variations Oberlack et al. (2000a) have recently shown that the probability of finding realizations apart from these two steady state solutions is indeed very small.

Chemical reactions that take place at the high temperatures on the upper branch of Figure 1.1 are nearly always fast compared to all turbulent time scales and, with the support of molecular diffusion, they concentrate in thin layers of a width that is typically smaller than the Kolmogorov scale. Except for density changes these layers cannot exert a feedback on the flow. Therefore they cannot influence the inertial range scaling. If these layers extinguish as the result of excessive heat loss, the temperature decreases such that chemistry becomes very slow and mixing can also be described by classical inertial range scaling.

In both situations, fast and slow chemistry, time and length scales of combustion are separated from those of turbulence in the inertial subrange. This *scale separation* is a specific feature of most practical applications of turbulent combustion.† It makes the mixing process in the inertial range independent of chemistry and simplifies modeling significantly. Almost all turbulent combustion models explicitly or implicitly assume scale separation.

As a general theme of this chapter, we will investigate whether the turbulence models to be discussed are based on the postulate of scale separation between turbulent and chemical time scales. In addition, it will be pointed out if a combustion model does not satisfy the postulate of Reynolds number independence in the large Reynolds number limit.

1.2 Statistical Description of Turbulent Flows

The aim of stochastic methods in turbulence is to describe the fluctuating velocity and scalar fields in terms of their statistical distributions. A convenient starting point for this description is the distribution function of a single variable, the velocity component u, for instance. The distribution function $F_u(U)$ of u is defined by the probability p of finding a value of $u < U$:

$$F_u(U) = p(u < U), \tag{1.1}$$

where U is the so-called sample space variable associated with the random stochastic variable u. The sample space of the random stochastic variable u consists of all possible realizations of u. The probability of finding a value of u in a certain interval $U_- < u < U_+$ is given by

$$p(U_- < u < U_+) = F_u(U_+) - F_u(U_-). \tag{1.2}$$

The probability density function (pdf) of u is now defined as

$$P_u(U) = \frac{dF_u(U)}{dU}. \tag{1.3}$$

It follows that $P_u(U)dU$ is the probability of finding u in the range $U \leq u < U + dU$. If the possible realizations of u range from $-\infty$ to $+\infty$, it follows that

$$\int_{-\infty}^{+\infty} P_u(U)\,dU = 1, \tag{1.4}$$

which states that the probability of finding the value u between $-\infty$ and $+\infty$ is certain (i.e., it has the probability unity). It also serves as a normalizing condition for P_u.

---

† A potential exception is the situation prior to ignition, where chemistry is neither slow enough nor fast enough to be separated from the turbulent time scales. We will discuss this situation in detail in Chapter 3, Section 3.12.

In turbulent flows the pdf of any stochastic variable depends, in principle, on the position x and on time t. These functional dependencies are expressed by the following notation:

$$P_u(U; x, t). \tag{1.5}$$

The semicolon used here indicates that P_u is a probability density in U-space and is a function of x and t. In stationary turbulent flows it does not depend on t and in homogeneous turbulent fields it does not depend on x. In the following, for simplicity of notation, we will not distinguish between the random stochastic variable u and the sample space variable U, dropping the index and writing the pdf as

$$P(u; x, t). \tag{1.6}$$

Once the pdf of a variable is known one may define its moments by

$$\overline{u(x, t)^n} = \int_{-\infty}^{+\infty} u^n P(u; x, t)\, du. \tag{1.7}$$

Here the overbar denotes the average or mean value, sometimes also called expectation, of u^n. The first moment ($n = 1$) is called the mean of u:

$$\bar{u}(x, t) = \int_{-\infty}^{+\infty} u\, P(u; x, t)\, du. \tag{1.8}$$

Similarly, the mean value of a function $g(u)$ can be calculated from

$$\bar{g}(x, t) = \int_{-\infty}^{+\infty} g(u) P(u; x, t)\, du. \tag{1.9}$$

Central moments are defined by

$$\overline{[u(x, t) - \overline{u(x, t)}]^n} = \int_{-\infty}^{+\infty} (u - \bar{u})^n P(u; x, t)\, du, \tag{1.10}$$

where the second central moment

$$\overline{[u(x, t) - \overline{u(x, t)}]^2} = \int_{-\infty}^{+\infty} (u - \bar{u})^2 P(u; x, t)\, du \tag{1.11}$$

is called the variance. If we split the random variable u into its mean and the fluctuations u' as

$$u(x, t) = \bar{u}(x, t) + u'(x, t), \tag{1.12}$$

where $\overline{u'} = 0$ by definition, the variance is found to be related to the first and second moment by

$$\overline{u'^2} = \overline{(u - \bar{u})^2} = \overline{u^2 - 2u\bar{u} + \bar{u}^2} = \overline{u^2} - \bar{u}^2. \tag{1.13}$$

Models for turbulent flows traditionally start from the Navier–Stokes equations to derive equations for the first and the second moments of the flow variables using (1.12). Since the three velocity components and the pressure depend on each other through the solutions of the Navier–Stokes equations they are correlated. To quantify these correlations it is convenient to introduce the joint probability density function of the random variables. For instance, the joint pdf of the velocity components u and v is written as

$$P(u, v; \boldsymbol{x}, t).$$

The pdf of u, for instance, may be obtained from the joint pdf by integration over all possible realizations of v,

$$P(u) = \int_{-\infty}^{+\infty} P(u, v)\,dv, \tag{1.14}$$

and is called the marginal pdf of u in this context. The correlation between u and v is given by

$$\overline{u'v'} = \int_{-\infty}^{+\infty}\int_{-\infty}^{+\infty} (u - \bar{u})(v - \bar{v}) P(u, v)\,du\,dv. \tag{1.15}$$

This can be illustrated by a so-called scatter plot (cf. Figure 1.2). If a series of instantaneous realizations of u and v are plotted as points in a graph of u and v, these points will scatter within a certain range. The means $\bar{u}$ and $\bar{v}$ are the average positions of the points in u and v directions, respectively. The correlation coefficient $\overline{u'v'}/\bar{u}\,\bar{v}$ is proportional to the slope of the average straight line through the data points.

A joint pdf of two independent variables can always be written as a product of a conditional pdf of one variable times the marginal pdf of the other, for

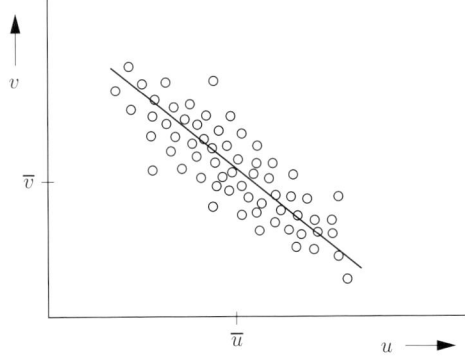

Figure 1.2. A scatter plot of two velocity components u and v illustrating the correlation coefficient.

example

$$P(u, v; \boldsymbol{x}, t) = P(u \mid v; \boldsymbol{x}, t) P(v; \boldsymbol{x}, t). \tag{1.16}$$

This is called Bayes' theorem. In this example the conditional pdf $P(u \mid v; \boldsymbol{x}, t)$ describes the probability density of u, conditioned at a fixed value of v. If u and v are not correlated they are called statistically independent. In that case the joint pdf is equal to the product of the marginal pdfs:

$$P(u, v; \boldsymbol{x}, t) = P(u; \boldsymbol{x}, t) P(v; \boldsymbol{x}, t). \tag{1.17}$$

By using this in (1.15) and integrating, we easily see that $\overline{u'v'}$ vanishes, if u and v are statistically independent. In turbulent shear flows $\overline{u'v'}$ is interpreted as a Reynolds shear stress, which is nonzero in general. The conditional pdf $P(u \mid v; \boldsymbol{x}, t)$ can be used to define conditional moments. For example, the conditional mean of u, conditioned at a fixed value of v, is given by

$$\langle u \mid v \rangle = \int_{-\infty}^{+\infty} u P(u \mid v) \, du. \tag{1.18}$$

In the following we will use angular brackets for conditional means only.

As a consequence of the nonlinearity of the Navier–Stokes equations several closure problems arise. These are not only related to correlations between velocity components among each other and the pressure, but also to correlations between velocity gradients and correlations between velocity gradients and pressure fluctuations. These appear in the equations for the second moments as dissipation terms and pressure–strain correlations, respectively. The statistical description of gradients requires information from adjacent points in physical space. Very important aspects in the statistical description of turbulent flows are therefore related to two-point correlations, which we will introduce in Section 1.4.

For flows with large density changes as occur in combustion, it is often convenient to introduce a density-weighted average $\tilde{u}$, called the Favre average, by splitting $u(\boldsymbol{x}, t)$ into $\tilde{u}(\boldsymbol{x}, t)$ and $u''(\boldsymbol{x}, t)$ as

$$u(\boldsymbol{x}, t) = \tilde{u}(\boldsymbol{x}, t) + u''(\boldsymbol{x}, t). \tag{1.19}$$

This averaging procedure is defined by requiring that the average of the product of u'' with the density ρ (rather than u'' itself) vanishes:

$$\overline{\rho u''} = 0. \tag{1.20}$$

The definition for $\tilde{u}$ may then be derived by multiplying (1.19) by the density ρ and averaging:

$$\overline{\rho u} = \overline{\rho \tilde{u}} + \overline{\rho u''} = \bar{\rho} \tilde{u}. \tag{1.21}$$

Here the average of the product $\rho \tilde{u}$ is equal to the product of the averages $\bar{\rho}$ and $\tilde{u}$, since $\tilde{u}$ is already an average defined by

$$\tilde{u} = \overline{\rho u}/\bar{\rho}. \tag{1.22}$$

This density-weighted average can be calculated, if simultaneous measurements of ρ and u are available. Then, by taking the average of the product ρu and dividing it by the average of ρ one obtains $\tilde{u}$. While such measurements are often difficult to obtain, Favre averaging has considerable advantages in simplifying the formulation of the averaged Navier–Stokes equations in variable density flows. In the momentum equations, but also in the balance equations for the temperature and the chemical species, the convective terms are dominant in high Reynolds number flows. Since these contain products of the dependent variables and the density, Favre averaging is the method of choice. For instance, the average of the product of the density ρ with the velocity components u and v would lead with conventional averages to four terms,

$$\overline{\rho u v} = \bar{\rho}\,\bar{u}\,\bar{v} + \bar{\rho}\overline{u'v'} + \overline{\rho'u'}\bar{v} + \overline{\rho'v'}\bar{u} + \overline{\rho'u'v'}. \tag{1.23}$$

Using Favre averages one writes

$$\begin{aligned}\rho u v &= \rho(\tilde{u} + \tilde{u}'')(\tilde{v} + v'') \\ &= \rho\tilde{u}\tilde{v} + \rho u''\tilde{v} + \rho v''\tilde{u} + \rho u''v''.\end{aligned} \tag{1.24}$$

Here fluctuations of the density do not appear. Taking the average leads to two terms only,

$$\overline{\rho u v} = \bar{\rho}\tilde{u}\tilde{v} + \bar{\rho}\widetilde{u''v''}. \tag{1.25}$$

This expression is much simpler than (1.23) and has formally the same structure as the conventional average of uv for constant density flows:

$$\overline{uv} = \bar{u}\bar{v} + \overline{u'v'}. \tag{1.26}$$

Difficulties arising with Favre averaging in the viscous and diffusive transport terms are of less importance since these terms are usually neglected in high Reynolds number turbulence.

The introduction of density-weighted averages requires the knowledge of the correlation between the density and the other variable of interest. A Favre pdf of u can be derived from the joint pdf $P(\rho, u)$ as

$$\bar{\rho}\tilde{P}(u) = \int_{\rho_{\min}}^{\rho_{\max}} \rho P(\rho, u)\,d\rho = \int_{\rho_{\min}}^{\rho_{\max}} \rho P(\rho\,|\,u)P(u)\,d\rho = \langle \rho\,|\,u\rangle P(u). \tag{1.27}$$

Multiplying both sides with u and integrating yields

$$\bar{\rho}\int_{-\infty}^{+\infty} u\tilde{P}(u)\,du = \int_{-\infty}^{+\infty} \langle \rho \mid u\rangle u P(u)\,du, \qquad (1.28)$$

which is equivalent to $\bar{\rho}\tilde{u} = \overline{\rho u}$. The Favre mean value of u therefore is defined as

$$\tilde{u} = \int_{-\infty}^{+\infty} u\tilde{P}(u)\,du. \qquad (1.29)$$

1.3 Navier–Stokes Equations and Turbulence Models

In the following we will first describe the classical approach to model turbulent flows. It is based on single point averages of the Navier–Stokes equations. These are commonly called Reynolds averaged Navier–Stokes equations (RANS). We will formally extend this formulation to nonconstant density by introducing Favre averages. In addition we will present the most simple model for turbulent flows, the k–ε model. Even though it certainly is the best compromise for engineering design using RANS, the predictive power of the k–ε model is, except for simple shear flows, often found to be disappointing. We will present it here, mainly to help us define turbulent length and time scales.

For nonconstant density flows the Navier–Stokes equations are written in conservative form:

Continuity

$$\frac{\partial \rho}{\partial t} + \nabla \cdot (\rho v) = 0, \qquad (1.30)$$

Momentum

$$\frac{\partial \rho v}{\partial t} + \nabla \cdot (\rho v v) = -\nabla p + \nabla \cdot \tau + \rho g. \qquad (1.31)$$

In (1.31) the two terms on the left-hand side (l.h.s.) represent the local rate of change and convection of momentum, respectively, while the first term on the right-hand side (r.h.s.) is the pressure gradient and the second term on the r.h.s. represents molecular transport due to viscosity. Here τ is the viscous stress tensor

$$\tau = \mu\left[2S - \frac{2}{3}\delta\nabla \cdot v\right] \qquad (1.32)$$

and

$$S = \frac{1}{2}(\nabla v + \nabla v^T) \qquad (1.33)$$

1.3 Navier–Stokes equations and turbulence models

is the rate of strain tensor, where ∇v^T is the transpose of the velocity gradient and μ is the dynamic viscosity. It is related to the kinematic viscosity ν as $\mu = \rho \nu$. The last term in (1.31) represents forces due to buoyancy.

Using Favre averaging on (1.30) and (1.31) one obtains

$$\frac{\partial \bar{\rho}}{\partial t} + \nabla \cdot (\bar{\rho} \tilde{v}) = 0, \qquad (1.34)$$

$$\frac{\partial \bar{\rho} \tilde{v}}{\partial t} + \nabla \cdot (\bar{\rho} \tilde{v} \tilde{v}) = -\nabla \bar{p} + \nabla \cdot \bar{\tau} - \nabla \cdot (\overline{\bar{\rho} v'' v''}) + \bar{\rho} \mathbf{g}. \qquad (1.35)$$

This equation is similar to (1.31) except for the third term on the l.h.s. containing the correlation $-\overline{\bar{\rho} v'' v''}$, which is called the Reynolds stress tensor.

The Reynolds stress tensor is unknown and represents the first closure problem for turbulence modeling. It is possible to derive equations for the six components of the Reynolds stress tensor. In these equations several terms appear that again are unclosed. Those so-called Reynolds stress models have been presented for nonconstant density flows, for example, by Jones (1994) and Jones and Kakhi (1996).

Although Reynolds stress models contain a more complete description of the physics, they are not yet widely used in turbulent combustion. Many industrial codes still rely on the k–ε model, which, by using an eddy viscosity, introduces the assumption of isotropy. It is known that turbulence becomes isotropic at the small scales, but this does not necessarily apply to the large scales at which the averaged quantities are defined. The k–ε model is based on equations where the turbulent transport is diffusive and therefore is more easily handled by numerical methods than the Reynolds stress equations. This is probably the most important reason for its wide use in many industrial codes.

An important simplification is obtained by introducing the eddy viscosity ν_t, which leads to the following expression for the Reynolds stress tensor:

$$-\bar{\rho} \widetilde{v'' v''} = \bar{\rho} \nu_t \left[2\tilde{S} - \frac{2}{3} \delta \nabla \cdot \tilde{v} \right] - \frac{2}{3} \delta \bar{\rho} \tilde{k}. \qquad (1.36)$$

Here δ is the tensorial Kronecker symbol δ_{ij} ($\delta_{ij} = 1$ for $i = j$ and $\delta_{ij} = 0$ for $i \neq j$) and ν_t is the kinematic eddy viscosity, which is related to the Favre average turbulent kinetic energy

$$\tilde{k} = \frac{1}{2} \widetilde{v'' \cdot v''} \qquad (1.37)$$

and its dissipation $\tilde{\varepsilon}$ by

$$\nu_t = c_\mu \frac{\tilde{k}^2}{\tilde{\varepsilon}}, \qquad c_\mu = 0.09. \qquad (1.38)$$

The introduction of the Favre averaged variables $\tilde{k}$ and $\tilde{\varepsilon}$ requires that modeled equations are available for these quantities. These equations are given here in their most simple form:

Turbulent kinetic energy

$$\bar{\rho}\frac{\partial \tilde{k}}{\partial t} + \bar{\rho}\tilde{v} \cdot \nabla \tilde{k} = \nabla \cdot \left(\frac{\bar{\rho}\nu_t}{\sigma_k}\nabla \tilde{k}\right) - \widetilde{\bar{\rho}v''v''} : \nabla \tilde{v} - \bar{\rho}\tilde{\varepsilon}, \qquad (1.39)$$

Turbulent dissipation

$$\bar{\rho}\frac{\partial \tilde{\varepsilon}}{\partial t} + \bar{\rho}\tilde{v} \cdot \nabla \tilde{\varepsilon} = \nabla \cdot \left(\bar{\rho}\frac{\nu_t}{\sigma_\varepsilon}\nabla \tilde{\varepsilon}\right) - c_{\varepsilon 1}\bar{\rho}\frac{\tilde{\varepsilon}}{\tilde{k}}\widetilde{v''v''} : \nabla \tilde{v} - c_{\varepsilon 2}\bar{\rho}\frac{\tilde{\varepsilon}^2}{\tilde{k}}. \qquad (1.40)$$

In these equations the two terms on the l.h.s. represent the local rate of change and convection, respectively. The first term on the r.h.s. represents the turbulent transport, the second one turbulent production, and the third one turbulent dissipation. As in the standard k–ε model, the constants $\sigma_k = 1.0$, $\sigma_\varepsilon = 1.3$, $c_{\varepsilon 1} = 1.44$, and $c_{\varepsilon 2} = 1.92$ are generally used. A more detailed discussion concerning additional terms in the Favre averaged turbulent kinetic energy equation may be found in Libby and Williams (1994).

It should be noted that for constant density flows the k-equation can be derived with few modeling assumptions quite systematically from the Navier–Stokes equations. From this derivation follows the definition of the viscous dissipation as

$$\varepsilon = \nu\overline{[\nabla v' + \nabla v'^T] : \nabla v'}. \qquad (1.41)$$

The ε-equation, however, cannot be derived in a systematic manner. The basis for the modeling of that equation are the equations for two-point correlations. Rotta (1972) has shown that by integrating the two-point correlation equations over the correlation coordinate r one can derive an equation for the integral length scale ℓ, which will be defined below. This leads to a k–ℓ-model. The ℓ-equation has been applied, for example, by Rodi and Spalding (1970) to turbulent jet flows. It is easily shown that from this model and from the algebraic relation between ℓ, k, and ε a balance equation for ε can be derived. A similar approach has recently been used by Oberlack (1997) to derive an equation for the dissipation tensor that is needed in Reynolds stress models.

The dissipation ε plays a fundamental role in turbulence theory, as will be shown in the next section. The eddy cascade hypothesis states that it is equal to the energy transfer rate from the large eddies to the smaller eddies and therefore is invariant within the inertial subrange of turbulence. By using this property for ε in the k-equation and by determining ε from an equation like (1.40) rather than from its definition (1.41) one obtains Reynolds number independent solutions for free shear flows where, owing to the absence of walls, the viscous stress tensor can be neglected compared to the Reynolds stress tensor. This is how inertial range invariance is built into turbulence models.

It may be counterintuitive to model dissipation, which is active at the small scales, by an equation that contains only quantities that are defined at the large integral scales (cf. Figure 1.5 below). However, it is only because inertial range invariance has been built into turbulence models that they reproduce the scaling laws that are experimentally observed. Based on the postulate formulated at the end of Section 1.1 the same must be claimed for turbulent combustion models in the large Reynolds number limit. Since combustion takes place at the small scales, inertial range invariant quantities must relate properties defined at the small scales to those defined at the large scales, at which the models are formulated.

1.4 Two-Point Velocity Correlations and Turbulent Scales

A characteristic feature of turbulent flows is the occurrence of eddies of different length scales. If a turbulent jet shown in Figure 1.3 enters with a high velocity into initially quiescent surroundings, the large velocity difference between the jet and the surroundings generates a shear layer instability, which, after a transition, becomes turbulent further downstream from the nozzle exit. The two shear layers merge into a fully developed turbulent jet. In order to characterize the distribution of eddy length scales at any position within the jet, one measures at point x and time t the axial velocity $u(x, t)$, and simultaneously at a second point $(x + r, t)$ with distance r apart from the first one, the velocity $u(x + r, t)$. Then the correlation between these two velocities is defined by the

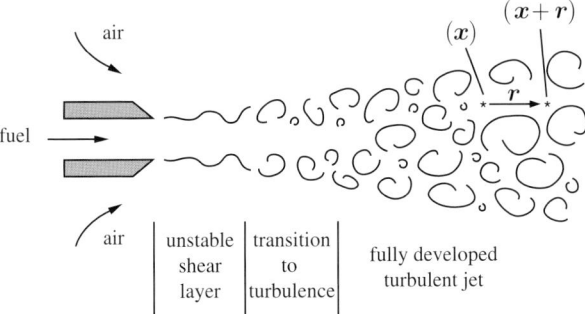

Figure 1.3. Schematic presentation of two-point correlation measurements in a turbulent jet.

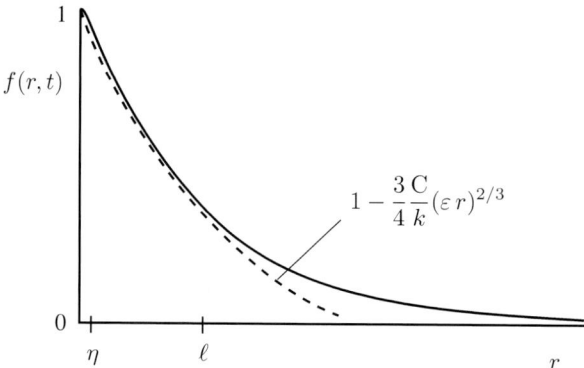

Figure 1.4. The normalized two-point velocity correlation for homogeneous isotropic turbulence as a function of the distance r between the two points.

average

$$R(x, r, t) = \overline{u'(x, t)u'(x + r, t)}. \tag{1.42}$$

For homogeneous isotropic turbulence the location x is arbitrary and r may be replaced by its absolute value $r = |r|$. For this case the normalized correlation

$$f(r, t) = R(r, t)/\overline{u'^2(t)} \tag{1.43}$$

is plotted schematically in Figure 1.4. It approaches unity for $r \to 0$ and decays slowly when the two points are only a very small distance r apart. With increasing distance it decreases continuously and may even take negative values. Very large eddies corresponding to large distances between the two points are rather seldom and therefore do not contribute much to the correlation.

Kolmogorov's 1941 theory for homogeneous isotropic turbulence assumes that there is a steady transfer of kinetic energy from the large scales to the small scales and that this energy is being consumed at the small scales by viscous dissipation. This is the eddy cascade hypothesis. By equating the energy transfer rate (kinetic energy per eddy turnover time) with the dissipation ε it follows that this quantity is independent of the size of the eddies within the inertial range. For the inertial subrange, extending from the integral scale ℓ to the Kolmogorov scale η, ε is the only dimensional quantity apart from the correlation coordinate r that is available for the scaling of $f(r, t)$. Since ε has the dimension [m²/s³], the second-order structure function defined by

$$F_2(r, t) = \overline{(u'(x, t) - u'(x + r, t))^2} = 2\overline{u'^2(t)}(1 - f(r, t)) \tag{1.44}$$

with the dimension [m²/s²] must therefore scale as

$$F_2(r, t) = C(\varepsilon r)^{2/3}, \tag{1.45}$$

1.4 Two-point velocity correlations and turbulent scales

where C is a universal constant called the Kolmogorov constant. In the case of homogeneous isotropic turbulence the velocity fluctuations in the three coordinate directions are equal to each other. The turbulent kinetic energy

$$k = \frac{1}{2}\overline{v' \cdot v'} \tag{1.46}$$

is then equal to $k = 3\overline{u'^2}/2$. Using this one obtains from (1.44) and (1.45)

$$f(r,t) = 1 - \frac{3}{4}\frac{C}{k}(\varepsilon r)^{2/3}, \tag{1.47}$$

which is also plotted in Figure 1.4.

There are eddies of a characteristic size containing most of the kinetic energy. At these eddies there remains a relatively large correlation $f(r, t)$ before it decays to zero. The length scale of these eddies is called the integral length scale ℓ and is defined by

$$\ell(t) = \int_0^\infty f(r,t)\,dr. \tag{1.48}$$

The integral length scale is also shown in Figure 1.4.

We denote the root-mean-square (r.m.s.) velocity fluctuation by

$$v' = \sqrt{2k/3}, \tag{1.49}$$

which represents the turnover velocity of integral scale eddies. The turnover time ℓ/v' of these eddies is then proportional to the integral time scale

$$\tau = \frac{k}{\varepsilon}. \tag{1.50}$$

For very small values of r only very small eddies fit into the distance between x and $x + r$. The motion of these small eddies is influenced by viscosity, which provides an additional dimensional quantity for scaling. Dimensional analysis then yields the Kolmogorov length scale

$$\eta = \left(\frac{\nu^3}{\varepsilon}\right)^{1/4}, \tag{1.51}$$

which is also shown in Figure 1.4.

The range of length scales between the integral scale and the Kolmogorov scale is called the inertial range. In addition to η a Kolmogorov time and a velocity scale may be defined as

$$t_\eta = \left(\frac{\nu}{\varepsilon}\right)^{1/2}, \quad v_\eta = (\nu\varepsilon)^{1/4}. \tag{1.52}$$

The Taylor length scale λ is an intermediate scale between the integral and the Kolmogorov scale. It is defined by replacing the average gradient in the definition of the dissipation (1.41) by v'/λ. This leads to the definition

$$\varepsilon = 15\nu \frac{v'^2}{\lambda^2}. \tag{1.53}$$

Here the factor 15 originates from considerations for isotropic homogeneous turbulence. Using (1.52) we see that λ is proportional to the product of the turnover velocity of the integral scale eddies and the Kolmogorov time:

$$\lambda = (15\nu \, v'^2/\varepsilon)^{1/2} \sim v' t_\eta. \tag{1.54}$$

Therefore λ may be interpreted as the distance that a large eddy convects a Kolmogorov eddy during its turnover time t_η. As a somewhat artificially defined intermediate scale it has no direct physical significance in turbulence or in turbulent combustion. We will see, however, that similar Taylor scales may be defined for nonreactive scalar fields, which are useful for the interpretation of mixing processes.

According to Kolmogorov's 1941 theory the energy transfer from the large eddies of size ℓ is equal to the dissipation of energy at the Kolmogorov scale η. Therefore we will relate ε directly to the turnover velocity and the length scale of the integral scale eddies,

$$\varepsilon \sim \frac{v'^3}{\ell}. \tag{1.55}$$

We now define a discrete sequence of eddies within the inertial subrange by

$$\ell_n = \frac{\ell}{2^n} \geq \eta, \quad n = 1, 2, \ldots. \tag{1.56}$$

Since ε is constant within the inertial subrange, dimensional analysis relates the turnover time t_n and the velocity difference v_n across the eddy ℓ_n to ε in that range as

$$\varepsilon \sim \frac{v_n^2}{t_n} \sim \frac{v_n^3}{\ell_n} \sim \frac{\ell_n^2}{t_n^3}. \tag{1.57}$$

This relation includes the integral scales and also holds for the Kolmogorov scales as

$$\varepsilon = \frac{v_\eta^2}{t_\eta} = \frac{v_\eta^3}{\eta}. \tag{1.58}$$

A Fourier transform of the isotropic two-point correlation function leads to a definition of the kinetic energy spectrum $E(k)$, which is the density of kinetic

1.4 Two-point velocity correlations and turbulent scales

energy per unit wavenumber k. Here, rather than presenting a formal derivation, we relate the wavenumber k to the inverse of the eddy size ℓ_n as

$$k = \ell_n^{-1}. \tag{1.59}$$

The kinetic energy v_n^2 at scale ℓ_n is then

$$v_n^2 \sim (\varepsilon \, \ell_n)^{2/3} = \varepsilon^{2/3} k^{-2/3} \tag{1.60}$$

and its density in wavenumber space is proportional to

$$E(k) = \frac{dv_n^2}{dk} \sim \varepsilon^{2/3} k^{-5/3}. \tag{1.61}$$

This is the well-known $k^{-5/3}$ law for the kinetic energy spectrum in the inertial subrange.

If the energy spectrum is measured in the entire wavenumber range one obtains the behavior shown schematically in a log–log plot in Figure 1.5. For small wavenumbers corresponding to large scale eddies the energy per unit wavenumber increases with a power law between k^2 and k^4. This range is not universal and is determined by large scale instabilities, which depend on the boundary conditions of the flow. The spectrum attains a maximum at a wavenumber that corresponds to the integral scale, since eddies of that scale contain most of the kinetic energy. For larger wavenumbers corresponding to the inertial subrange the energy spectrum decreases following the $k^{-5/3}$ law. There is a cutoff at the Kolmogorov scale η. Beyond this cutoff, in the range called

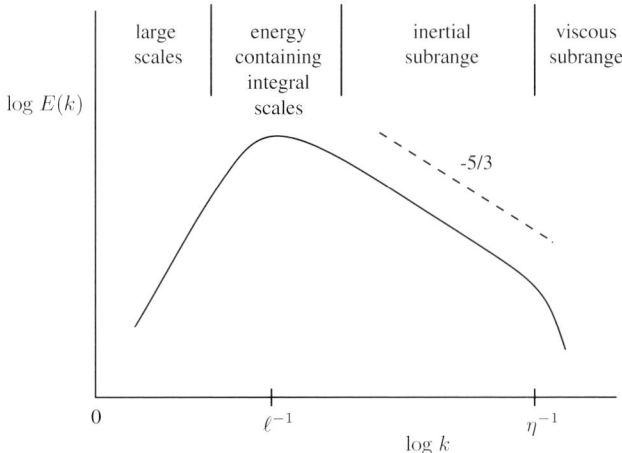

Figure 1.5. Schematic representation of the turbulent kinetic energy spectrum as a function of the wavenumber k.

the viscous subrange, the energy per unit wavenumber decreases exponentially owing to viscous effects.

In one-point averages the energy containing eddies at the integral length scale contribute the most to the kinetic energy. Therefore RANS averaged mean quantities essentially represent averages over regions in physical space that are of the order of the integral scale. This was meant by the statement at the end of Section 1.3 that RANS averages are defined at the large scales. In Large Eddy Simulations (LES), to be discussed in Section 1.14, filtering over smaller regions than the integral length scale leads to different mean values and, in particular, to smaller variances.

1.5 Balance Equations for Reactive Scalars

Combustion is the conversion of chemical bond energy contained in fossil fuels into heat by chemical reactions. The basis for any combustion model is the continuum formulation of the balance equations for energy and the chemical species. We will not derive these equations here but refer to Williams (1985a) for more details. We consider a mixture of n chemically reacting species and start with the balance equations for the mass fraction of species i,

$$\rho \frac{\partial Y_i}{\partial t} + \rho \boldsymbol{v} \cdot \nabla Y_i = -\nabla \cdot \boldsymbol{j}_i + \omega_i, \qquad (1.62)$$

where $i = 1, 2, \ldots, n$. In these equations the terms on the l.h.s. represent the local rate of change and convection. The diffusive flux in the first term on the r.h.s. is denoted by $\boldsymbol{j}_i$ and the last term ω_i is the chemical source term.

The molecular transport processes that cause the diffusive fluxes are quite complicated. A full description may be found in Williams (1985a). Since in models of turbulent combustion molecular transport is less important than turbulent transport, it is useful to consider simplified versions of the diffusive fluxes; the most elementary is the binary flux approximation

$$\boldsymbol{j}_i = -\rho D_i \nabla Y_i, \qquad (1.63)$$

where D_i is the binary diffusion coefficient, or mass diffusivity, of species i with respect to an abundant species, for instance N_2. It should be noted, however, that in a multicomponent system this approximation violates mass conservation, if nonequal diffusivities D_i are used, since the sum of all n fluxes has to vanish and the sum of all mass fractions is unity. Equation (1.63) is introduced here mainly for the ease of notation, but it must not be used in laminar flame calculations.

1.5 Balance equations for reactive scalars

For simplicity it will also be assumed that all mass diffusivities D_i are proportional to the thermal diffusivity denoted by

$$D = \lambda/\rho\, c_p \tag{1.64}$$

such that the Lewis numbers

$$Le_i = \lambda/(\rho\, c_p D_i) = D/D_i \tag{1.65}$$

are constant. In these equations λ is the thermal conductivity and c_p is the heat capacity at constant pressure of the mixture.

Before going into the definition of the chemical source term ω_i to be presented in the next section, we want to consider the energy balance in a chemically reacting system. The enthalpy h is the mass-weighted sum of the specific enthalpies h_i of species i:

$$h = \sum_{i=1}^{n} Y_i h_i. \tag{1.66}$$

For an ideal gas h_i depends only on the temperature T:

$$h_i = h_{i,\text{ref}} + \int_{T_\text{ref}}^{T} c_{p_i}(T)\, dT. \tag{1.67}$$

Here c_{pi} is the specific heat capacity of species i at constant pressure and T is the temperature in Kelvins. The chemical bond energy is essentially contained in the reference enthalpies $h_{i,\text{ref}}$. Reference enthalpies of H_2, O_2, N_2, and solid carbon are in general chosen as zero, while those of combustion products such as CO_2 and H_2O are negative. These values as well as polynomial fits for the temperature dependence of c_{p_i} are documented, for instance, for many species used in combustion calculations in Burcat (1984). Finally the specific heat capacity at constant pressure of the mixture is

$$c_p = \sum_{i=1}^{n} Y_i c_{p_i}. \tag{1.68}$$

A balance equation for the enthalpy can be derived from the first law of thermodynamics as (cf. Williams, 1985a)

$$\rho \frac{\partial h}{\partial t} + \rho \boldsymbol{v} \cdot \nabla h = \frac{\partial p}{\partial t} + \boldsymbol{v} \cdot \nabla p - \nabla \cdot \boldsymbol{j}_q + q_R. \tag{1.69}$$

Here the terms on the l.h.s. represent the local rate of change and convection of enthalpy. We have neglected the term that describes frictional heating because it is small for low speed flows. The local and convective change of pressure is important for acoustic interactions and pressure waves. We will not consider the term $\boldsymbol{v} \cdot \nabla p$ any further since we are interested in the small Mach number limit only. The transient pressure term $\partial p/\partial t$ must be retained in applications for reciprocating engines but can be neglected in open flames where the pressure is approximately constant and equal to the static pressure. The heat flux j_q includes the effect of enthalpy transport by the diffusive fluxes j_i:

$$j_q = -\lambda \nabla T + \sum_{i=1}^{n} h_i j_i. \qquad (1.70)$$

Finally, the last term in (1.69) represents heat transfer due to radiation and must be retained in furnace combustion and whenever strongly temperature dependent processes, such as NO_x formation, are to be considered.

The static pressure is obtained from the thermal equation of state for a mixture of ideal gases

$$p = \rho \frac{RT}{W}. \qquad (1.71)$$

Here R is the universal gas constant and W is the mean molecular weight given by

$$W = \left(\sum_{i=1}^{n} \frac{Y_i}{W_i} \right)^{-1}. \qquad (1.72)$$

The molecular weight of species i is denoted by W_i. For completeness we note that mole fractions X_i can be converted into mass fractions Y_i via

$$Y_i = \frac{W_i}{W} X_i. \qquad (1.73)$$

We now want to simplify the enthalpy equation. Differentiating (1.66) one obtains

$$dh = c_p dT + \sum_{i=1}^{n} h_i dY_i, \qquad (1.74)$$

where (1.67) and (1.68) have been used. If (1.70), (1.74), and (1.63) are inserted into the enthalpy equation (1.69) with the term $v \cdot \nabla p$ removed, it takes the form

$$\rho \frac{\partial h}{\partial t} + \rho v \cdot \nabla h = \frac{\partial p}{\partial t} + \nabla \cdot \left(\frac{\lambda}{c_p} \nabla h \right) + q_R$$
$$- \sum_{i=1}^{n} h_i \nabla \cdot \left[\left(\frac{\lambda}{c_p} - \rho D_i \right) \nabla Y_i \right]. \qquad (1.75)$$

It is immediately seen that the last term disappears, if all Lewis numbers are assumed equal to unity. If, in addition, unsteady pressure changes and radiation heat transfer can be neglected, the enthalpy equation contains no source terms.

1.5 Balance equations for reactive scalars

It then has a form similar to the equation for a conserved scalar such as the mixture fraction that will be introduced in Chapter 3.

Another form of the energy equation can be written in terms of the temperature. Inserting (1.62), (1.70), and (1.74) into (1.69) one obtains

$$\rho c_p \frac{\partial T}{\partial t} + \rho c_p \boldsymbol{v} \cdot \nabla T = \frac{\partial p}{\partial t} + \nabla \cdot (\lambda \nabla T) - \sum_{i=1}^{n} c_{p_i} \boldsymbol{j}_i \cdot \nabla T - \sum_{i=1}^{n} h_i \omega_i + q_R. \tag{1.76}$$

If, for simplicity, the specific heat capacities $c_{p,i}$ are all assumed equal and constant, the pressure is constant, and the heat transfer due to radiation is neglected, the temperature equation becomes

$$\rho \frac{\partial T}{\partial t} + \rho \boldsymbol{v} \cdot \nabla T = \nabla \cdot (\rho D \nabla T) + \omega_T. \tag{1.77}$$

Here (1.64) was used and the heat release due to chemical reactions is written as

$$\omega_T = -\frac{1}{c_p} \sum_{i=1}^{n} h_i \omega_i. \tag{1.78}$$

This form of the temperature equation resembles that for the mass fractions of species i, which becomes with the binary diffusion approximation (1.63)

$$\rho \frac{\partial Y_i}{\partial t} + \rho \boldsymbol{v} \cdot \nabla Y_i = \nabla \cdot (\rho D_i \nabla Y_i) + \omega_i. \tag{1.79}$$

If, in addition, a one-step reaction and equal diffusivities ($D_i = D$) are assumed, coupling relations between the temperature and the species mass fractions can be derived (cf. Williams, 1985a). These assumptions are often used in mathematical analyses of combustion problems.

In the following we will use the term "reactive scalars" for the mass fraction of all chemical species and temperature and introduce the vector

$$\boldsymbol{\psi} = (Y_1, Y_2, \ldots, Y_n, T). \tag{1.80}$$

Here n is the number of reactive species. For simplicity of notation, the balance equation for the reactive scalar ψ_i will be written

$$\rho \frac{\partial \psi_i}{\partial t} + \rho \boldsymbol{v} \cdot \nabla \psi_i = \nabla \cdot (\rho D_i \nabla \psi_i) + \omega_i, \tag{1.81}$$

where $i = 1, 2, \ldots, n+1$. The diffusivity $D_i (i = 1, 2, \ldots, n)$ is the mass diffusivity for the species and the thermal diffusivity D is D_{n+1}. Similarly, ω_{n+1} is defined as ω_T. The chemical source term will also be written as

$$\omega_i = \rho S_i. \tag{1.82}$$

1.6 Chemical Reaction Rates and Multistep Asymptotics

Over the past thirty years, knowledge about gas phase reactions relevant to combustion has continuously improved. Most of the rate data for pure hydrocarbon fuels up to n-heptane and iso-octane and some aromatic fuels such as toluol are sufficiently accurate to predict ignition delay times, laminar burning velocities, and extinction strain rates (cf. the recent review by Lindstedt, 1998). Many data collections are available in the combustion literature (cf., for instance, Baulch et al., 1994). Many authors have selected reactions and their rate data and have combined them into tables called elementary reaction mechanisms. For example, a mechanism for methane oxidation comprising 277 reactions among 49 chemical species, called GRI-Mech 2.11, has been compiled by the Gas Research Institute (cf. Bowman et al., 1999). Kinetic data for C_1–C_3 hydrocarbons have been evaluated by Leung and Lindstedt (1995) for the purpose of calculating laminar diffusion flames. A comprehensive mechanism for hydrocarbon and methanol combustion may also be found in Warnatz et al. (1996).

The rate of reaction k in a mechanism containing r chemical reactions is

$$w_k = k_{fk} \prod_{j=1}^{n} \left(\frac{\rho Y_j}{W_j} \right)^{v'_{jk}} - k_{bk} \prod_{j=1}^{n} \left(\frac{\rho Y_j}{W_j} \right)^{v''_{jk}}. \tag{1.83}$$

Here k_{fk} and k_{bk} are the rate coefficients of the forward and backward reaction, respectively. In general they are temperature dependent and may also depend on pressure. The exponents v'_{jk} and v''_{jk} are the stoichiometric coefficients of reaction k in forward and backward direction, respectively.

The chemical source term ω_i, which is the mass of species i produced per unit volume and unit time, is the sum over all reactions in the mechanism:

$$\omega_i = W_i \sum_{k=1}^{r} v_{ik} w_k, \tag{1.84}$$

where $v_{ik} = v''_{ik} - v'_{ik}$. The sum over all source terms vanishes:

$$\sum_{i=1}^{n} \omega_i = 0. \tag{1.85}$$

Using (1.84) the heat release rate in (1.76) may be transformed to

$$-\sum_{i=1}^{n} h_i \omega_i = -\sum_{k=1}^{r} \sum_{i=1}^{n} v_{ik} W_i h_i w_k = \sum_{k=1}^{r} Q_k w_k, \tag{1.86}$$

where

$$Q_k = -\sum_{i=1}^{n} v_{ik} W_i h_i \tag{1.87}$$

is the heat of reaction of reaction k. It follows that the heat release rate (1.78) may also be expressed as the sum over all chemical reactions,

$$\omega_T = \frac{1}{c_p} \sum_{k=1}^{r} Q_k w_k. \qquad (1.88)$$

If elementary reaction mechanisms are used, the chemical source term ω_i contains the contributions from many fast reactions. This leads to a system of stiff nonlinear equations, a straightforward integration of which in the context of turbulent combustion models is often deemed to be prohibitive. Therefore the need arises to simplify the kinetic mechanism without however, losing the more important part of the chemical information contained in it. Several methods have been developed in the past to reduce the computational costs of chemistry calculations. Since hydrocarbon chemistry proceeds by chain reactions, an evident choice is to introduce the Quasi-Steady-State Assumption (QSSA) for intermediate species, as proposed for methane–air combustion by Peters (1985). This method has extensively been used for laminar flame studies and flamelet calculations (cf. Peters and Rogg, 1993). It has the advantage that the fast reactions depleting the quasi-steady state intermediate species are eliminated, while only the slow rate-determining reactions remain and determine the rates of global reactions. The resulting global reaction mechanism is considerably smaller, but formally similar to an elementary mechanism.

It has been argued that the disadvantage of the method is that a certain insight into the reaction mechanism is required to enable one to choose the steady state species in advance. Such insight can be gained by numerically solving one-dimensional flame problems or even homogeneous reactor problems with the elementary mechanism, thereby identifying the species that remain at very low concentrations. By setting these chemical species into steady state, the fast consumption reactions can be eliminated from the system of governing differential equations, which also reduces the stiffness of the system.

When the Quasi-Steady-State Assumption is employed, the accuracy and the numerical stability can be improved by using less steady state assumptions than could potentially be applied. Rather than the 4-step global mechanism for methane that will be discussed below, a 6-step or even a 10-step mechanism should be used in numerical computations, keeping in the case of the 6-step mechanism the radicals OH and O, and in case of the 10-step mechanism, in addition the species CH_3, C_2H_6, C_2H_4, and C_2H_2 in non–steady state. This has the advantage that fewer of the algebraic steady state relations must be truncated, which sometimes leads to numerical stiffness. For higher hydrocarbons, such as n-heptane, for example, where low temperature ignition chemistry is of interest, the Quasi-Steady-State Assumption leads to 14 global reactions, reduced from a skeletal mechanism of 98 reactions (cf. Pitsch and Peters, 1998b).

A method that reduces the number of independent variables to a minimum while still maintaining a high accuracy is the method of Intrinsic Low-Dimensional Manifolds (ILDM) by Maas and Pope (1992a,b). This method is particularly suited for calculating the chemical kinetics of a large number of Lagrangian particles in the context of models based on the pdf transport equation. As with the method of Computational Singular Perturbations (CSP) by Lam and Goussis (1988), the ILDM method uses the fact that many of the chemical time scales involving intermediates in the reaction chains are fast and thereby not rate determining. By suppressing the fast modes one is able to identify those chemical species that are in steady state and those reactions that are in chemical equilibrium. Then the chemical state of the system depends on a much lower number of variables. These variables represent combinations of species concentrations that change along the trajectory in reactive scalar space. The method uses multidimensional tables to store the chemical source terms needed to calculate changes of composition associated with the Lagrangian particles (cf. (1.128) below). For reasons of storage and accuracy of interpolation, these tables should not exceed two or three dimensions. This limits the ILDM method to chemical systems with relatively few degrees of freedom. Other methods used in the context of the pdf transport equation will be presented below. A review of mathematical tools for the reduction of reaction mechanisms has recently been given by Tomlin et al. (1997).

In the past, investigations focusing on the fluid dynamical aspects of systems with combustion often have used the model of a one-step global reaction between fuel and oxygen, yielding the products P,

$$\nu_F F + \nu_{O_2} O_2 \to \nu_P P, \tag{1.89}$$

with the reaction rate written as

$$w = A \left(\frac{\rho Y_F}{W_F} \right)^{n_F} \left(\frac{\rho Y_{O_2}}{W_{O_2}} \right)^{n_{O_2}} \exp\left(-\frac{E}{RT} \right), \tag{1.90}$$

where A is the preexponential factor, n_F and n_{O_2} are reaction orders, and E is the activation energy. All these quantities are chosen empirically or are simply assigned. The chemical source terms in the temperature and the species equation are

$$\omega_T = \frac{Q}{c_p} w, \quad \omega_i = \nu_i W_i w. \tag{1.91}$$

Asymptotic studies exploiting the limit of a large activation energy are presented, for example, in Williams (1985a). Peters and Williams (1987) have

1.6 Chemical reaction rates and multistep asymptotics

shown that the large activation energy assumption represents the temperature sensitivity of the overall combustion process very well, although it can not be derived from elementary kinetics.

In the subsequent chapters we often will refer to results obtained with the four-step reduced mechanism for methane–air combustion by Peters (1985) and by Peters and Williams (1987). This mechanism was obtained by introducing steady state and partial equilibrium assumptions into the first 17 reactions of the elementary mechanism given in Smooke (1991). Details of the reduction procedure may be found, for instance, in Peters (1988). The four global steps are

$$
\begin{aligned}
&\text{I} && CH_4 + 2H + H_2O = CO + 4H_2, \\
&\text{II} && CO + H_2O = CO_2 + H_2, \\
&\text{III} && H + H + M = H_2 + M, \\
&\text{IV} && O_2 + 3H_2 = 2H + 2H_2O.
\end{aligned} \quad (1.92)
$$

The principle rates that govern these reactions are

$$ w_I = w_{11}, \quad w_{II} = w_9, \quad w_{III} = w_5, \quad w_{IV} = w_1, \quad (1.93) $$

which correspond to the elementary reactions

$$
\begin{aligned}
&11 && CH_4 + H \rightarrow CH_3 + H_2, \\
&9 && CO + OH \rightleftharpoons CO_2 + H, \\
&5 && H + O_2 + M \rightarrow HO_2 + M, \\
&1 && H + O_2 \rightleftharpoons OH + O,
\end{aligned} \quad (1.94)
$$

respectively. The one-sided arrow indicates that only the forward rates of reaction 11 and 5 are used, while for reaction 9 and 1 both the forward and the backward rates are retained. Since OH and O appear in the rates of these reactions, we need to express these radicals in terms of the remaining species in the four-step mechanism. We will use the partial equilibrium assumption of the reactions 2 and 3:

$$
\begin{aligned}
&2 && O + H_2 = OH + H, \\
&3 && OH + H_2 = H + H_2O,
\end{aligned} \quad (1.95)
$$

which read

$$
\begin{aligned}
[O] &= \frac{[H][OH]}{K_2[H_2]}, \\
[OH] &= \frac{[H_2O][H]}{K_3[H_2]}.
\end{aligned} \quad (1.96)
$$

This leads to the following reaction rates of the global steps I–IV:

$$w_\mathrm{I} = k_{11}[\mathrm{CH_4}][\mathrm{H}],$$
$$w_\mathrm{II} = \frac{k_{9f}}{K_3}\frac{[\mathrm{H}]}{[\mathrm{H_2}]}\left\{[\mathrm{CO}][\mathrm{H_2O}] - \frac{1}{K_\mathrm{II}}[\mathrm{CO_2}][\mathrm{H_2}]\right\},$$
$$w_\mathrm{III} = k_5[\mathrm{H}][\mathrm{O_2}][\mathrm{M}], \qquad (1.97)$$
$$w_\mathrm{IV} = k_{1f}\frac{[\mathrm{H}]}{[\mathrm{H_2}]^3}\left\{[\mathrm{O_2}][\mathrm{H_2}]^3 - \frac{1}{K_\mathrm{IV}}[\mathrm{H}]^2[\mathrm{H_2O}]^2\right\}.$$

This mechanism is explicit in terms of the concentrations $[X_i] = \rho Y_i/W_i$ of species appearing in the four-step mechanism. In (1.96) and (1.97) K_2, K_3, K_II, and K_IV are equilibrium constants of the elementary reactions 2 and 3 and the global reactions II and IV, respectively. It is worth noting that kinetic rate data of only four elementary reactions, namely those in (1.94), enter into the global mechanism.

Peters and Williams (1987) go one step further and assume steady state of the radical H. Adding in (1.92) reaction IV to I and III, and canceling the H, leads to the three steps

$$\begin{array}{ll} \mathrm{I'} & \mathrm{CH_4 + O_2 = CO + H_2 + H_2O,} \\ \mathrm{II'} & \mathrm{CO + H_2O = CO_2 + H_2,} \\ \mathrm{III'} & \mathrm{O_2 + 2H_2 = 2H_2O,} \end{array} \qquad (1.98)$$

which are governed by the rates w_I, w_II, and w_III, respectively. A further approximation that will reduce the three-step mechanism effectively to a two-step mechanism is the assumption of partial equilibrium of reaction II. This has been used by Peters and Williams (1987) and in a number of subsequent papers. Later on Bui-Pham et al. (1992) introduced the steady state assumption for $\mathrm{H_2}$, instead. Adding one half of reaction III′ to I′ and II′ and canceling the $\mathrm{H_2}$ reduces (1.98) to a two-step mechanism:

$$\begin{array}{ll} \mathrm{I''} & \mathrm{CH_4 + \frac{3}{2}O_2 = CO + 2H_2O,} \\ \mathrm{II''} & \mathrm{CO + \frac{1}{2}O_2 = CO_2.} \end{array} \qquad (1.99)$$

This shows further improvements in the prediction of burning velocities of lean-to-stoichiometric methane flames. Similar analyses have been performed for many hydrocarbon flames up to n-heptane and iso-octane and also for hydrogen flames (cf. Seshadri et al., 1997, Pitsch et al., 1996a, and Seshadri et al., 1994, respectively). A review of multistep asymptotic analyses, also called rate-ratio asymptotics, has been given by Seshadri (1996).

The schematic representation in Figure 1.6, taken from Peters (1997), shows that the structure of a premixed methane–air flame contains several zones. There is a chemically inert preheat zone of order unity followed by several

1.6 Chemical reaction rates and multistep asymptotics

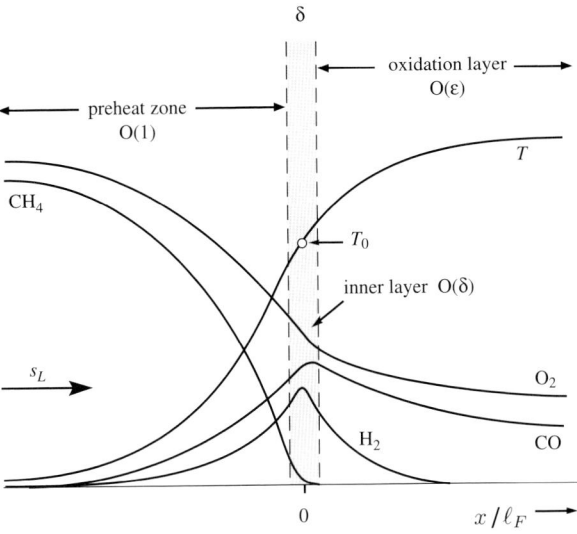

Figure 1.6. Schematic illustration of the structure of a premixed methane–air flame. From Peters (1997). (Reprinted with permission by the author.)

reaction layers; the first one is the fuel consumption layer of order δ where the fuel is consumed and the radicals are depleted by chain-breaking reactions. Another important layer is the oxidation layer of order ε, which is also shown in Figure 1.6. The fuel consumption layer has also been called the inner layer (cf. Seshadri and Peters, 1990) and the temperature in this layer the inner layer temperature, T_0.[†] This layer is responsible for keeping the reaction process alive. The rate-determining reaction occurring in this layer is reaction 11, which is sensitive to the temperature and the presence of H radicals. As this reaction also consumes radicals, as illustrated by the global reaction I, the entire flame structure will be disrupted if the structure of the inner layer is affected by turbulence. This would happen if turbulent eddies penetrate into it and by turbulent mixing, which would enhance heat conduction and diffusion of radicals out of it. Since the smallest turbulent eddy is the Kolmogorov eddy of size η, scale separation between chemistry and turbulence requires that the inner layer be thinner than η.

Let us estimate the thickness of the inner layer. According to $(1.97)_1$ the product $k_{11}[H]$ is the inverse of the chemical time scale t_δ of the fuel consumption reaction. From the rate given for reaction 11 on page 23 in Smooke (1991)

[†] The inner layer temperature T_0 also corresponds to the crossover temperature between chain-branching and chain-breaking reactions (cf. Peters and Williams, 1987 and Peters, 1997).

one obtains for k_{11} at 1,500 K the value of 3.93×10^{12} cm^3/(mol s). Taking for the mole fraction of the H radical the value of 0.0077 from page 55 of Smooke (1991), one obtains at the same temperature and atmospheric pressure the concentration [H] = 5.61×10^{-8} mol/cm^3. This leads to a chemical time scale t_δ of 4.5 μs. Calculating the diffusion coefficient defined as $D = (\lambda/c_p)_0/\rho_u$ by (2.7) in Chapter 2 with the formula for λ/c_p from page 11 of Smooke (1991) at $T_0 = 1,500$ K and $T_u = 300$ K one obtains $D = 7.2 \times 10^{-5}$m^2/s. The diffusive length scale corresponding to the thickness of the fuel consumption layer then is $\ell_\delta = (Dt_\delta)^{1/2} = 0.018$ mm. This length scale decreases with pressure and with temperature.

Another important length scale in laminar premixed flames is the flame thickness ℓ_F, which will be defined by (2.5) in Chapter 2. It has been used to nondimensionalize the x coordinate in Figure 1.6. For a stoichiometric methane flame at 1 atm it is estimated by Peters (1991) as $\ell_F = 0.175$ mm. The thickness of the inner layer is a fraction δ of the flame thickness,

$$\ell_\delta = \delta \ell_F, \tag{1.100}$$

which leads to a value of $\delta = 0.1$ for the example considered above. A diagram in Peters (1991) shows that δ varies from values of approximately $\delta = 0.1$ at atmospheric pressure to $\delta = 0.03$ at pressures around 30 atm.

In Chapter 2, dealing with premixed combustion, we will compare the thickness ℓ_δ to the Kolmogorov scale η to evaluate the validity of the assumption of *scale separation*, on which most of the combustion models are based. In a regime called the thin reaction zones regime, we will identify the inner layer with the thin reaction zone. We purposely do not use the thickness

$$\ell_\varepsilon = \varepsilon \ell_F \tag{1.101}$$

of the oxidation layer, which is typically three times larger than ℓ_δ (cf. Peters, 1991). Perturbations of that layer will affect the oxidation of CO to CO_2, but they will not have major consequences for the inner layer, because the feedback from the downstream oxidation layer on the inner layer is rather weak (cf. Peters, 1997). Kolmogorov eddies that have crossed the inner layer will grow because of the increase of the kinematic viscosity in (1.51) as the temperature increases. Furthermore, two-dimensional (2D) simulations by Hilka et al. (1996), using detailed chemistry for premixed methane flames, show that the vorticity is strongly attenuated as it passes through the fuel consumption layer. The same conclusion has been drawn from experimental investigations by Mueller et al. (1998).

The situation is different in nonpremixed combustion, where turbulent eddies can penetrate from either side into the reaction zone. We will therefore base the

1.7 Moment Methods for Reactive Scalars

estimate of the validity of scale separation in Chapter 3 on a comparison of the oxidation layer thickness ℓ_ε with the Kolmogorov scale η.

1.7 Moment Methods for Reactive Scalars

Favre averaged equations for the mean and the variance of the reactive scalars can be derived by splitting $\psi_i(\boldsymbol{x}, t)$ into a Favre mean and a fluctuation:

$$\psi_i(\boldsymbol{x}, t) = \tilde{\psi}_i(\boldsymbol{x}, t) + \psi_i''(\boldsymbol{x}, t). \tag{1.102}$$

When this is introduced into (1.81) one obtains after averaging

$$\bar{\rho}\frac{\partial \tilde{\psi}_i}{\partial t} + \bar{\rho}\tilde{\boldsymbol{v}} \cdot \nabla \tilde{\psi}_i = \nabla \cdot \overline{(\rho D_i \nabla \psi_i)} - \nabla \cdot (\bar{\rho}\widetilde{\boldsymbol{v}'' \psi_i''}) + \bar{\rho}\tilde{S}_i. \tag{1.103}$$

In this equation the terms on the l.h.s. are closed, while those on the r.h.s. must be modeled. In high Reynolds number flows the molecular transport term containing the molecular diffusivity D_i is small and can be neglected. Closure is required for the second term on the r.h.s., the turbulent transport term, and for the last term, the mean chemical source term.

The modeling of the mean chemical source term has often been considered to be the main problem of moment methods in turbulent combustion. To discuss the difficulties associated with the closure of this term, we assume that coupling relations exist between the chemical species and the temperature. As noted before, such coupling relations can easily be derived for the case of a one-step reaction and equal diffusivities. With this assumption we consider the following form of the heat release rate:

$$\omega_T(T) = \rho\, S_T(T) = \rho B (T_b - T) \exp\left(-\frac{E}{RT}\right). \tag{1.104}$$

Here B contains the frequency factor and the heat of reaction, T_b is the adiabatic flame temperature, E the activation energy, and R the universal gas constant. Introducing $T = \tilde{T} + T''$ into (1.104) we can expand the argument of the exponential term around $\tilde{T}$ for small T'' as

$$\frac{E}{RT} = \frac{E}{R\tilde{T}} - \frac{ET''}{R\tilde{T}^2}. \tag{1.105}$$

If the expansion is also introduced into the preexponential term, the quantity S_T becomes

$$S_T(T) = S_T(\tilde{T}) \left(1 - \frac{T''}{T_b - \tilde{T}}\right) \exp\left(\frac{ET''}{R\tilde{T}^2}\right). \tag{1.106}$$

Typically, the grouping $E/R\tilde{T}$ is of the order of 10 in the reaction zone of a flame and the absolute value of $T''/\tilde{T}$ varies between 0.1 and 0.3. Therefore the exponential term in (1.106) causes enhanced fluctuations of the chemical source term around its mean value evaluated with the mean temperature $\tilde{T}$. If one expands the exponential term into a series for small values of the argument and takes averages, one needs to consider a large number of higher moments before the exponential term will converge.

1.8 Dissipation and Scalar Transport of Nonreacting and Linearly Reacting Scalars

Another unresolved problem of moment methods is the closure of the turbulent transport term $\widetilde{v''\psi_i''}$ in (1.103). It is general practice in turbulent combustion to employ the gradient transport assumption not only for nonreacting but also for reactive scalars. The scalar flux then takes the form

$$\widetilde{v''\psi_i''} = -D_t \nabla \tilde{\psi}_i. \qquad (1.107)$$

Here D_t is a turbulent diffusivity, which is modeled by analogy to the eddy viscosity as

$$D_t = \frac{\nu_t}{Sc_t}, \qquad (1.108)$$

where Sc_t is a turbulent Schmidt number. We want to show that the gradient transport assumption is unacceptable for reactive scalars.

To this end we first must derive an equation for the variance $\widetilde{\psi_i''^2}$. By subtracting (1.103) from (1.81) after both have been divided by ρ and $\bar{\rho}$, respectively, an equation for the fluctuation ψ_i'' is obtained:

$$\frac{\partial \psi_i''}{\partial t} + (\tilde{v} + v'') \cdot \nabla \psi_i'' + v'' \cdot \nabla \tilde{\psi}_i$$
$$= \frac{1}{\rho} \nabla \cdot (\rho D_i \nabla \psi_i) - \frac{1}{\bar{\rho}} \nabla \cdot \overline{\left(\rho D_i \nabla \psi_i \right)} + \nabla \cdot (\widetilde{\bar{\rho} v'' \psi_i''}) + S_i''. \qquad (1.109)$$

Here the continuity equation was used and $S_i'' = S_i - \tilde{S}_i$ describes fluctuation of the chemical source term. If derivatives of ρ and D_i and their mean values are neglected for simplicity, the first two terms on the r.h.s. of (1.109) can be combined to obtain a term proportional to $D_i \nabla^2 \psi_i''$. Introducing this and multiplying (1.109) by $2\rho \psi_i''$ one obtains an equation for $\widetilde{\psi_i''^2}$. With the use of

1.8 Dissipation and scalar transport of nonreacting and linearly

the continuity equation and averaging one obtains

$$\bar{\rho}\frac{\partial \widetilde{\psi_i''^2}}{\partial t} + \bar{\rho}\tilde{v} \cdot \nabla \widetilde{\psi_i''^2} = -\nabla \cdot \left(\overline{\rho v'' \psi_i''^2}\right)$$
$$+ 2\bar{\rho}(\widetilde{-v''\psi_i''}) \cdot \nabla \tilde{\psi}_i - \bar{\rho}\tilde{\chi}_i + 2\overline{\rho\psi_i'' S_i''}. \qquad (1.110)$$

As before, the terms on the l.h.s. describe the local rate of change and convection. The first term on the r.h.s. is the turbulent transport term. The second term on the r.h.s. accounts for the production of scalar fluctuations. The mean molecular transport term has been neglected for simplicity but the molecular diffusivity still appears in the dissipation term. The Favre scalar dissipation rate is defined as

$$\tilde{\chi}_i = 2 D_i \widetilde{(\nabla \psi_i'')^2}. \qquad (1.111)$$

Finally, the last term in (1.110) contains the covariance of the reactive scalar with the chemical source term.

A integral scalar time scale can be defined by

$$\tau_i = \frac{\widetilde{\psi_i''^2}}{\tilde{\chi}_i}. \qquad (1.112)$$

In the nonreacting case it is often set proportional to the flow time $\tau = \tilde{k}/\tilde{\varepsilon}$,

$$\tau = c_\chi \tau_i, \qquad (1.113)$$

where the constant of proportionality c_χ is of order unity but its value depends on the length of the inertial range and additional assumptions about the spectrum. A value $c_\chi = 2.0$ is often used (cf. Section 3.7). Combining (1.112) and (1.113) leads to the model

$$\tilde{\chi}_i = c_\chi \frac{\tilde{\varepsilon}}{\tilde{k}} \widetilde{\psi_i''^2}. \qquad (1.114)$$

Now, using the assumption "production = dissipation" in (1.110) we are able to justify the gradient flux approximation (1.107) for nonreacting scalars. Setting

$$2(\widetilde{-v''\psi_i''}) \cdot \nabla \tilde{\psi}_i = c_\chi \frac{\tilde{\varepsilon}}{\tilde{k}} \widetilde{\psi_i''^2} \qquad (1.115)$$

and multiplying both sides by the turbulent diffusivity D_t, which is proportional to $\tilde{k}^2/\tilde{\varepsilon}$, one obtains

$$D_t(\widetilde{-v''\psi_i''}) \cdot \nabla \tilde{\psi}_i \sim c_\chi \tilde{k} \widetilde{\psi_i''^2}. \qquad (1.116)$$

The r.h.s. of (1.116) has the same order of magnitude and dimension as $(\widetilde{-v''\psi_i''})^2$. Assuming isotropy and thereby a proportionality between these two quantities one obtains the gradient transport assumption written as

$$(-\widetilde{v''\psi_i''}) \sim c_\chi^{-1} D_t \nabla \tilde{\psi}_i. \tag{1.117}$$

This model is valid for nonreactive scalars only.

We now want to analyze the influence of chemistry on the scalar time scale and the scalar flux. For illustration purpose and mathematical convenience the following linear reaction rate is considered:

$$\omega_i = -\rho B \psi_i, \tag{1.118}$$

where B is the frequency factor, which is inversely proportional to a chemical time. By considering linear chemistry and the case of isotropic turbulence with a stationary spectrum, Corrsin (1961) has identified a critical wavenumber k_c equal to $(B^3/\varepsilon)^{1/2}$, at which the chemical time is equal to the turnover time of an eddy of size k_c^{-1} within the inertial range. Larger eddies have a larger turnover time and therefore chemistry will be able to reduce scalar fluctuations at large length scales. Smaller eddies, having a smaller turnover time, will not be affected by chemistry. The decrease of scalar fluctuations at the large scales influences the entire spectrum, also reducing the scalar fluctuations at the small scales. Since the latter determine the scalar dissipation rate χ_i, the scalar time scale τ_i increases, as compared to the nonreacting case. Therefore chemistry interacts with turbulence, as long as k_c falls into the inertial range, and the assumption of scale separation is no longer satisfied. As a consequence, the flow to scalar time ratio τ/τ_i depends on chemistry. For the linear reaction rate, Corrsin (1961) has carried the analysis to the point where τ/τ_i can be evaluated as a function of $B\tau$ (which may be interpreted as a Damköhler number). Equations (28) and (57) of his paper can be combined to obtain Figure 1.7. The large Reynolds number limit has been employed and the small wavenumber cut-off was chosen to obtain $\tau/\tau_i = 2.0$ in the limit $B\tau \to 0$. For increasing values of $B\tau$ the flow to scalar time ratio decreases rapidly.

If for a linearly reacting scalar the balance of production, dissipation, and reaction in the variance Equation (1.110) is used, one obtains by a similar procedure as before instead of (1.116)

$$D_t(-\widetilde{v''\psi_i''}) \cdot \nabla \tilde{\psi}_i \sim (\tau/\tau_i + 2B\tau) \tilde{k} \widetilde{\psi_i''^2}. \tag{1.119}$$

Using the same arguments as those that led to (1.117) we can derive the following modified gradient transport approximation:

$$(-\widetilde{v''\psi_i''}) \sim \frac{D_t}{\tau/\tau_i + 2B\tau} \nabla \tilde{\psi}_i, \tag{1.120}$$

where D_t corresponds to the turbulent diffusivity used in the nonreacting case. Only if $B\tau$ is small and τ/τ_i is close to the value for the nonreacting case

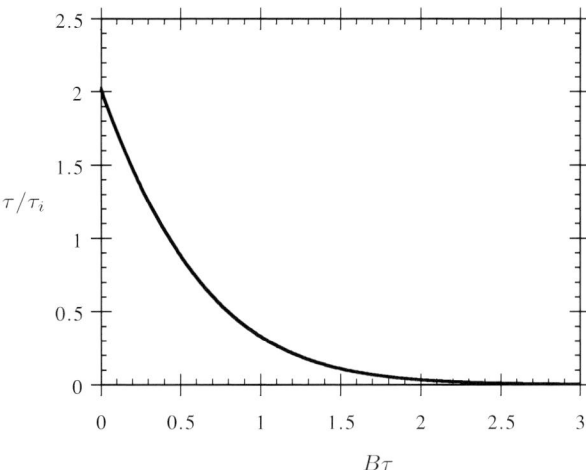

Figure 1.7. The ratio of the flow time to the scalar time for linear chemistry, evaluated from Corrsin (1961).

does the turbulent diffusivity remain of the same order of magnitude. For large values of B the scalar flux decreases to zero, owing to the second term in the dominator on the r.h.s. of (1.120). Physically this behavior results from the change of scalar fluctuations brought about by chemical reactions in the energy containing eddies, with a corresponding change of $\widetilde{v''\psi_i''}$.

Whether the simple analysis presented here remains valid for nonlinear chemistry must be questioned. Nevertheless, for linearly reacting scalars, the turbulent diffusivity is expected to be different and most likely to be smaller than in the nonreacting case. This has also been found by Elperin et al. (1998), who discussed turbulent transport in the framework of irreversible thermodynamics. For premixed turbulent flames, where chemistry is concentrated in thin layers, recent experimental data by Gagnepain et al. (1998) show that $\tau/\tau_i = 2.0$ in the unburnt gas, but it increases in the burnt gas. These results differ from Corrsin's prediction for the linear reaction rate. Further investigations on the interaction between chemistry and turbulence in the inertial range are certainly required.

1.9 The Eddy-Break-Up and the Eddy Dissipation Models

An early attempt to provide a closure for the chemical source term is due to Spalding (1971) who argued that since turbulent mixing may be viewed as a cascade process from the integral down to the molecular scales, the cascade process also controls the chemical reactions as long as mixing rather than chemistry is the rate-determining process. This model was called the Eddy-Break-Up Model (EBU). The turbulent mean reaction rate of products was expressed as

$$\overline{\omega_p} = \rho C_{\text{EBU}} \frac{\varepsilon}{k} \left(\overline{Y_p''^2}\right)^{1/2}, \qquad (1.121)$$

where $\overline{Y_p''^2}$ is the variance of the product mass fraction and C_{EBU} is the Eddy-Break-Up constant.

This model has been modified by Magnussen and Hjertager (1977) who replaced $(\overline{Y_p''^2})^{1/2}$ simply by the mean mass fraction of the deficient species (fuel for lean or oxygen for rich mixtures) calling it the Eddy Dissipation Model (EDM). The model takes the minimum of three rates, those defined with the mean fuel mass fraction

$$\overline{\omega_F} = \bar{\rho} A \overline{Y_F} \frac{\varepsilon}{k}, \qquad (1.122)$$

with the mean oxidizer mass fraction

$$\overline{\omega_{O_2}} = \bar{\rho} \frac{A \overline{Y_{O_2}}}{\nu} \frac{\varepsilon}{k}, \qquad (1.123)$$

and with the product mass fraction

$$\overline{\omega_P} = \bar{\rho} \frac{A \cdot B}{(1+\nu)} \overline{Y_P} \frac{\varepsilon}{k}, \qquad (1.124)$$

in order to calculate the mean chemical source term. In (1.122)–(1.124) A and B are modeling constants and ν is the stoichiometric oxygen to fuel mass ratio to be defined in (3.5) in Chapter 3.

The Eddy-Break-Up Model and its modifications are based on intuitive arguments. The main idea is to replace the chemical time scale of an assumed one-step reaction by the turbulent time scale $\tau = k/\varepsilon$. Thereby the model eliminates the influence of chemical kinetics, representing the fast chemistry limit only. When these models are used in computational fluid dynamics (CFD) calculations, it turns out that the constants C_{EBU} or A and B must be "tuned" within a wide range to obtain reasonable results for a particular problem.

The Eddy-Break-Up Model was formulated primarily for premixed combustion. For nonpremixed combustion, we will show in Chapter 3 that by a linear combination of fuel and oxidizer mass fractions, the mixture fraction can be introduced to eliminate the chemical reaction rate. To describe nonpremixed combustion, one needs to know the probability density function (pdf) of the mixture fraction Z at a point x and time t. The concept of a pdf for nonpremixed combustion was already introduced by Hawthorne et al. (1949). Later on, an expression for the mean turbulent reaction rate for nonpremixed combustion was derived in a rigorous way by Bilger (1976) who showed that in the fast chemistry limit the mean fuel consumption rate may be expressed as

$$\overline{\omega_F} = -\frac{1}{2}\bar{\rho}\frac{Y_{F,1}}{1-Z_{st}}\tilde{\chi}_{st}\tilde{P}(Z_{st}). \tag{1.125}$$

Here $Y_{F,1}$ is the mass fraction of fuel in the fuel stream, and $\tilde{\chi}_{st}$ is the scalar dissipation rate and $\tilde{P}(Z_{st})$ is the probability density function of the mixture fraction, both conditioned at the stoichiometric value $Z = Z_{st}$. We will define these quantities in Chapter 3.

1.10 The Pdf Transport Equation Model

Similar to moment methods, models based on a pdf transport equation for the velocity and the reactive scalars are usually formulated for one-point statistics. Within that framework, however, they represent a very general statistical description of turbulent reacting flows, applicable to premixed, nonpremixed, and partially premixed combustion. A joint pdf transport equation for the velocity and the reactive scalars can be derived, which is equivalent to an infinite hierarchy of one-point moment equations for these quantities. Inclusion of the gradients of the velocity and the reactive scalars would provide information at neighboring points and thereby would solve some of the problems associated with the modeling of viscous and scalar dissipation to be discussed below. Pope (1990) proposes the use of a transport equation for the joint pdf of velocity, viscous dissipation, and reactive scalars. This equation does not include scalar gradients and therefore contains no information about the mixing time scale $\tau_i \sim \widetilde{\psi_i''^2}/\tilde{\chi}_i$. Dopazo (1994) therefore advocates the use of a transport equation for the joint statistics of velocity, velocity gradient, reactive scalars, and their gradients. In this equation in addition to convection and chemical reaction being closed, so also is the term that describes the straining and rotation of scalar gradients, a mechanism that is believed to be essential in turbulent reacting flows. The closure problem is, however, shifted to the mixing of scalar gradients. This formulation has not yet been applied to flows with combustion.

For simplicity, we will consider here the transport equation for the joint pdf of velocity and reactive scalars only. If we denote the set of reactive scalars such as the temperature and the mass fraction of reacting species by the vector ψ, then $P(v, \psi; x, t)dvd\psi$ is the probability of finding at point x and time t the velocity components and the reactive scalars within the interval $v - dv/2 < v < v + dv/2$ and $\psi - d\psi/2 < \psi < \psi + d\psi/2$, respectively.

There are several ways to derive a transport equation for the probability density $P(v, \psi; x, t)$ (cf., for instance, Lundgren, 1967 and O'Brien, 1980).

We refer here to the presentation in Pope (1985, cf. also Pope 2000), but we write the convective terms in conservative form:

$$\frac{\partial(\rho P)}{\partial t} + \nabla \cdot (\rho v P) + (\rho \mathbf{g} - \nabla \bar{p}) \cdot \nabla_v P + \sum_{i=1}^{n} \frac{\partial}{\partial \psi_i}[\omega_i P]$$

$$= \nabla_v \cdot [\langle -\nabla \cdot \boldsymbol{\tau} + \nabla p' \,|\, v, \psi\rangle P] - \sum_{i=1}^{n} \frac{\partial}{\partial \psi_i}[\langle \nabla \cdot (\rho D \nabla \psi_i)\rangle \,|\, v, \psi\rangle P]. \tag{1.126}$$

In deriving this equation, the equations for all reactive scalars, including that for temperature, have been cast into the form (1.81), for simplicity. The symbol ∇_v denotes the divergence operator with respect to the three components of velocity. The angular brackets denote conditional averages, conditioned with respect to fixed values of v and ψ. For simplicity of presentation we do not use different symbols for the random variables describing the stochastic fields and the corresponding sample space variables, which are the independent variables in the pdf equation.

The first two terms on the l.h.s. of (1.126) are the local rate of change and convection of the probability density function in physical space. The third term represents transport in velocity space by gravity and the mean pressure gradient. The last term on the l.h.s. contains the chemical source terms. All these terms are in closed form, since they are local in physical space. Note that the mean pressure gradient does not present a closure problem, since the pressure is calculated independently of the pdf equation using the mean velocity field. For chemical reacting flows it is of particular interest that the chemical source terms can be treated exactly for arbitrarily complex chemical kinetics. It has often been argued that in this respect the transported pdf formulation has a considerable advantage compared to other formulations.

However, on the r.h.s. of the transport equation there are two terms that contain gradients of quantities conditioned on the values of velocity and composition. Therefore, if gradients are not included as sample space variables in the pdf equation, these terms occur in unclosed form and have to be modeled. The first unclosed term on the r.h.s. describes transport of the probability density function in velocity space induced by the viscous stresses and the fluctuating pressure gradient. The second term represents transport in reactive scalar space by molecular fluxes. This term represents molecular mixing.

The predictive capability of pdf methods for turbulent combustion depends on the quality of the models that can be constructed for the unclosed terms. For slow chemistry Hůlek and Lindstedt (1998) have obtained very good agreement with experimental data using the joint scalar-velocity formulation. However, when chemistry is fast, mixing and reaction take place in thin layers where molecular transport and the chemical source term balance each other. Therefore, the closed chemical source term and the unclosed molecular mixing term, being leading order terms in a asymptotic description of the flame structure, are closely linked to each other. Pope and Anand (1984) have illustrated this for the case of premixed turbulent combustion by comparing a standard pdf closure for the molecular mixing term with a formulation where the molecular diffusion term was combined with the chemical source term to define a modified reaction rate. They call the former distributed combustion and the latter flamelet combustion and find considerable differences in the Damköhler number dependence of the turbulent burning velocity normalized with the turbulent intensity.

From a numerical point of view, the most apparent property of the pdf transport equation is its high dimensionality. Finite-volume and finite-difference techniques are not very attractive for this type of problem, as memory requirements increase roughly exponentially with dimensionality. Therefore, virtually all numerical implementations of pdf methods for turbulent reactive flows employ Monte-Carlo simulation techniques (cf. Pope, 1981, 1985). The advantage of Monte-Carlo methods is that their memory requirements depend only linearly on the dimensionality of the problem. Monte-Carlo methods employ a large number, N, of particles. In the Lagrangian algorithm (Pope, 1985) the particles are not bound to grid nodes. Instead, each particle has its own position and moves through the computational domain with its own instantaneous velocity. The particles should be considered as different realizations of the turbulent reactive flow problem under investigation. The state of the particle is described by its position and velocity, and by the values of the reactive scalar that it represents as a function of time. These particles should not be confused with real fluid elements, which behave similarly in a number of respects.

In the following we will present the solution strategy based on the concepts presented by Pope (1990) and Correa and Pope (1992). The main aspect of the solution algorithm is the method by which Lagrangian particles are tracked in the physical domain. The latter is subdivided by a computational grid, on which mean quantities are calculated. The simplest method to estimate local means is to compute cell averages. In addition, mean quantities such as $\tilde{k}$ and $\tilde{\varepsilon}$, which are needed in the Lagrangian equations, are calculated on the grid using a finite-volume method. Therefore the method used by Correa and Pope (1992) is a hybrid numerical method consisting of a Lagrangian Monte-Carlo method and an Eulerian finite-volume method.

The main drawback of Monte-Carlo methods is that they suffer from a statistical error that decreases only slowly with the number of particles N_{pc} per cell: The error is proportional to $1/\sqrt{N_{pc}}$. For an acceptable numerical accuracy far more than a hundred particles must be present in each cell. For industrial CFD problems, which require large numerical grids consisting of typically several hundred thousand grid cells, this leads to very large numbers of particles unless special conditions, such as stationarity of the mean flow, can be exploited. Recently Xu and Pope (1998) have quantified the different errors that occur in Monte-Carlo methods for turbulent reactive flows.

The Lagrangian motion of a particle j is given by the equation

$$\frac{dx^{(j)}}{dt} = v^{(j)}(x), \tag{1.127}$$

where the velocity vector $v^{(j)}$ describing the motion of the particle is a stochastic quantity. According to the theory of stochastic differential equations the evolution equation of the spatial distribution of particles is represented by the first two terms on the l.h.s. of (1.126). Similarly, the change of the value ψ_i of the reactive scalars of the Lagrangian particle j, given by

$$\rho^{(j)} \frac{d\psi_i^{(j)}}{dt} = \omega_i^{(j)}, \tag{1.128}$$

is represented by the local term and the last term on the l.h.s. of (1.126). This illustrates how the solution of only time-dependent Lagrangian equations for particles simulates the solution of the transported pdf equation.

In the Lagrangian simulation, typically the method of fractional steps (cf. Pope, 1990) is used. This method is based on the observation that the various terms describing the time evolution of the pdf in (1.126) are additive. Therefore the processes in physical, velocity, and reactive scalar space may be treated sequentially rather than simultaneously. The method of fractional steps proceeds as follows: For a sufficiently small time step Δt, the motion of particle j may be approximated by

$$x^{(j)}(t + \Delta t) = x^{(j)}(t) + v^{(j)} \Delta t \tag{1.129}$$

and the change in composition by

$$\psi_i^{(j)}(t + \Delta t) = \psi_i^{(j)}(t) + \frac{\omega_i^{(j)}}{\rho^{(j)}} \Delta t. \tag{1.130}$$

The method of fractional steps for Lagrangian particles therefore requires that the remaining closed term on the l.h.s. in (1.126) is represented by increments of velocity during a time step Δt. The third term on the l.h.s. yields

$$v^{(j)}(t + \Delta t) = v^{(j)}(t) + \left(g - \frac{1}{\rho} \nabla \bar{p}\right)^{(j)} \Delta t. \tag{1.131}$$

Now we will consider the unclosed terms on the r.h.s. of (1.126). Pope (1994a) has proposed a stochastic Lagrangian equation to model the velocity increment due to the combined effect of viscous stresses and the fluctuating pressure gradient, the first term on the r.h.s. of (1.126). A simplified Langevin model leads to the velocity change

$$v^{(j)}(t + \Delta t) = v^{(j)}(t) - \left(\frac{1}{2} + \frac{3}{4}C_0\right)\frac{\tilde{\varepsilon}}{\tilde{k}}(v^{(j)} - \bar{v})\Delta t + \sqrt{C_0\tilde{\varepsilon}}\,W^{(j)}. \tag{1.132}$$

Here the first term of the velocity increment is a drift term, describing a relaxation toward the local mean velocity $\bar{v}$ with characteristic time scale $\tilde{k}/\tilde{\varepsilon}$. The constant $C_0 = 2.1$ was determined from diffusion measurements in grid turbulence. The last term represents a stochastic diffusion term where the vector W represents an isotropic Wiener process. Pope (1994b) describes the relationship between Langevin models and second-order closure models for turbulent flows.

For chemically reacting flows the last term in (1.126), the molecular mixing term, is the most difficult to model. There are several models proposed in the literature. Following the concise presentation in Nooren (1998) these models will be discussed next.

Interaction by Exchange with the Mean (IEM)

This simplest mixing model was first proposed by Villermaux and Devillon (1972). It is also known as the Linear Mean Square Estimation Model (LMSE) (cf. Dopazo, 1975, O'Brien, 1980, and Borghi, 1988). In this model, the scalar values of individual particles are subject to relaxation toward the mean value:

$$\frac{d\psi_i^{(j)}}{dt} = -\frac{\psi_i^{(j)} - \tilde{\psi}_i^{(j)}}{\tau_i}. \tag{1.133}$$

By construction, the IEM model leaves the mean unchanged and produces the correct decay of the variance for nonreacting scalars. For inert scalar mixing simulations using $\tau_i = \tau/2.0$, Wouters et al. (1998) show that the model is reasonably successful at the prediction of the pdf shapes in an inhomogeneous turbulent jet. For reactive scalars (1.133) would require that τ_i is known as a function of chemistry as in the example in Section 1.8. Borghi and Gonzalez (1986) discuss these matters and propose that τ_i should follow from a turbulent time scale distribution that could be parameterized by the integral time τ as well as the Kolmogorov time t_η. The latter would, however, introduce the viscosity into the combustion model and thereby violate the requirement of Reynolds number independence. In most calculations, the influence of chemistry on τ_i is ignored and τ_i is set proportional to the integral time scale τ. This implicitly introduces scale separation of chemistry and turbulence in the inertial range and thereby satisfies the postulate formulated at the end of Section 1.1. Because of its simplicity of implementation the IEM model is widely used in combustion simulations.

Coalescence–Dispersion (C–D)

In coalescence–dispersion (C–D) models, also known as particle-interaction models, mixing takes place in particle pairs. In the basic version of the model, which is due to Curl (1963), two particles p and q, selected randomly from the ensemble of N_{pc} particles in the cell, mix with a certain probability p_{mix} during a time step Δt. After mixing, the particles take new scalar values, equal to the mean of the two values before mixing. If the scalar values of particle j before and after mixing are denoted by $\psi_i^{(j)}$ and $\psi_i^{(j)}(t + \Delta t)$, respectively, the model reads for two particles p and q

$$\psi_i^{(p)}(t + \Delta t) = \psi_i^{(q)}(t + \Delta t) = \frac{1}{2}\left[\psi_i^{(p)}(t) + \psi_i^{(q)}(t)\right]. \tag{1.134}$$

For $\Delta t \to 0$, the desired variance decay is obtained if p_{mix} is chosen as

$$p_{\text{mix}} = C_D \frac{\tilde{\varepsilon}}{\tilde{k}} N_{pc} \Delta t. \tag{1.135}$$

A drawback of this version of the model is that the new scalar values can only take values from a limited set. This effect can be illustrated for the case of binary mixing in homogeneous turbulence. The unmixed state at time $t = 0$ is represented by a two δ-function pdf. At later times the model produces pdfs consisting of series of δ-functions. Longer mixing times merely increase the number of δ-functions in the pdf and do not remove its discrete character. This unphysical behavior is remedied in the modified Curl model by Janicka et al. (1979), in which the degree of mixing is a uniformly distributed random variable. In order to obtain the correct variance decay, the probability of mixing p_{mix} has to be increased by a factor $3/2$ compared to that in Equation (1.135). As in the IEM model the turbulent time $\tau = \tilde{k}/\tilde{\varepsilon}$ is used instead of τ_i as the mixing time scale. Since τ is independent of chemical reactions the model also assumes scale separation of chemistry and turbulence.

In homogeneous turbulence, the modified C–D model gives a relaxation of an arbitrary initial pdf to a bell-shaped distribution. In this respect, the model is superior to the IEM model. However, the normalized higher moments produced by the model do not approach the values for a Gaussian distribution but become infinite instead (cf. Pope, 1982). The deviations from the Gaussian shape are

1.10 The pdf transport equation model

found primarily in the tails of the distribution. Just as the IEM model, C–D models are nonlocal in composition space.

Mapping Closure

The two mixing models introduced above have at least two shortcomings: They cannot predict the relaxation of an arbitrary initial pdf to a Gaussian pdf in homogeneous turbulence and they are nonlocal in reactive scalar space. Both of these shortcomings can, at least partially, be attributed to the fact that both models lack a sound physical basis. Mapping closures, introduced by Chen et al. (1989), have a sounder basis and overcome these shortcomings. In mapping closure models, the scalar field is mapped to a Gaussian reference field. This was extensively studied at first by Gao (1991a,b). For a detailed description of the model and its implementation, the reader is refered to Pope (1991) and to Dopazo (1994). The mapping closure is local in reactive scalar space and gives excellent agreement with data from direct numerical simulations for the case of a single decaying nonreacting scalar in homogeneous turbulence. An extension to multiscalar mixing was formulated by Girimaji (1993).

Euclidean Minimum Spanning Trees (EMST)

Nonlocalness in reactive scalar space has recently been remedied by a model based on Euclidean Minimum Spanning Trees (EMST) by Subramaniam and Pope (1998). In nonpremixed combustion with fast chemistry, for instance, mixing based on the IEM or the C–D model can result in the transition of cold fuel and oxidizer particles across the reaction zone without these being subjected to the fast reaction rates occurring in that zone. This was demonstrated by Norris and Pope (1991) and is also discussed by Jones and Kakhi (1996). It violates the physical principle that mixing of adjacent material in physical space is equivalent to mixing of neighboring particles in reactive scalar space.

The basic idea behind the EMST model is that mixing of scalar particles should be governed by its close neighborhood in reactive scalar space. This model can be viewed as a multiscalar mixing model in which the interaction between adjacent particles resembles that of the mapping closure.

Chemistry Models

It was pointed out that Lagrangian particles are used to solve the pdf transport equation. In reacting flow calculations each particle carries the information about changes in chemical composition and temperature (or enthalpy) along its trajectory. These changes are due to chemistry and mixing and are calculated by solving a system of zero-dimensional time-dependent equations. If, for example, the IEM model is used to model the mixing process, these equations are of the type

$$\rho^{(j)} \frac{d\psi_i^{(j)}}{dt} = -\rho^{(j)} \frac{(\psi_i^{(j)} - \tilde{\psi}_i)}{\tau_i} + \omega_i^{(j)}. \tag{1.136}$$

Given the very large number of particles to be used, there is a need to reduce the computational costs for the calculation of the chemistry. In the ILDM method by Maas and Pope (1992a,b), for instance, it is common practice to determine and to tabulate the reactive scalar increments due to chemical reactions beforehand as a function of the degrees of freedom. A look-up table is generated from which this information can be retrieved using multilinear interpolation. This method is feasible, if the number of degrees of freedom is two or three, but since storage and retrieval time grows exponentially with the dimension of the table, this method can only be applied to small reaction systems.

To overcome these limitations, an algorithm called In Situ Adaptive Tabulation (ISAT) has recently been developed by Yang and Pope (1998) and Pope (1997). Here the tabulation is done in situ during the combustion calculation with the consequence that only those regions in reactive scalar space will be tabulated which have been accessed during the successive calculations. Test calculations for a detailed methane–air mechanism with 40 reactions between 16 species show a speed-up factor of three orders of magnitude compared to the direct numerical integration of the species equations.

The idea behind the ISAT algorithm is that in combustion calculations only a very small fraction of the multidimensional reactive scalar space is really accessed. This is reminiscent of the discussion in Section 1.1 with respect to the S-shaped curve. Pope (1997) provides an illustration by considering a steady, laminar two-dimensional flame involving 50 chemical species. The realizable region in reactive scalar space is a 50-dimensional manifold. But the accessed region is just a two-dimensional manifold obtained from mapping the solution in two-dimensional physical space into reactive scalar space. This points to the properties of laminar flamelet models, which will be presented in the following.

1.11 The Laminar Flamelet Concept

The view of a turbulent diffusion flame as consisting of an ensemble of stretched laminar flamelets is due to Williams (1975). Later, Liew et al. (1981) proposed using profiles taken from laminar diffusion flames to calculate means and variances in turbulent flames. Flamelet equations based on the mixture fraction as

1.11 The laminar flamelet concept

independent variable, using the scalar dissipation rate for the mixing process, were independently derived by Peters (1980) and Kuznetsov (1982). A first review of diffusion flamelet models was given by Peters (1984). For premixed and diffusion flames the flamelet concept was reviewed by Peters (1986) and Bray and Peters (1994).

Flamelets are thin reactive-diffusive layers embedded within an otherwise nonreacting turbulent flow field. Once ignition has taken place, chemistry accelerates as the temperature increases. When the temperature reaches values that are in the vicinity of the close-to-equilibrium branch in Figure 1.1, the reactions that determine fuel consumption become very fast. For methane combustion, for example, the time scale of the rate-determining reaction in the fuel consumption layer was estimated at the end of Section 1.6. Since the chemical time scale of this reaction is short, chemistry is most active within a thin layer, namely the fuel consumption or inner layer. If this layer is thin compared to the size of a Kolmogorov eddy, it is embedded within the quasi-laminar flow field of such an eddy and the assumption of a laminar flamelet structure is justified. If, on the contrary, turbulence is so intense that Kolmogorov eddies become smaller than the inner layer and can penetrate into it, they are able to destroy its structure. Under these conditions the entire flame is likely to extinguish.

The location of the inner layer defines the flame surface. In contrast to moment methods or methods based on a pdf transport equation, statistical considerations in the flamelet concept focus on the location of the flame surface and not on the reactive scalars themselves. That location is defined as an iso-surface of a nonreacting scalar quantity, for which a suitable field equation is derived. For nonpremixed combustion the mixture fraction Z is that scalar quantity; for premixed combustion the scalar G will be introduced. Once equations that describe the statistical distributions of Z and G are solved, the profiles of the reactive scalars normal to the surface are calculated using flamelet equations. These profiles are assumed to be attached to the flame surface and are convected with it in the turbulent flow field. Therefore the statistical moments of the reactive scalars can be obtained from the statistical distribution of the scalar quantities Z and G.

Since the scalar quantities Z and G are nonreacting, their field equations do not contain a chemical source term. Therefore classical turbulent modeling assumptions used for nonreacting scalars can be applied. Using, for instance, the presumed shape pdf approach, to be introduced in Section 2.15 for premixed and in Section 3.8 for nonpremixed turbulent combustion, the probability density function of these scalar quantities can be constructed, once equations for their means and variances have been solved as part of a RANS or LES calculation.

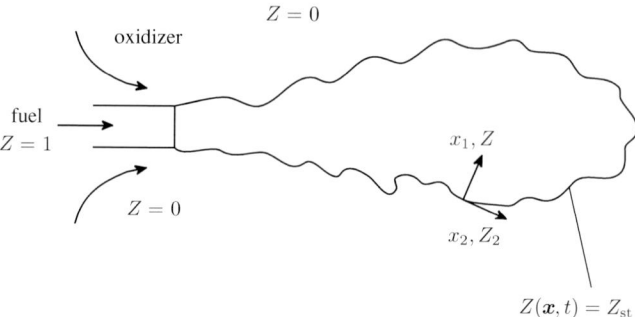

Figure 1.8. Surface of stoichiometric mixture in a turbulent jet diffusion flame.

With the pdf for Z or G known, one can determine the probability of finding the flame surface at any point x and time t in the flow field. The presumed shape pdf approach can then be used to calculate mean values of all reactive scalars and mean reaction rates, once the flamelet equations have been solved.

Flamelet equations describe the reactive-diffusive structure in the vicinity of the flame surface as a function of Z or G. For nonpremixed combustion the steady state formulation of these equations has been derived by Peters (1980) and Kuznetsov (1982), and the unsteady formulation by Peters (1984). An analogous formulation for premixed combustion has been presented by Peters (1993). The derivation is based on two steps: In the first step a coordinate transformation, applied at the flame surface, is introduced. This is illustrated in Figure 1.8 for a turbulent jet diffusion flame, where, for simplicity, it is assumed that the fuel consumption layer is located at stoichiometric conditions such that the flame surface is defined as the surface of stoichiometric mixture $Z(x, t) = Z_{st}$. In Figure 1.8 the coordinate x_1, defined as being locally normal to the flame surface, is replaced by the new independent variable Z, whereas the tangential coordinates x_2 and x_3 are the same as the new coordinates Z_2 and Z_3, respectively. In the second step, based on the normalized thickness of the reaction zone as a small parameter, asymptotic arguments are used to show that derivatives of the reactive scalars in tangential directions are negligible compared to those in normal direction, expressed now in terms of the new independent variable Z. This is a classical boundary layer argument for thin layers: Since the temperature, for instance, is nearly constant along the flame surface, $Z(x, t) = Z_{st}$, gradients along the surface are expected to be small compared to those normal to it. The same argument can be put forward for the mass fractions of chemical species.

A more general derivation of the flamelet equations will be given in the forthcoming Chapters 2 and 3. It will be based on a two-scale asymptotic

1.11 The laminar flamelet concept

expansion similar to that used by Keller and Peters (1994) for premixed combustion. In addition, the nonreacting scalar fields outside of the instantaneous reactive-diffusive structure will be analyzed by considering filtered fields of the reactive scalars and the mixture fraction (or the scalar G). Introducing the normalized thickness of the diffusive layers in the vicinity of the flame surface as a small parameter in an asymptotic formulation, we can then relate the filtered reactive scalars to the filtered mixture fraction (or the scalar G) as in the flamelet equations, except that the chemical source term is missing. The one-dimensional equations for the nonreacting outer fields can be matched to (or simply combined with) the flamelet equations valid in the thin layer. This leads to one-dimensional equations in terms of Z or G in the entire domain of definition of these scalar variables.

Nonpremixed Combustion

Let us first present the flamelet model for nonpremixed combustion. As a starting point, it is necessary to formulate the field equation for the mixture fraction that determines the location of the flame surface. As shown in Chapter 3, this equation is

$$\rho \frac{\partial Z}{\partial t} + \rho \boldsymbol{v} \cdot \nabla Z = \nabla \cdot (\rho D \nabla Z). \tag{1.137}$$

Once the solution of (1.137) is known in the entire flow field, the flame surface is defined as the surface of stoichiometric mixture, which is obtained by setting

$$Z(\boldsymbol{x}, t) = Z_{st} \tag{1.138}$$

as shown in Figure 1.8. It will also be shown in Chapter 3 that in the vicinity of that surface the reactive-diffusive structure can be described by the flamelet equations

$$\rho \frac{\partial \psi_i}{\partial t} = \frac{\rho}{Le_i} \frac{\chi}{2} \frac{\partial^2 \psi_i}{\partial Z^2} + \omega_i. \tag{1.139}$$

In these equations the instantaneous scalar dissipation rate defined as

$$\chi = 2D |\nabla Z|^2 \tag{1.140}$$

has been introduced. At the flame surface, it takes the value χ_{st}. If χ is a function of Z, as discussed in Section 3.4, this functional dependence can be parameterized by χ_{st}. It acts as an external parameter that is imposed on the flamelet structure by the mixture fraction field.

The scalar dissipation rate plays a very important role in flamelet models for nonpremixed turbulent combustion. It has the dimension of an inverse time and therefore represents the inverse of a diffusion time scale. It also can be thought of as a diffusivity in mixture fraction space, where D is multiplied by the square of the gradient $|\nabla Z|$. This gradient is amplified, for instance, by compressive strain (cf. Gibson et al., 1988) imposed by turbulence on the mixture fraction field. As the viscous dissipation rate ε the mean scalar dissipation rate is inertial range invariant. It also must be modeled. Then neither a molecular diffusivity nor a molecular viscosity appears in the flamelet equations, and the resulting turbulent combustion model does not introduce an additional Reynolds number dependence, a requirement that has been stated at the end of Section 1.1.

Equation (1.139) shows that ψ_i depends on the mixture fraction Z, on the scalar dissipation rate χ_{st}, and on time t. Referring to the comment by Pope (1997) quoted at the end of the last section, we note that the flamelet concept for nonpremixed combustion presumes that the reactive scalars represent a three-dimensional manifold. By the use of (1.139) it is implied that the reactive scalars are constant along iso-mixture fraction surfaces at a given time and a prescribed functional form of the scalar dissipation rate. Thereby the fields of the reactive scalars are aligned to that of the mixture fraction and are transported together with it by the flow field.

The flamelet equations (1.139) are valid for burning flamelets where the temperature is sufficiently high such that radicals are present and the fuel consumption layer becomes thin, but also for nonreacting layers where the reaction rates vanish. During rapid transitions between these two states, which correspond to ignition and extinction events discussed in the context of Figure 1.1, the unsteady term balances the reaction rate and the diffusion term.

In principle, both the mixture fraction Z and the scalar dissipation rate χ are fluctuating quantities and their statistical distribution needs to be considered, if one wants to calculate statistical moments of the reactive scalars (cf. Peters, 1984). If the joint pdf $\tilde{P}(Z, \chi_{st})$ is known, and the steady state flamelet equations are solved to obtain ψ_i as a function of Z and χ_{st}, the Favre mean of ψ_i can be obtained from

$$\tilde{\psi}_i(\boldsymbol{x}, t) = \int_0^1 \int_0^\infty \psi_i(Z, \chi_{st}) \tilde{P}(Z, \chi_{st}; \boldsymbol{x}, t) \, d\chi_{st} \, dZ. \qquad (1.141)$$

This has sometimes been called the Stretched Laminar Flamelet Model (SLFM). For further reading see Peters (1984) and Lentini (1994).

If, for reasons to be discussed in Section 3.13, the unsteady term in the flamelet equation must be retained, joint statistics of Z and χ_{st} become impractical. Then, in order to reduce the dimension of the statistics, it is useful

1.11 The laminar flamelet concept

to average the flamelet equations over all realizations of χ_{st}, thereby obtaining equations for conditional mean values.

In a one-dimensional problem, the scalar dissipation rate can be determined as a function of the mixture fraction. In Section 3.4, we will present such a functional dependence on Z, showing that it is the same for two very different situations, the counterflow configuration and the unsteady mixing layer. If the Favre average over all realizations of $\chi(Z)$ is taken, conditioned at fixed Z and t, and the scalar dissipation rate χ in (1.139) is replaced by the conditional Favre mean scalar dissipation rate $\tilde{\chi}_Z$ defined by

$$\tilde{\chi}_Z = \frac{\langle \rho \chi \mid Z \rangle}{\langle \rho \mid Z \rangle}, \qquad (1.142)$$

the flamelet equations for the conditional mean values take the form

$$\rho \frac{\partial \psi_i}{\partial t} = \frac{\rho}{Le_i} \frac{\tilde{\chi}_Z}{2} \frac{\partial^2 \psi_i}{\partial Z^2} + \omega_i. \qquad (1.143)$$

Such an average over the distribution of scalar dissipation rates, however, is unable to account for those ignition and extinction events that are triggered by small and large values of χ, respectively. For extinction, this has been discussed in Peters (1984). There, by presuming the pdf $\tilde{P}(\chi_{st})$ of χ_{st} as a lognormal distribution, the fraction of burning flamelets p_b was calculated as the probability that $\chi_{st} < \chi_q$, where χ_q is the scalar dissipation rate at quenching:

$$p_b = \int_0^{\chi_q} \tilde{P}(\chi_{st}) \, d\chi_{st}. \qquad (1.144)$$

With $\psi_i(Z, \tilde{\chi}_Z, t)$ obtained from solving (1.143), Favre mean values $\tilde{\psi}_i$ can be obtained at any point x and time t in the flow field by

$$\tilde{\psi}_i(x, t) = \int_0^1 \psi_i(Z, \tilde{\chi}_Z, t) \tilde{P}(Z; x, t) \, dZ. \qquad (1.145)$$

This formulation is the same as the one used in the concept of Conditional Moment Closure to be presented in the next section.

The unsteady flamelet equations have been used to simulate ignition, combustion and pollutant formation in diesel engines. These applications will be discussed in Chapter 3. They have also been used, for instance, by Mauss et al. (1990) to calculate extinction and ignition in piloted turbulent jet diffusion flames. There the time t in (1.139) was interpreted as a Lagrangian residence time in the jet. We will call this model the Lagrangian Flamelet Model (LFM). More recent applications of this model to jet flames are due to Pitsch et al. (1998) and Pitsch (1999).

An alternative way to use the unsteady flamelet equations is to imagine marker particles to which the flamelets are thought to be attached. These particles would follow the flow and encounter different histories of scalar dissipation rates in different regions of the flow. Particle trajectories must not necessarily be solved by a Lagrangian procedure as in the pdf transport equation model. Barths et al. (1998) have used a modeled Eulerian convective-diffusive equation to calculate the probability of finding such particles. Typically up to ten different particles are introduced corresponding to ten different flamelet histories. We will call this the Eulerian Particle Flamelet Model (EPFM) and discuss it in more detail in Section 3.15. Using an Eulerian description for all field equations makes the method consistent with the Eulerian framework in which flamelet models are usually formulated.

Instead of flamelet profiles the equilibrium solution or, in the case of a one-step irreversible reaction, the Burke–Schumann solution, both presented in Section 3.2, can be used as a first step in a model of nonpremixed turbulent combustion. These solutions are independent of the scalar dissipation rate. Using those solutions together with the presumed pdf approach is called the Conserved Scalar Equilibrium Model (CSEM). The flamelet model may be regarded as an extension of the equilibrium model to finite rate chemistry. It is important to note that arbitrary complex chemistry can be used, if needed, in solving the flamelet equations.

Although the presumed shape pdf approach is the most appropriate method to be used together with flamelet equations, there is valuable information contained in the pdf transport equation (1.126) written for the mixture fraction as the only sample space variable. In that case the chemical source term disappears and the turbulent transport terms in physical space must be modeled. Furthermore, the molecular mixing term may be split into a molecular diffusion term and a term containing the conditional mean scalar dissipation rate (cf. O'Brien, 1980). Neglecting the molecular diffusion term, the equation for the Favre pdf of the mixture fraction becomes

$$\bar{\rho}\frac{\partial \tilde{P}}{\partial t} + \bar{\rho}\tilde{v} \cdot \nabla \tilde{P} = \nabla \cdot (\bar{\rho} D_t \nabla \tilde{P}) - \frac{1}{2}\frac{\partial^2}{\partial Z^2}(\langle \rho \chi \mid Z \rangle P). \qquad (1.146)$$

Here the gradient transport approximation for the turbulent flux of $\tilde{P}$ has been introduced. This equation shows that there is a close link between $\tilde{P}(Z)$ and the conditional mean scalar dissipation rate $\tilde{\chi}_Z$. Using (1.143) and (1.27) the last term in (1.146) may also be written as

$$-\frac{1}{2}\bar{\rho}\frac{\partial^2}{\partial Z^2}(\tilde{\chi}_Z \tilde{P}). \qquad (1.147)$$

Rather than calculating $\tilde{P}(Z)$ from (1.146) one can use that equation to calculate $\tilde{\chi}_Z$ by inserting $\tilde{P}(Z)$ obtained from the presumed shape pdf approach. This was first done by Janicka and Peters (1982) for a turbulent jet diffusion flame by inserting a presumed beta function pdf for $\tilde{P}(Z)$. Later, Girimaji (1992) used the same procedure and presented analytical solutions for the case of isotropic turbulence.

At this stage it is worth comparing the flamelet formulation for nonpremixed combustion with moment methods for reactive scalars. If one multiplies (1.146) by $\psi_i(Z, \tilde{\chi}_Z, t)$, obtained from the solution of the flamelet equations (1.143), one may move ψ_i underneath all spatial derivatives and integrate over Z to obtain

$$\bar{\rho}\int_0^1 \psi_i \frac{\partial \tilde{P}}{\partial t}\,dZ + \bar{\rho}\tilde{v}\cdot\nabla\tilde{\psi}_i = \nabla\cdot(\bar{\rho}D_t\nabla\tilde{\psi}_i) - \frac{\bar{\rho}}{2}\int_0^1 \psi_i \frac{\partial^2}{\partial Z^2}(\tilde{\chi}_Z\tilde{P}(Z))\,dZ. \quad (1.148)$$

The last term in this equation is transformed by applying partial integration twice:

$$\frac{\bar{\rho}}{2}\int_0^1 \psi_i \frac{\partial^2}{\partial Z^2}(\tilde{\chi}_Z\tilde{P}(Z))\,dZ = \bar{\rho}\int_0^1 \frac{\tilde{\chi}_Z}{2}\frac{\partial^2 \psi_i}{\partial Z^2}\tilde{P}(Z)\,dZ. \quad (1.149)$$

Here it has been assumed that the value and the derivative of $\tilde{\chi}_Z\tilde{P}(Z)$ are zero at both limits of the integral (cf. Janicka and Peters, 1982 and Girimaji, 1992). If in (1.143) the Lewis number is assumed equal to unity, the flamelet equations can be introduced into the last term of (1.149) to show that the two integrals in (1.148) can be combined to obtain the equation for the Favre mean of ψ_i as

$$\bar{\rho}\frac{\partial\tilde{\psi}_i}{\partial t} + \bar{\rho}\tilde{v}\cdot\nabla\tilde{\psi}_i = \nabla\cdot(\rho D_t\nabla\tilde{\psi}_i) + \bar{\omega}_i. \quad (1.150)$$

This is the same as (1.103) if the molecular diffusion term is neglected in that equation and the gradient flux approximation (1.107) is introduced. This indicates that in nonpremixed combustion, turbulent transport of the reactive scalar is solely due to turbulent transport of the mixture fraction field. It supports a model, commonly used in nonpremixed combustion, where the balance equations for the mean reactive scalars are solved with the mean chemical source term obtained from the presumed shape pdf approach using the solution of the flamelet equations.

There are certain additional aspects of flamelet models in nonpremixed combustion that will be discussed in Chapter 3: Differential diffusion effects, multiple unsteady flamelets that account for multiple histories of scalar dissipation rates and thereby model a distribution of χ, steady versus unsteady flamelet

Premixed Combustion

For premixed combustion, flamelet models are either based on the progress variable c or on the scalar G. The progress variable c is defined as a normalized temperature or normalized product mass fraction

$$c = \frac{T - T_u}{T_b - T_u} \quad \text{or} \quad c = \frac{Y_P}{Y_{P,b}}, \tag{1.151}$$

which implies a one-step reaction $A \to P$ and a corresponding heat release raising the temperature from T_u to T_b. In flamelet models based on the progress variable the flame structure is assumed to be infinitely thin and no intermediate values of temperature between T_u and T_b are resolved. This corresponds to the fast chemistry limit. The progress variable therefore is a step function that separates unburnt mixture and burnt gas in a given flow field.

The classical model for premixed turbulent combustion, the Bray–Moss–Libby (BML) model, was initiated by Bray and Moss (1977) by assuming the pdf of the progress variable c to be a two delta function distribution. This assumption only allows for entries at $c = 0$ and $c = 1$ in a turbulent premixed flame, but it illustrates important features, such as countergradient diffusion of the progress variable. Let us consider the equation for the Favre mean progress variable $\tilde{c}$:

$$\bar{\rho}\frac{\partial \tilde{c}}{\partial t} + \bar{\rho}\boldsymbol{v} \cdot \nabla \tilde{c} + \nabla \cdot (\widetilde{\bar{\rho} v'' c''}) = \bar{\omega}_c, \tag{1.152}$$

where the molecular diffusion term has been neglected. This equation requires the modeling of the turbulent transport term $\widetilde{v'' c''}$ and the mean reaction term $\bar{\omega}_c$. Libby and Bray (1981) and Bray et al. (1981) have shown that the gradient transport assumption (1.107) is not applicable to $\widetilde{v'' c''}$. This is due to gas expansion effects at the flame surface and is called countergradient diffusion. A more detailed analysis will be presented in Section 2.4. Countergradient diffusion has been found in many experiments and in many one-dimensional numerical analyses. However, there is no model available that could be used in three-dimensional calculations solving (1.152) with countergradient diffusion included.

Models for the mean reaction rate by Bray et al. (1984a) and Bray and Libby (1986) focus on a time series of step function events of the progress variable. This makes the mean source term proportional to the flamelet crossing

1.11 The laminar flamelet concept

frequency. Further modeling, discussed in more detail in Bray and Libby (1994), then leads to the expression

$$\overline{\omega_c} = \rho_u s_L I_0 \Sigma, \quad (1.153)$$

where s_L is the laminar burning velocity, I_0 is a stretch factor, and Σ is the flame surface density (flame surface per unit volume). The flame surface density may either be expressed by an algebraic model being proportional to

$$\Sigma \sim \frac{\bar{c}(1-\bar{c})}{\hat{L}_y}, \quad (1.154)$$

where $\bar{c}$ is the mean progress variable and $\hat{L}_y$ is the crossing length scale (cf. Bray and Libby, 1994) or via a modeled transport equation. The latter is based on an exact but unclosed formulation by Pope (1988) and Candel and Poinsot (1990), for which several modeling strategies have been proposed. An empirically based formulation by Darabiha et al. (1987) has been generalized by Candel et al. (1990) to also include nonpremixed combustion. This is called the Coherent Flame Model (CFM), which is attributed to Marble and Broadwell (1977). A comparison of the performance of different formulations of the model for one-dimensional turbulent flames was made by Duclos et al. (1993). Later modeling based on DNS data has led Trouvé and Poinsot (1994) to the following formulation of an equation for the flame surface density:

$$\frac{\partial \Sigma}{\partial t} + \nabla \cdot (\tilde{v}\Sigma) = \nabla \cdot (D_t \nabla \Sigma) + C_1 \frac{\varepsilon}{k}\Sigma - C_2 s_L \frac{\Sigma^2}{1-\bar{c}}. \quad (1.155)$$

The terms on the l.h.s. represent the local rate of change and convection; the first term on the r.h.s. represents turbulent diffusion, the second term production by flame stretch, and the last term flame surface annihilation. The stretch term is proportional to the inverse of the integral time scale $\tau = k/\varepsilon$, which is to be evaluated in the unburnt gas.

Inspection of (1.153) and (1.155), with the stretch factor set equal to unity, shows that no chemical time scale enters into the mean reaction rate. In fact, if (1.155) is multiplied by s_L, one obtains an equation for the product $s_L \Sigma$ that is independent of s_L. Therefore the mean progress variable also becomes independent of s_L and, except for the stretch term, which is rather empirical, the influence of chemistry on $\bar{c}$ disappears. The rate-determining time scale is the integral time $\tau = k/\varepsilon$, which indicates that the model describes the fast chemistry limit.

Some of the formulations of the Coherent Flame Model discussed by Duclos et al. (1993) introduce the inverse of the Kolmogorov time scale rather than of the integral time scale in modeling the production term of (1.155) (cf. Cant et al., 1990 and Mantel and Borghi, 1994). This builds a Reynolds number dependence into those formulations, which violates the requirement stated at the end of Section 1.1 that combustion models should be Reynolds number independent in the limit of large Reynolds numbers.

More recently, flamelet models for premixed combustion have been based on the level set approach using the scalar G rather than on the progress variable. An equation defining G will be derived in Chapter 2 from the kinematic balance among the flow velocity, the burning velocity normal to the flame front, and the flame front propagation velocity. By introducing the normal vector $\boldsymbol{n}$ to the front as

$$\boldsymbol{n} = -\frac{\nabla G}{|\nabla G|} \tag{1.156}$$

one obtains the G-equation

$$\frac{\partial G}{\partial t} + \boldsymbol{v} \cdot \nabla G = s_L |\nabla G|. \tag{1.157}$$

This equation is only defined at the premixed flame surface, which is the iso-scalar surface

$$G(\boldsymbol{x}, t) = G_0, \tag{1.158}$$

where G_0 is arbitrary but fixed. Viewed as a field equation, the G-equation plays a similar role in premixed combustion as the balance Equation (1.137) for the mixture fraction plays in nonpremixed combustion.

The level set approach based on the G-equation makes it possible to derive flamelet equations in an analogous manner to (1.139) with G as the independent variable. To leading order one obtains the steady state equations

$$\rho s_L |\nabla G| \frac{d\psi_i}{dG} = \frac{d}{dG}\left(\rho D_i |\nabla G|^2 \frac{d\psi_i}{dG}\right) + \omega_i. \tag{1.159}$$

By introducing the flame normal coordinate

$$x_n = \frac{G - G_0}{|\nabla G|} \tag{1.160}$$

one may convert this into a system of equations describing a one-dimensional steady premixed flame:

$$\rho s_L \frac{d\psi_i}{dx_n} = \frac{d}{dx_n}\left(\rho D_i \frac{d\psi_i}{dx_n}\right) + \omega_i. \tag{1.161}$$

As for nonpremixed combustion, complex chemistry can be used in solving the flamelet equations (1.159) or (1.161). Therefore the flamelet model based on the G-equation is able to account for nonequilibrium effects.

The G-equation (1.157) is generalized in Chapter 2 to enable one to describe two different regimes of premixed turbulent combustion. When Reynolds averaging and modeling is introduced, one obtains equations for the Favre mean $\tilde{G}(\boldsymbol{x}, t)$ and the variance $\widetilde{G''^2}(\boldsymbol{x}, t)$, as well as an additional equation for the flame surface area ratio $\bar{\sigma} = \overline{|\nabla G|}$. In these equations turbulent fluxes normal to the mean turbulent flame front do not appear. This allows the flamelet model based on the level set approach to circumvent modeling of countergradient diffusion. Together with the Reynolds averaged Navier–Stokes equations for the flow field, the Reynolds stress equations (or that for the turbulent kinetic energy k), and the ε-equation, the three equations for $\tilde{G}$, $\widetilde{G''^2}$, and $\bar{\sigma}$ form a complete set to determine premixed turbulent flame propagation. The mean fields of the reactive scalars and the density can then be calculated by solving the flamelet equations and using the presumed shape pdf approach. There are many analogies with the flamelet approach for nonpremixed combustion but also some mathematical subtleties that must be considered.

For partially premixed combustion the two flamelet concepts for premixed and nonpremixed combustion will be combined to derive a model for the turbulent flame propagation in layered mixtures. The resulting model will be presented in Chapter 4.

Flamelet models explicitly assume scale separation between the small scales at which reaction occurs and the larger scales in the inertial subrange of turbulence. Since mixing is modeled by inertial range invariant scalar dissipation rates they do not introduce a Reynolds number dependence in the large Reynolds number limit.

1.12 The Concept of Conditional Moment Closure

Klimenko (1990) and Bilger (1993) independently proposed an interesting concept for nonpremixed turbulent combustion, called Conditional Moment Closure (CMC). The suggestion is that, rather than considering conventional averages, one should condition the reactive scalars on the mixture fraction. Klimenko (1990) has emphasized that turbulent diffusion in mixture fraction space can be modeled more rigorously than in physical space. Bilger (1993) based his derivation on the observation that most of the fluctuations of the reactive scalars can be associated with fluctuations of the mixture fraction. These alternative views have been compared and reviewed by Klimenko and Bilger (2000) where an extension to premixed turbulent combustion, with conditioning on the progress variable, is also proposed.

Rather than considering flame surface statistics and the laminar reactive-diffusive structure attached to the flame surface as in the flamelet model, the

CMC model is based on conditional moments at a fixed location x and time t within the flow field. Using the conditional probability density function

$$P(\psi_i \mid Z; x, t) = \frac{P(\psi_i, Z; x, t)}{P(Z; x, t)} \qquad (1.162)$$

the first conditional moment of the reactive scalars is defined as

$$Q_i(Z; x, t) = \langle \psi_i \mid Z \rangle = \int_0^1 \psi_i \, P(\psi_i \mid Z; x, t) \, d\psi_i. \qquad (1.163)$$

Because of conditioning, the quantity Q_i is a function of not only x and t, but also of Z. Klimenko (1990) derives an equation for the first conditional moment by starting from the joint pdf transport equation, while Bilger (1993) decomposes the reactive scalar into a conditional mean and a conditional fluctuation y_i:

$$\psi_i(x, t) = Q_i(Z; x, t) + y_i(Z; x, t). \qquad (1.164)$$

Similar to classical moment methods, he introduces (1.164) into the governing equations (1.81) for the reactive scalars. When the conditional average of the resulting equation is taken one obtains

$$\langle \rho \mid Z \rangle \frac{\partial Q_i}{\partial t} + \langle \rho \mid Z \rangle \tilde{v}_Z \cdot \nabla Q_i = \langle \rho \mid Z \rangle \tilde{\chi}_Z \frac{\partial^2 Q_i}{\partial Z^2} + \langle \omega_i \mid Z \rangle. \qquad (1.165)$$

Some additional terms that account for diffusive fluxes and conditional turbulent transport in physical space are not written down here for convenience. It is argued that conditional turbulent transport may be modeled by the gradient flux approximation in a similar way as for unconditional moments. It is neglected in all known applications of the CMC model.

The Favre conditional velocity $\tilde{v}_Z$, the Favre conditional scalar dissipation rate $\tilde{\chi}_Z$, and the conditional chemical source term are unclosed. In all the known applications the conditional velocity is replaced by the unconditional Favre mean velocity. For calculating the conditional scalar dissipation rate the procedure based on the pdf transport equation presented in the previous section is advocated. Likewise the presumed shape pdf approach is used to calculate unconditional mean values of the reactive scalars and the mean reaction rate.

In CMC modeling, higher moments of the chemical source term are often assumed to be negligible. This leads to the closure of the chemical source term:

$$\langle \omega_i \mid Z \rangle = \omega_i(\langle \psi_i \mid Z \rangle). \qquad (1.166)$$

This assumption is found to be valid, for instance, by comparison with DNS data (cf. Mell et al., 1994 or Swaminathan and Bilger, 1999). We will call this the Conditional First Moment Closure (CFMC).

Equations for higher moments can also be formulated in the CMC concept. For a second-order closure one obtains for n reactive scalars $n(n+1)/2$ differential equations for variances and covariances. For the problem of predicting NO formation in turbulent jet diffusion flames to be discussed in Chapter 3, Section 3.14, Kronenburg et al. (1998) have modeled and solved a transport equation for the conditional variance of the temperature. In a recent analysis of DNS data Swaminathan and Bilger (1998) have examined the different terms in the conditional variance equation for the fuel mass fraction of a two-step mechanism. They plot the correlation coefficients relating the different terms in the equations to known quantities as a function of the mixture fraction. However, no closure of the variance equation is proposed.

In choosing the mixture fraction (rather than another quantity such as the temperature, for instance) as the particular scalar upon which all reactive scalars are conditioned, CMC for nonpremixed combustion undeniably follows the flamelet concept. The CMC derivation has the advantage that it clearly identifies $\tilde{\chi}_Z$ as a conditional scalar dissipation rate. The derivation of the flamelet equations, in contrast, allows for a statistical distribution of χ_{st}, which is important for ignition and extinction phenomena.

Conditional First Moment Closure has been used either in homogeneous flows or in boundary layer flows (cf. Klimenko, 1995). In the latter only the axial component of velocity is retained. In the homogeneous flow case the first moment CMC equations are identical to the flamelet Equations (1.143) while in the boundary layer case the formulation is very similar to the Lagrangian Flamelet Model. This similarity is at first surprising, because the derivation of both models appears to be quite different. Since the same equations result from different procedures, logic requires that the underlying assumptions, either explicit or implicit, must be the same and that the equations describe the same physics. The turbulent mixing model in CMC, for instance, being based on the scalar dissipation rate in the same way as in the flamelet model, implies the assumption of scale separation. For CMC to go beyond the flamelet approach, not only as a concept but as an alternative combustion model, it seems necessary that the potential benefit of higher moment closure be convincingly demonstrated.

1.13 The Linear Eddy Model

Another approach to account for nonequilibrium chemistry in turbulent combustion is the Linear Eddy Model (LEM), which was formulated for nonreacting flows by Kerstein (1988a,b, 1989, 1990, 1991) and was extended to reactive scalars by Kerstein (1992a,b). Linear eddy modeling is a method of simulating

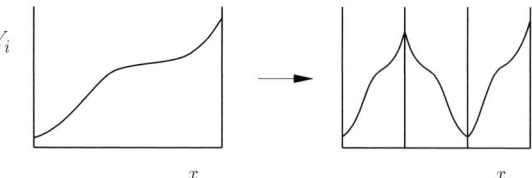

Figure 1.9. Illustration of a triplet map used in the Linear Eddy Model to account for rearrangement caused by turbulent mixing of the 1D scalar fields.

molecular mixing on a one-dimensional (1D) domain embedded in a turbulent flow. The method involves two concurrent processes to determine the scalar $\psi_i(x, t)$: The first process describes the evolution of the reactive scalar field in one dimension. It is governed by the system of parabolic equations

$$\frac{\partial(\rho\psi_i)}{\partial t} = \frac{\partial}{\partial x}\left(\rho D_i \frac{\partial \psi_i}{\partial x}\right) + \omega_i, \qquad (1.167)$$

which must be solved numerically. The second process consists of a stochastic sequence of instantaneous, statistically independent "rearrangement events." Both processes are performed at the finest scales of fluid-property variations in physical space, which makes this method computationally very expensive.

Each rearrangement event may be viewed as representing the effect of an individual eddy on the scalar field. They are prescribed, for instance, by the triplet map shown in Figure 1.9 (cf. Kerstein, 1991). Here the one-dimensional scalar field within a chosen segment is compressed by a factor of three. Three adjacent copies of the new scalar field, with the middle copy mirror-inverted, are then used to replace the original scalar field. The map causes an increase of the scalar gradient, thereby simulating the effect of shear or compressive strain imposed by eddies of any size within the inertial range on the scalar field. Two quantities govern each event: the eddy size and the location within the one-dimensional domain. The eddy size is determined randomly from a pdf of eddy sizes that are modeled using inertial range scaling. The location within the one-dimensional domain is chosen randomly from a uniform distribution. The overall event rate is chosen to match the turbulent diffusivity of the flow being simulated. The method is suited for both premixed and nonpremixed turbulent combustion.

The Linear Eddy Model has been applied by McMurtry et al. (1992) to hydrogen–air combustion and by Menon and Kerstein (1992) on the G-equation. In the latter paper the linear relation between the turbulence intensity and the turbulent burning velocity is reproduced as expected for the corrugated flamelets regime (cf. Chapter 2).

Menon and Calhoon (1996) have used the LEM formulation as a subgrid model together with large eddy simulations and applied it to a shear layer in which the simple irreversible one-step reaction $F + O_2 \to P$ with small heat release occurs. The LEM results are compared with experimental data in the near field of a reacting shear layer. The small Reynolds number dependence of the normalized product thickness in the experimental data was used to determine the calibration coefficient used to rescale the subgrid stirring and diffusion time scales. Smith and Menon (1996) extended this model in order to study freely propagating premixed turbulent flames. They varied the Lewis number between 0.8 and 1.2 and compared the turbulent burning velocity with that from DNS data, obtaining similar trends. This model was further extended by Smith and Menon (1997) to also include the 4-step model (1.92) for methane combustion. Comparisons with experimental burning velocities and with the G-equation simulations by Menon and Kerstein (1992) were made.

Recently, the ideas behind the Linear Eddy Model have been used by Kerstein (1999) to include modeling of the velocity field. This model, called One-Dimensional Turbulence (ODT) is a self-contained formulation that does not require assumptions about the energy cascade process but rather reproduces it by one-dimensional simulations of a localized shear layer structure and appropriately scaled rearrangement events.

1.14 Combustion Models Used in Large Eddy Simulation

Turbulence models based on Reynolds averaged Navier–Stokes equations (RANS) employ turbulent transport approximations with an effective turbulent viscosity that is by orders of magnitude larger than the molecular viscosity. In particular if steady state versions of these equations are used, this tends to suppress large scale instabilities, which occur in flows with combustion even more frequently than in nonreacting flows. If those instabilities are to be resolved in numerical simulations, it is necessary to resort to more advanced, but computationally more expensive methods such as direct numerical simulation (DNS) or large eddy simulation (LES). As noted in the preface, DNS is still out of reach as a method to predict turbulent flows with combustion for practical engineering applications for many years to come.

Large eddy simulation, in contrast, does not intend to numerically resolve all turbulent length scales, but only a fraction of the larger energy-containing scales within the inertial subrange. Modeling is then applied to represent the smaller unresolved scales, which contain only a small fraction of the turbulent kinetic energy. Therefore the computed flows are usually less sensitive to modeling assumptions. The distinction between the resolved large scales and the modeled small scales is made by the grid resolution that can be afforded. The model for the smaller scales is called the subgrid model. In deriving the basic LES equations, the Navier–Stokes equations are spatially filtered with a filter of size Δ, which is of the size of the grid cell (or a multiple thereof) in order to remove the direct effect of the small scale fluctuations (cf. Ghosal and Moin, 1995). These show up indirectly through nonlinear terms in the subgrid-scale stress tensor as subgrid-scale Reynolds stresses, Leonard stresses, and subgrid-scale cross stresses. The latter two contributions result from the fact that, unlike with the traditional Reynolds averages, a second filtering changes an already filtered field. In a similar way, after filtering the equations for nonreacting scalars such as the mixture fraction, one has to model the filtered scalar flux vectors that contain subgrid scalar fluxes, Leonard fluxes, and subgrid-scale cross fluxes.

Review papers have been written, for instance, by Lesieur and Métais (1996) and by Moin (1997). Often the gradient transport assumption employing the Smagorinsky model is introduced for the Reynolds stress tensor and the scalar fluxes. The Smagorinsky model for the subgrid-scale stress tensor takes the form

$$\bar{\tau} = -2\bar{\rho}C_s\Delta^2|\bar{S}|\bar{S}, \qquad (1.168)$$

where $\bar{S}$ is the filtered rate of strain tensor and C_s is the Smagorinsky coefficient. The Smagorinsky model is a generalization of Prandtl's mixing length model. It requires C_s to be positive and to be known in advance. It has been found, however, that to obtain satisfactory results, largely different values of C_s, ranging from 0.01 to 0.3, depending on the flow and grid resolution, must be chosen.

As an example for the scalar fluxes, the subgrid-scale heat flux vector in a compressible nonreacting flow can be modeled as (cf. Moin et al., 1991)

$$\bar{j}_q = -\bar{\rho}\frac{C_s\Delta^2|\bar{S}|}{Pr_t}\nabla\bar{T}. \qquad (1.169)$$

Here Pr_t is a subgrid-scale turbulent Prandtl number. Some of the more recent approaches to modeling the additional stresses are discussed by Speziale (1998).

A breakthrough in subgrid modeling is the introduction of a method called dynamic modeling by Germano et al. (1991). In the dynamic subgrid-scale model, a test filter $\hat{\Delta}$ is introduced in addition to the grid filter Δ. Often the test filter is set to two times the grid filter. A variable Smagorinsky coefficient can then be calculated, which depends on the filtered stresses and fluxes, as well as on those that are resolved at the test filter level. The Smagorinsky coefficient may thereby take positive or negative values. A positive value implies that energy flows from the resolved to the subgrid scales, which is in agreement with the concept of the cascade hypothesis, while a negative coefficient implies an inverse cascade or "backscatter". It has been argued by Piomelli et al. (1991) that this back scatter has a physical basis, but with the original formulation of the dynamic model, where up to 30% negative values of C_s were calculated, it leads to undesirable numerical instabilities. Various remedies to resolve this problem have been proposed (cf. Germano et al., 1991, Piomelli and Liu, 1995, and Meneveau et al., 1996). The dynamic model was extended by Moin et al. (1991) to scalar transport where it served to determine the subgrid-scale turbulent Prandtl number.

When finite-difference methods are used in LES of inhomogeneous flows, different numerical discretization errors may inflict serious damage on simulations. Among these are the truncation error due to the representation of derivatives by finite differences and the aliasing error that arises when nonlinear terms are represented on a discrete grid. A systematic study of this issue has been undertaken by Kravchenko and Moin (1997) showing that both aliasing and truncation errors of low-order schemes can degrade LES computations. Of these, the aliasing error was found to be the source of the most serious problems, because it interfered with the energy conserving nature of the scheme. Mathematical and physical constraints of LES have recently been reviewed by Ghosal (1999).

For combustion simulations with LES, in addition to accurate resolution, a reliable subgrid model for scalar fields is of particular importance. Cook and Riley (1994) have proposed a scale similarity assumption for the subgrid variance of the mixture fraction:

$$\overline{Z'^2} = \overline{Z^2} - \bar{Z}^2 = c_Z(\widehat{\overline{Z^2}} - \hat{\bar{Z}}^2). \tag{1.170}$$

Here the bar denotes filtering at the grid filter and the hat denotes filtering using the test filter. The hypothesis behind scale similarity is that the largest unresolved scales have a structure similar to the smallest of the resolved scales. A theoretical estimate for c_Z is given by Jiménez et al. (1997), who predict that it depends on the exponent of the scalar spectrum function in the large Reynolds limit. However, Cook (1997) derives a method for calculating the coefficient of scale similarity models that he shows to depend on the grid size of the test filter and the Reynolds number.

An alternative approach assumes that production equals dissipation in the subgrid variance equation. Pierce and Moin (1998a) use this to calculate the constant C in the resulting expression for the subgrid-scale Favre variance $\widetilde{Z''^2}$

of the mixture fraction,

$$\widetilde{Z''^2} = C\Delta^2|\nabla \tilde{Z}|^2, \qquad (1.171)$$

by a dynamic modeling approach. Here $|\nabla \tilde{Z}|$ is the gradient of the resolved Favre mean mixture fraction. Comparing this dynamic model with the scale similarity model, Pierce and Moin (1998b) found remarkable differences in the large eddy simulation of a swirling, confined, coaxial jet flame.

Germano et al. (1997) extend the scale similarity model to reacting scalars and postulate that the subgrid variance of a scalar or the co-variance of two different scalars can be calculated from known quantities at the two filter levels. They find, however, that the similarity constant depends not only on the mesh size but also on the chemical time scale. This is because filtering over thin reaction zones underestimates the unresolved chemical source term. As the grid filter decreases to values of the order of the reaction zone thickness, an increasing part of the reaction rate occurs between the two filter levels. Therefore the similarity constant corresponding to c_Z in (1.170) decreases with the mesh size (cf. Vervisch and Veynante, 2000).

Certain formulations of subgrid models profitably take guidance from closure procedures that have been successful on the RANS level. Cook and Riley (1994) have proposed a presumed shape beta function pdf formulation, called Large Eddy Probability Density Function (LEPDF), for the mixture fraction. It uses the filtered mean and the subgrid variance taken from (1.170) to achieve closure for nonpremixed combustion. For the reactive scalars they apply the Conserved Scalar Equilibrium Model. Later on, Cook et al. (1997) extend this by using profiles for the reactive scalars from a steady state flamelet model based on a single one-step reaction. This model was found to be reasonably accurate compared to DNS data of homogeneous, isotropic decaying turbulence. Cook and Riley (1998) performed further a priori testing (comparison with DNS data) of this model by varying the activation energy of the one-step model. This shows a better agreement with DNS data than models using equilibrium chemistry models or a closure using filtered mean reactive scalars for the reaction rate.

De Bruyn Kops et al. (1998) performed a full LES calculation of the filtered mean and variance of the mixture fraction, determined by using their balance equations, a presumed shape for the scalar dissipation rate, and a subgrid-scale presumed shape pdf. A table was constructed by integrating steady state flamelet profiles. The table is parameterized in terms of the filtered mean and the variance of the mixture fraction and the mean scalar dissipation rate. The model accurately reproduces the spatial average of the filtered species concentration obtained from DNS data.

1.14 Combustion models used in large eddy simulation

The large eddy presumed shape beta function LEPDF model by Cook and Riley (1994) has been tested by Jiménez et al. (1997) together with other assumptions, such as the lognormal pdf for scalar gradients and the scale similarity assumption. They find that presuming a beta function pdf yields excellent agreement with DNS data not only for higher moments of temperature in the fast chemistry limit, but even for the pdf of the mixture fraction itself. DesJardin and Frankel (1998) have performed a priori and a posteriori assessments of several subgrid-scale combustion models. They compared LES calculations using a Scale Similarity Filtered Reactive Rate Model (SSFRRM) and a Conserved Scalar Equilibrium Model. They find that the SSFRRM model provides the best agreement with the DNS data. A model based on a transport equation for the subgrid-scale pdf, called Filtered Density Function (FDF), was developed by Colucci et al. (1998) and compared to presumed subgrid pdfs. Dynamic modeling has been applied by Réveillon and Vervisch (1997, 1998) to test a closure of the mixing term in the subgrid pdf equation for the mixture fraction. Similarly, Cook and Bushe (2000) use a priori testing of a subgrid-scale model for the scalar dissipation rate.

There are so far only a few large eddy simulations of large scale turbulent nonpremixed flames. Pierce and Moin (1998b) calculate a swirling, confined, coaxial jet flame based on the presumed beta function subgrid LES approach including heat release in the fast chemistry limit and show convincing comparisons with experimental data. Branley and Jones (1997) compute a hydrogen jet diffusion flame using the Smagorinsky model with $C_s = 0.1$, a subgrid-scale Prandtl number of 0.7, the presumed shape beta function subgrid pdf model with the variance calculated from (1.171) with $C = 0.13$, and an equilibrium model for hydrogen–air combustion. Forkel and Janicka (1999) perform a large eddy simulation of the diluted hydrogen–air diffusion flame documented by Tacke et al. (1998). They use the dynamic procedure, clipping negative values of the Smagorinsky constant. Chemical equilibrium profiles are tabulated as a function of the mixture fraction. It is found that inlet conditions must be well defined to describe the flow and the mixture fraction field close to the nozzle accurately enough. In a recent paper Branley and Jones (1999) calculate a swirling methane flame using dynamic modeling for the velocity and a unity Prandtl number for the scalar flux. The conserved scalar pdf is again obtained from the presumed shape beta function subgrid pdf model with the variance obtained from (1.171) with $C = 0.09$. The chemistry model is based on a single flamelet profile computed at a strain rate of 200/s. The results show good qualitative agreement with measurements.

A recent simulation of a turbulent jet diffusion flame by Pitsch and Steiner (2000) based on the Lagrangian Flamelet Model is shown in Figure 1.10. Only

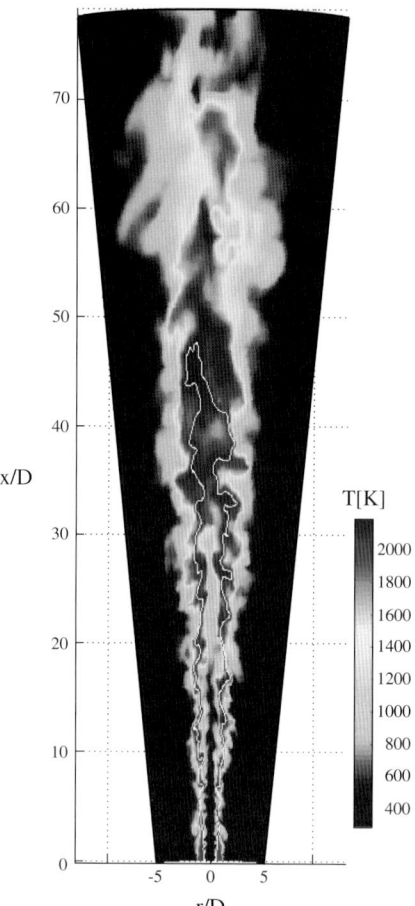

Figure 1.10 (see Color Plate I, Left). Large eddy simulations of a turbulent jet diffusion flame by Pitsch and Steiner (2000). (Reprinted with permission by H. Pitsch and H. Steiner.)

the conservation equations for mass, momentum, and the mixture fraction were solved in the LES. The subgrid stresses and scalar fluxes are determined using the dynamic model indicating that a constant Schmidt number approximation with a value of $Sc_t = 0.4$ is appropriate for LES. The mixture fraction variance is evaluated by (1.171). The mean scalar dissipation rate, including resolved and subgrid contributions, is calculated as proposed by Girimaji and Zhou (1996). In this study the shape of the scalar dissipation rate conditioned on the mixture fraction has not been presumed but is given by a model based on the assumption of a homogeneous distribution of this quantity in radial direction. Comparison of the computational results with experimental data shows very good agreement for temperature, major species, as well as OH and NO mass fractions.

There have been very few attempts to address premixed turbulent combustion by LES. Im et al. (1997) have performed calculations based on the G-equation, where the dynamic procedure was applied to several presumed expressions of the subgrid-scale burning velocity as a function of the subgrid turbulence intensity. Boger et al. (1998) used DNS data of a premixed turbulent flame in a homogeneous and isotropic flow field to test the flame surface density concept for LES. They find agreement with (1.154) for the modeling of Σ. As in RANS models, countergradient diffusion is also found in their LES results, but the effect is lower because part of the phenomenon is incorporated in the motion of the resolved flow structures.

1.15 Summary of Turbulent Combustion Models

In this chapter we have discussed a variety of combustion models that are being used by different groups in the combustion community. These models are schematically represented in Table 1.1 with respect to their capability to treat premixed or nonpremixed combustion, or both, and with respect to assumptions about the chemistry. Not shown in this table are the large differences in computational time required by the different models.

Some models, such as the Pdf Transport Equation Model and the Linear Eddy Model, can be applied to premixed as well as to nonpremixed combustion. This is because these models are less specific in their choice of the turbulent mixing model, such as for instance the flamelet models, where mixing is controlled by transport in either mixture fraction space or G-space in the respective flamelet equations. Other models, such as the BML model or the Conserved Scalar Equilibrium Model, are formulated only for premixed or for nonpremixed combustion, respectively. The CMC model is closely related to the flamelet model for nonpremixed combustion and shares its properties with respect to mixing. The Coherent Flame Model is claimed to be valid for both premixed and nonpremixed combustion, but most of its applications are in premixed combustion. The Eddy Break-Up Model and the Eddy Dissipation Model were not included in Table 1.1 because their physical content is rather limited.

While the BML model, the Coherent Flame Model, and the Conserved Scalar Equilibrium Model are valid for infinitely fast chemistry only, the other models are formulated such that finite rate chemistry can be used, involving as many reactions as are thought to be necessary. In the Pdf Transport Equation Model, however, restrictions are imposed by the computational time required to tabulate

Table 1.1. *Classification of turbulent combustion models in terms of chemistry and mixing*

	Premixed Combustion	Nonpremixed Combustion
Infinitely fast chemistry	Bray–Moss–Libby Model Coherent Flame Model	Conserved Scalar Equilibrium Model
Finite rate chemistry	Pdf Transport Equation Model	
	Flamelet Model Based on the G-Equation	Flamelet Model Based on Mixture Fraction Conditional Moment Closure
	Linear Eddy Model	

and retrieve reaction rates for a large number of chemical species. Similarly, solutions of a system of parabolic equations at each grid point in the Linear Eddy Model will become increasingly expensive for large reaction mechanisms.

It seems that most of the calculations using complex chemistry in turbulent combustion have been done using flamelet models. Here the system of parabolic equations needs to be calculated only once in a typical application. If multiple flamelets are calculated, as in the Eulerian Particle Flamelet Model, the computational time needed for chemistry calculations needs to be multiplied by the number of flamelets. If the Conditional First Moment Closure is applied, it has, for obvious reasons, the same computational time requirements.

It is worth comparing the flamelet equations (1.139) with the Lagrangian IEM equation (1.136) used in the PdfTransport Equation Model and the one-dimensional equation (1.167) of the Linear Eddy Model. It is clear that the unsteady terms and the chemical source terms are the same, while the mixing term is different in those three formulations. In all mixing models used in the Pdf Transport Equation Model, Lagrangian particles mix in reactive scalar space at a rate that is the same for all reactive and nonreactive scalars. In the flamelet equation (1.139) the second derivative in mixture fraction space drives the mixing, while gradients along iso-mixture fraction surfaces are assumed to be negligible. In the LEM model mixing occurs in a one-dimensional domain in physical space.

1.15 Summary of turbulent combustion models

The scalar dissipation rate appearing in the flamelet formulation may be interpreted as a diffusivity in mixture fraction space, which contains effects of compressive strain and shear on the scalar field (cf. Peters, 1984). Since the scalar dissipation rate is inertial range invariant it is independent of the eddy size. The rearrangement events of the LEM model simulate this numerically by modeling the event rate according to inertial rate scaling, thereby allowing the LEM model to account for the interaction between chemistry and turbulence in the inertial range. It is the only model that does not assume scale separation and therefore has the potential of testing the postulate formulated at the end of Section 1.1. The One-Dimensional Turbulence Model is even able to consider situations where the Reynolds number is not very large and to test the assumption of Reynolds number independence in the large Reynolds number limit.

There is an interesting analogy between Lagrangian particles in the Pdf Transport Equation Model and the Eulerian Particle Flamelet Model. Both methods capture residence time effects that are important for slow chemistry, but they use different numerical formulations.

In the future, large eddy simulation will certainly be the method of choice to eliminate inaccuracies of RANS models in turbulent combustion. In particular, combustion in reciprocating engines is an ideal candidate for large eddy simulation since the flow is three dimensional and time dependent, thereby reducing the advantages that RANS models have for flows that are steady in the mean. But LES is also well suited for combustion simulations in gas turbines and burners, as the work of Pierce and Moin (1998b) has shown. In those flows combustion induces large scale instabilities that are not easily accounted for by RANS models. In most applications of LES a flamelet model combined with the presumed shape subgrid pdf approach is probably the best choice, because of its ease of implementation but also because of its capability of incorporating rather complex chemistry at reasonable computational costs. Its use requires, however, that the assumption of scale separation between chemistry and turbulence in the inertial range is valid.

2

Premixed turbulent combustion

2.1 Introduction

Premixed combustion requires that fuel and oxidizer be completely mixed before combustion is allowed to take place. Examples of practical applications are spark-ignition engines, lean-burn gas turbines, and household burners. In all three cases fuel and air are mixed before they enter into the combustion chamber. Such a premixing is only possible at sufficiently low temperatures where the chain-branching mechanism that drives the reaction chain in hydrogen and hydrocarbon oxidation is unable to compete with the effect of three-body chain-breaking reactions. Under such low temperature conditions combustion reactions are said to be "frozen." At ambient pressures the crossover from chain-branching to chain-breaking happens when the temperature decreases to values lower than approximately 1,000 K for hydrogen flames or lower than approximately 1,300 K for hydrocarbon flames (cf. Peters, 1997). The frozen state is metastable, because a sufficiently strong heat source, a spark for example, can raise the temperature beyond the crossover temperature and initiate combustion.

Once fuel and oxidizer have homogeneously been mixed and a heat source is supplied it becomes possible for a flame front to propagate through the mixture. This will happen if the fuel-to-air ratio lies between the flammability limits: Flammable mixtures range typically from approximately $\phi = 0.5$ to $\phi = 1.5$, where ϕ is the fuel-air-equivalence ratio defined by (3.12) in Chapter 3. Owing to the temperature sensitivity of the reaction rates the gas behind the flame front rapidly approaches the burnt gas state close to chemical equilibrium, while the mixture in front of the flame typically remains in the unburnt state. Therefore, the combustion system on the whole contains two stable states, the unburnt (index u) and the burnt gas state (index b). These two states correspond to the

2.1 Introduction

lower and the upper branch in Figure 1.1, respectively. In premixed combustion both states exist in the system at the same time; they are spatially separated by the flame front where the transition from one to the other takes place.

The most important application of premixed turbulent combustion is to spark-ignition engines. In a homogeneous charge spark-ignition engine, fuel is injected into the intake manifold where it mixes with the intake air. When this mixture enters into the cylinder, it mixes further with the remaining burnt gas from the previous cycle during the subsequent compression. At approximately 40–20 degrees crank angle before top dead center (TDC) the mixture is nearly entirely homogeneous. It is ignited by a spark forming a laminar flame kernel at first, which rapidly becomes turbulent. This kernel develops into a turbulent flame, which, owing to the low mean velocity in the cylinder, grows nearly spherically until it reaches the combustion chamber walls. Figure 2.1 shows Schlieren photographs of flame propagation in a disk-shaped combustion chamber of a 1.6 liter transparent Volkswagen engine at 2,000 rpm. The piston is equipped with a quartz window that allows one to observe the combustion process. The two large semi circles in Figure 2.1 correspond to the valves of the engine; the larger one on the r.h.s. is the inlet valve; the smaller one on the l.h.s. is the exhaust valve. The spark plug is located in the dark semi circle on the top of the pictures in Figure 2.1. In this series of pictures, ignition had occurred at approximately 40 degrees before TDC. At 22 degrees before TDC the flame kernel has grown to a few millimeters. It then develops further until at 14 degrees before TDC large turbulent structures become visible. The corrugated flame front is located within the bright regions where large density gradients occur. At 4 degrees before TDC there appears a dark region behind the front corresponding to the burnt gas region. At TDC the flame has propagated across most of the visible part of the combustion chamber.

As the burnout of the charge must be completed within a crank angle range up to 40 degrees after TDC, it is clear that the propagation velocity through the charge is a very important quantity that needs to be known. Combustion models that are unable to predict this quantity clearly are incomplete and of little interest for practical applications.

In this chapter we will first introduce the concept of the burning velocity for laminar and turbulent combustion and describe some experimental devices by which the latter can be measured. This implies that the turbulent burning velocity is a well-defined quantity, a problem that will be addressed in Section 2.12 below. We will classify in Section 2.4 the modes of premixed turbulent combustion in terms of a regime diagram. This will guide us through the remaining sections of this chapter. A review of the BML model and the Coherent Flame Model, the most commonly used premixed turbulent combustion models, will

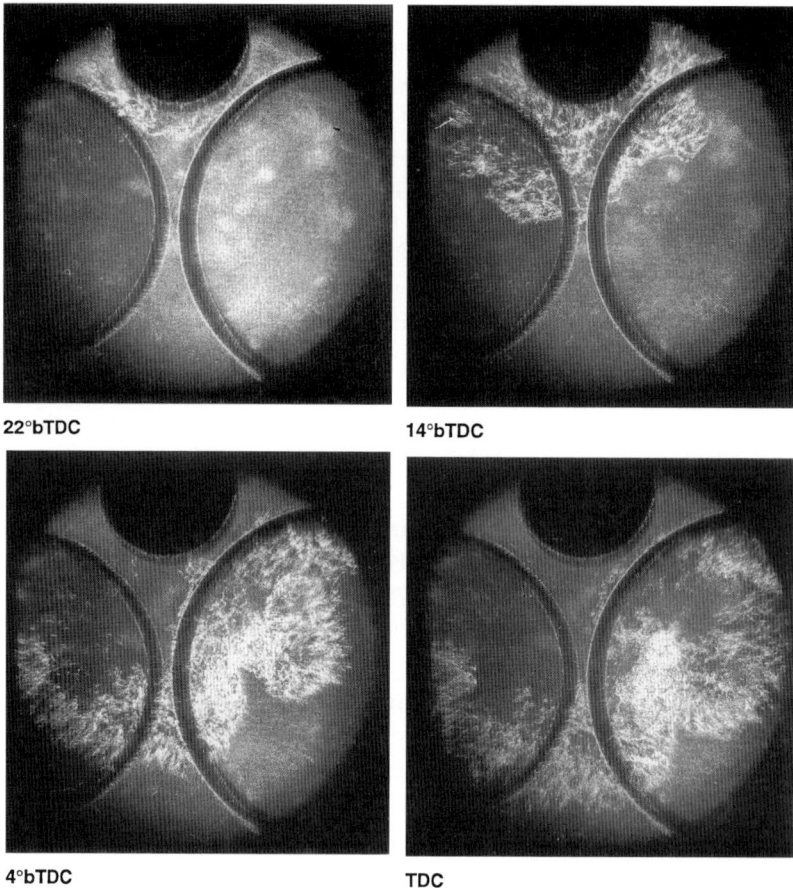

Figure 2.1. Schlieren photographs of turbulent flame propagation, viewed through the transparent piston of a gasoline engine. The spark plug is located at the top of the pictures. The large semicircles are the edges of the valves.

then be presented. The level set approach based on the G-equation will be introduced in Sections 2.5–2.7, and its turbulent modeling in Section 2.8 and 2.9. Since many readers are probably not familiar with the level set approach, various example calculations are presented. The central part of this chapter comprises Sections 2.10–2.12, where Damköhler's scaling laws of the turbulent burning velocity for large scale and small scale turbulence are rederived and are combined into a single expression. Sections 2.13–2.15 are devoted to flamelet equations and the presumed shape pdf approach, which are needed to predict the details of the reactive-diffusive flamelet structure and the calculation of mean scalar quantities. A literature review of premixed turbulent combustion

2.2 Laminar and Turbulent Burning Velocities

calculation models and applications is presented in Section 2.16 and an example calculation of a turbulent Bunsen flame, based on the equations derived in Sections 2.8–2.15, is presented in Section 2.17. Section 2.18 concludes this chapter.

The most important quantity in premixed combustion is the velocity at which the flame front propagates normal to itself and relative to the flow into the unburnt mixture. This velocity is called the laminar burning velocity s_L. It is a thermo-chemical transport property that depends primarily on the fuel-to-air equivalence ratio ϕ, the temperature in the unburnt mixture, and the pressure. It has been measured for various fuels over a wide range of these parameters (cf. Law, 1993). It also can be calculated numerically using elementary or reduced reaction mechanisms and molecular transport properties. For that purpose one considers a planar steady state flame configuration normal to the x direction with the unburnt mixture at $x \to -\infty$, and the burnt gas at $x \to +\infty$. The one-dimensional balance equations for continuity, mass fractions of the chemical species, and energy following from (1.30), (1.62), and (1.76) are

$$\frac{\partial(\rho u)}{\partial x} = 0, \tag{2.1}$$

$$\rho u \frac{\partial Y_i}{\partial x} = -\frac{\partial j_i}{\partial x} + \omega_i, \tag{2.2}$$

$$c_p \rho u \frac{\partial T}{\partial x} = \frac{\partial}{\partial x}\left(\lambda \frac{\partial T}{\partial x}\right) - \sum_{i=1}^{n} c_{p,i} j_i \frac{\partial T}{\partial x} - \sum_{i=1}^{n} h_i \omega_i + q_R. \tag{2.3}$$

The continuity equation may be integrated to show that the mass flow rate through the flame is constant. This defines the burning velocity s_L^0:

$$(\rho u)_{-\infty} = \left(\rho s_L^0\right)_u. \tag{2.4}$$

Here the suffix 0 indicates that the flame is planar and the flow is one dimensional. The momentum equation has been used in the limit of small Mach numbers to obtain the condition $p =$ constant throughout the flame (cf. Williams, 1985a, p. 143). Solving (2.2) and (2.3) with prescribed values for $Y_{i,u}$ and T_u and zero gradient or equilibrium boundary conditions downstream yields the burning velocity as an eigenvalue of the problem. Results of numerical calculations of laminar burning velocities for hydrogen, methanol, and hydrocarbon fuels up to propane may be found in Peters and Rogg (1993). The influence of heat losses due to radiation have been analyzed, for instance, by Kennel et al. (1990). Numerical and asymptotic analyses based on reduced chemical mechanisms are

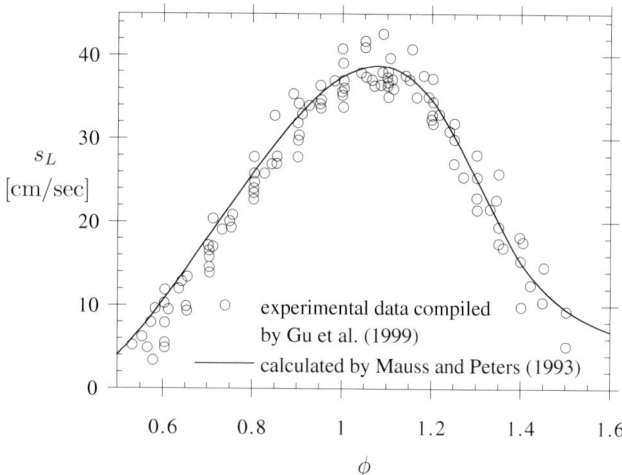

Figure 2.2. Burning velocities calculated by Mauss and Peters (1993) with a detailed mechanism, containing up to C_2-hydrocarbons, compared with data compiled by Gu et al. (2000) for atmospheric methane–air-flames.

reviewed by Seshadri and Williams (1994) and Seshadri (1996). As an example, burning velocities of methane–air-flames calculated by Mauss and Peters (1993), based on an elementary mechanism containing species up to C_2-hydrocarbons, are compared with a recent compilation of experimental data by Gu et al. (2000) and are shown as a function of the equivalence ratio ϕ in Figure 2.2.

The very existence of a property with dimension [m/s] introduces new scaling laws. The analysis in this chapter will emphasize the consequences that result from the laminar burning velocity as an additional property in a fluid-dynamical system. An immediate consequence of the existence of a velocity scale in a diffusive medium is the existence of a length scale. Length scales in laminar flames have been discussed by Peters (1991). A suitable definition of a flame thickness of a premixed flame is

$$\ell_F = \frac{(\lambda/c_p)_0}{(\rho s_L^0)_u} \qquad (2.5)$$

(cf. Göttgens et al., 1992). Here the heat conductivity λ and the heat capacity c_p are evaluated at the inner layer temperature T_0 while, as a consequence of (2.4), the product of the density ρ and the laminar burning velocity s_L^0 is evaluated in the unburnt gas. If the definition $\ell_F = D/s_L^0$ relating the flame thickness ℓ_F to the diffusivity D and the burning velocity is used, as in (2.23) below, and s_L^0 is taken as $s_{L,u}^0$, (2.5) defines the diffusivity D as

$$D = s_{L,u}^0 \ell_F = (\lambda/c_p)_0/\rho_u. \qquad (2.6)$$

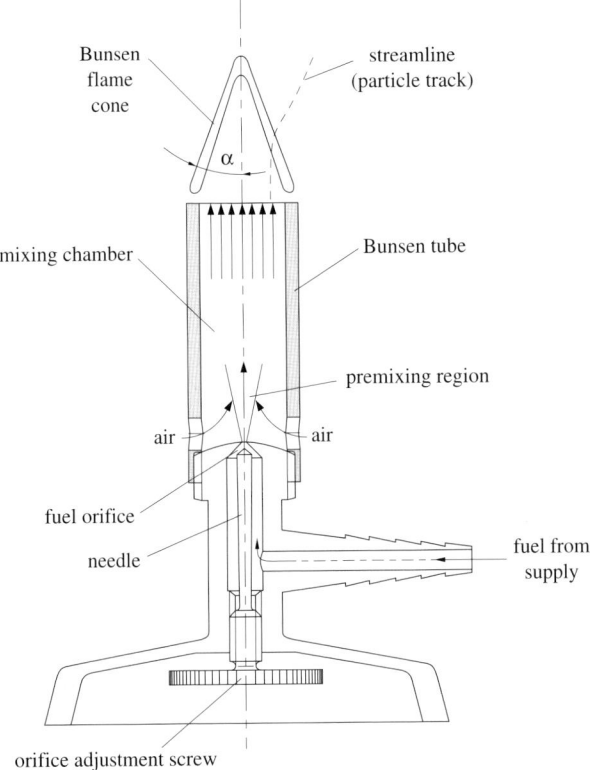

Figure 2.3. The design of a classical Bunsen burner showing the premixing in the Bunsen tube and the Bunsen cone at the exit of the burner.

To address the various issues and physical phenomena associated with flame propagation, we first consider a classical experimental device, the Bunsen burner shown in Figure 2.3. Gaseous fuel from the fuel supply enters through an orifice into the mixing chamber, into which air is entrained from the outside through adjustable openings. The opening area of the fuel orifice may be adjusted by moving the needle through an adjustment screw into the orifice, thereby allowing the velocity of the jet entering into the mixing chamber to be varied and the entrainment of the air and the mixing to be optimized. The mixing chamber must be long enough to generate a fully premixed gas.

Laminar or turbulent steady premixed flames can be established on a Bunsen burner. If the velocity of the mixture is sufficiently large, the flow inside the Bunsen tube becomes turbulent. Turbulence may also be generated by a turbulence grid at the upper end of the mixing chamber, for example. If the velocity of the flow issuing from the tube is larger than the laminar or the turbulent

2. Premixed turbulent combustion

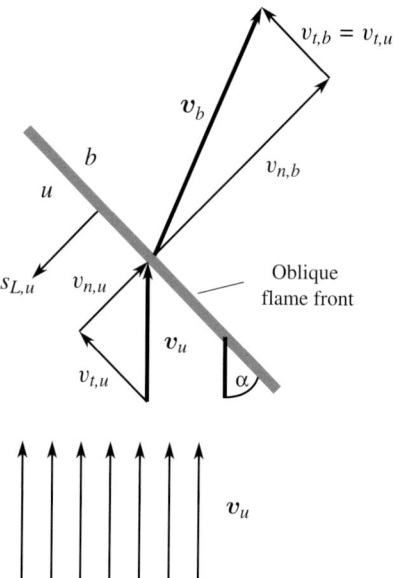

Figure 2.4. Kinematic balance for a steady oblique flame.

burning velocity s_L or s_T, respectively, a flame cone is established. The angle α of the cone is a measure for the laminar or turbulent burning velocity, as will be shown in the following.

The kinematic balance between the flow velocity and the burning velocity is illustrated for a steady oblique flame in Figure 2.4. The oncoming flow velocity vector v_u of the unburnt mixture is split into a component $v_{t,u}$ that is tangential to the flame front and a component $v_{n,u}$ normal to the flame front. Owing to gas expansion within the flame front the normal velocity component $v_{n,b}$ on the burnt gas side is larger than $v_{n,u}$, since, because of continuity, the mass flow ρv_n in normal direction through the flame must be the same in the unburnt mixture and in the burnt gas:

$$(\rho v_n)_u = (\rho v_n)_b, \qquad (2.7)$$

while the density decreases. Therefore

$$v_{n,b} = v_{n,u} \frac{\rho_u}{\rho_b}. \qquad (2.8)$$

The tangential velocity component v_t is not affected by the gas expansion and remains the same:

$$v_{t,b} = v_{t,u}. \qquad (2.9)$$

2.2 Laminar and turbulent burning velocities

Vector addition of the velocity components in the burnt gas in Figure 2.4 then leads to v_b, which points into a direction that is deflected from the direction of v_u in the unburnt mixture.

Finally, since the flame front is stationary in this example, the burning velocity with respect to the unburnt mixture must be equal to the flow velocity of the unburnt mixture normal to the front. For a laminar flow the laminar burning velocity is obtained from the kinematic balance

$$s_{L,u}^0 = v_{n,u}. \tag{2.10}$$

Similarly, for a turbulent flow the turbulent burning velocity $s_{T,u}$ is equal to the mean normal velocity,

$$s_{T,u}^0 = \bar{v}_{n,u}, \tag{2.11}$$

indicating that the turbulent burning velocity is an averaged quantity. With the angle of the Bunsen flame cone in Figure 2.4 denoted by α, the normal velocity is

$$v_{n,u} = |v_u| \sin \alpha, \tag{2.12}$$

for both laminar and turbulent flow. Therefore the laminar and the turbulent burning velocities with respect to the unburnt gas are

$$\begin{aligned} s_{L,u}^0 &= |v_u| \sin \alpha, \\ s_{T,u}^0 &= |\bar{v}_u| \sin \alpha. \end{aligned} \tag{2.13}$$

This allows one to determine the burning velocity experimentally by measuring the flow velocity and the cone angle α. It will be shown below that for a given fuel–air mixture the turbulent burning velocity depends on the turbulence intensity v' and the integral length scale ℓ. In general, not only the mean flow velocity $\bar{v}_u$ but also v' and ℓ depend on the radial direction in a turbulent Bunsen flow. Therefore the flame angle and also the flame brush thickness, to be introduced below, varies with radial distance.

A particular phenomenon occurs at the flame tip of a Bunsen flame. If the tip is closed, which is in general the case for turbulent flames in the mean (but not necessarily for laminar flames), the burning velocity at the tip, being opposite and therefore equal to the flow velocity, is a factor $1/\sin \alpha$ larger than the burning velocity through the oblique part of the cone. This local increase of the burning velocity is caused by the merging of the flame fronts leading to an enhanced burnout. Finally, it is shown in Figure 2.3 that the flame is detached from the rim of the burner. This is caused by heat loss to the burner, which,

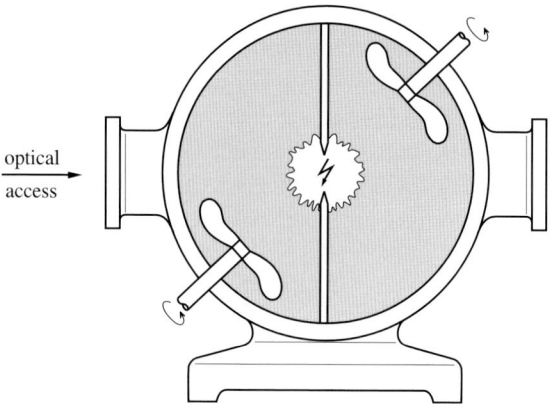

Figure 2.5. Spherical flame propagation in a combustion vessel.

in the region very close to the rim, leads to temperatures at which combustion cannot be sustained.

Another example of an experimental device for measuring burning velocities is the combustion vessel (Figure 2.5) within which a flame is initiated by a central spark. The spherical flame propagation that follows may optically be detected through quartz windows and the flame propagation velocity dr_f/dt may be recorded. This set-up is designed to generate a nonstationary motion of the flame front. It can be used to measure laminar and turbulent burning velocities. In the latter case a turbulent flow field must be generated before spark ignition. Using four mutually opposed high speed fans, a nearly homogeneous isotropic turbulence field has been generated by Andrews et al. (1975), Abdel-Gayed and Bradley (1977), Abdel-Gayed and Bradley (1981), and in many subsequent studies by the Leeds group (cf. Bradley, 1992). One or two flame kernels were initiated by electrical sparks. The flame kernel development and the turbulent burning velocity are measured by high speed cinematography. This configuration is very suitable for studying the unsteady flame development at early times.

To investigate the structure of steady turbulent flames and their turbulent burning velocities, one would like to generate a steady planar turbulent flame in a sufficiently isotropic turbulent flow field. Three different configurations, namely the conical Bunsen flame, the rod-stabilized V-flame, and the turbulent flame stabilized in the stagnation flow in front of a disk, have been compared by Cheng and Shepherd (1991). They find that, while the measured turbulent burning velocities are consistent in these configurations, the turbulent transport processes are geometry dependent.

2.2 Laminar and turbulent burning velocities

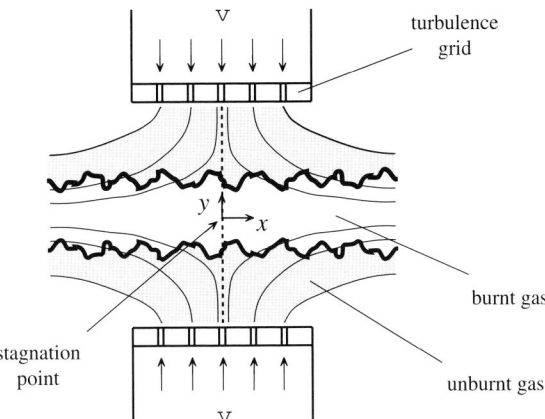

Figure 2.6. Two turbulent counterflow premixed flames.

A modification of the stagnation flow geometry is the counterflow configuration shown in Figure 2.6, where two planar flames are stabilized in a divergent axially symmetric flow between two opposed ducts. There are a number of experimental and theoretical papers using this configuration to study the structure of turbulent premixed flames (cf. Kostiuk et al., 1993a,b, Bray et al., 1991, and Wu and Bray, 1997). In this configuration, which we will consider for the laminar case in an example in Section 2.5, the component of the velocity in the y direction decreases from the values at the exits of the ducts to zero at the stagnation point but is approximately independent of the x direction. Therefore the two flame fronts are approximately normal to the y direction. Disadvantages of this configuration are the existence of a considerable mean strain and the fact that the two flame fronts may interact with each other.

To avoid these influences on the turbulent flame structure, while maintaining a turbulent flow normal to the main flow direction, a new type of burner, called the "weak swirl burner" has been designed by Bédat and Cheng (1995). This device is shown schematically in Figure 2.7. It consists of a flame tube, which, in addition to the main premixed flow, has four tangential air inlets that generate a circumferential velocity component. This velocity component, however, is restricted only to a small annular region of the outlet flow at the perimeter close to the burner rim but leaves the center core flow unchanged. The weak swirl generates a slightly diverging flow close to the burner rim that stabilizes the downward propagating flame at the vertical position where the mean flow velocity equals the turbulent burning velocity. Turbulent quantities such as v' and ℓ are nearly independent of the radial direction. This setup produces stable premixed turbulent flames for a wide range of premixtures and turbulence

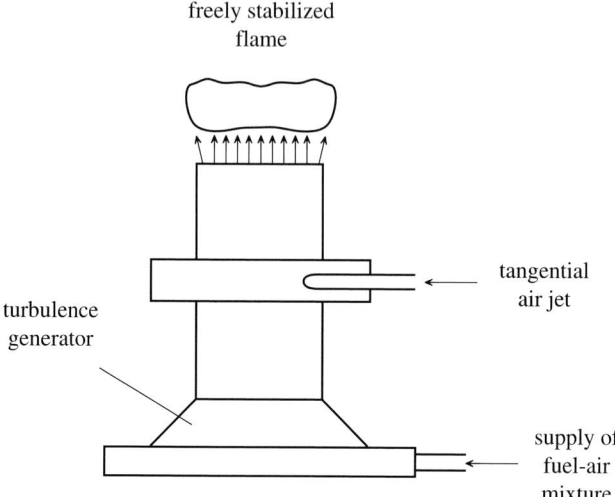

Figure 2.7. The weak swirl burner by Bédat and Cheng (1995) with a freely stabilized premixed flame normal to the mean flow.

intensities. Detailed measurements of the velocity field were reported by Cheng (1995). Turbulent burning velocities and scalar fields were recently measured by Plessing et al. (2000).

To illustrate the kinematic balance among flame propagation velocity, flow velocity, and burning velocity the case of an unsteady radially propagating flame will be discussed next, using the laminar notation, for simplicity. We will ignore the initial flame development, where influences from the spark still prevail, and consider a well-established flame at times, where the increase of pressure (and thereby of the temperature due to adiabatic compression) within the vessel is still negligible. Effects of flame curvature will not be considered here but in the second example in Section 2.5 below. For the unsteady case, the propagation velocity dr_f/dt of the flame front results from an imbalance of the flow velocity and the burning velocity, here written with respect to the unburnt mixture:

$$\frac{dr_f}{dt} = v_u + s_{L,u}^0. \tag{2.14}$$

This can also be written with respect to the burnt gas as

$$\frac{dr_f}{dt} = v_b + s_{L,b}^0. \tag{2.15}$$

2.2 Laminar and turbulent burning velocities

In a moving frame of reference attached to the flame the balance of the mass flow rate through the front is

$$\rho_u \left(v_u - \frac{dr_f}{dt} \right) = \rho_b \left(v_b - \frac{dr_f}{dt} \right). \quad (2.16)$$

In the present example of a laminar spherical flame the flow velocity v_b in the burnt gas behind the flame is zero because of symmetry. With the gas expansion parameter defined as

$$\gamma = \frac{\rho_u - \rho_b}{\rho_u}, \quad (2.17)$$

$v_b = 0$, and (2.16) this leads to

$$\gamma \frac{dr_f}{dt} = v_u. \quad (2.18)$$

Using (2.14) the velocity of the unburnt mixture at the front is calculated as

$$v_u = \frac{\gamma}{1-\gamma} s^0_{L,u}. \quad (2.19)$$

This velocity is induced by gas expansion within the flame front. Introducing (2.19) into (2.18) we see that the propagation dr_f/dt velocity is related to the burning velocity $s^0_{L,u}$ by

$$(1-\gamma) \frac{dr_f}{dt} = s^0_{L,u}. \quad (2.20)$$

Measuring the propagation velocity dr_f/dt then allows us to determine $s_{L,u}$. Furthermore, from (2.15) it follows with $v_b = 0$ that

$$\frac{dr_f}{dt} = s^0_{L,b}. \quad (2.21)$$

Comparing (2.20) and (2.21) shows that the burning velocity with respect to the burnt gas is a factor $\rho_u/\rho_b = 1/(1-\gamma)$ larger than that with respect to the unburnt gas,

$$s^0_{L,b} = \frac{s^0_{L,u}}{1-\gamma}. \quad (2.22)$$

In the following we will use the notation $(\rho s^0_L) = \rho_u s^0_{L,u} = \rho_b s^0_{L,b}$ for the mass flow rate through a laminar flame front, where the brackets indicate that the mass flow rate is a constant. The notation $(\bar{\rho} s^0_T)$ has the equivalent meaning for a turbulent flame front. The burning velocities s^0_L or s^0_T appearing in the text indicate that the location, at which s^0_L or s^0_T are to be evaluated, is not specified. For general considerations, such as regime diagrams and scaling laws it is implicitly understood that s^0_L or s^0_T should be evaluated with respect to the unburnt mixture. For simplicity of notation, we will also remove the suffix 0.

2.3 Regimes in Premixed Turbulent Combustion

Diagrams defining regimes of premixed turbulent combustion in terms of velocity and length scale ratios have been proposed by Borghi (1985), Peters (1986), Abdel-Gayed and Bradley (1989), Poinsot et al. (1990), and many others. For scaling purposes it is useful to assume equal diffusivities for all reactive scalars, to take a Schmidt number $Sc = \nu/D$ of unity, and to define the flame thickness ℓ_F and the flame time t_F as

$$\ell_F = \frac{D}{s_L}, \qquad t_F = \frac{D}{s_L^2}. \tag{2.23}$$

Then, using the turbulent intensity v' and the turbulent length scale ℓ introduced in Chapter 1, we define the turbulent Reynolds number as

$$Re = \frac{v'\ell}{s_L \ell_F} \tag{2.24}$$

and the turbulent Damköhler number as

$$Da = \frac{s_L \ell}{v' \ell_F}. \tag{2.25}$$

Furthermore, with the Kolmogorov time, length, and velocity scales defined in Chapter 1, we introduce two turbulent Karlovitz numbers; the first one defined as

$$Ka = \frac{t_F}{t_\eta} = \frac{\ell_F^2}{\eta^2} = \frac{v_\eta^2}{s_L^2} \tag{2.26}$$

measures the ratios of the flame scales in terms of the Kolmogorov scales. Using the definitions (1.51)–(1.52) with $\nu = D$ and (1.55) taken as equality we see that (2.24)–(2.26) can be combined to show that

$$Re = Da^2 Ka^2. \tag{2.27}$$

Referring to the discussion about the appropriate reaction zone thickness in premixed flames at the end of Section 1.6, one may introduce a second Karlovitz number

$$Ka_\delta = \frac{\ell_\delta^2}{\eta^2} = \delta^2 Ka, \tag{2.28}$$

where (1.100) has been used.

In the following we will discuss a new regime diagram, Figure 2.8, for premixed turbulent combustion (cf. Peters, 1999). Using (2.23) and (2.24) and the definition of the Kolmogorov length scale (1.51), where, for scaling purposes,

2.3 Regimes in premixed turbulent combustion

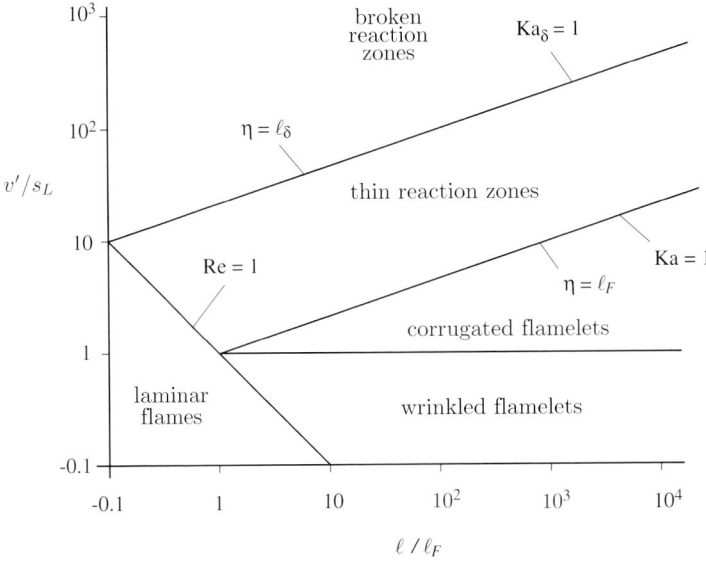

Figure 2.8. Regime diagram for premixed turbulent combustion.

ε is set equal to v'^3/ℓ, the ratios v'/s_L and ℓ/ℓ_F may be expressed in terms of the two nondimensional numbers Re and Ka as

$$\frac{v'}{s_L} = Re \left(\frac{\ell}{\ell_F}\right)^{-1}$$

$$= Ka^{2/3} \left(\frac{\ell}{\ell_F}\right)^{1/3}. \tag{2.29}$$

The lines $Re = 1$ and $Ka = 1$ represent boundaries between different regimes of premixed turbulent combustion in Figure 2.8. Other boundaries of interest are the line $v'/s_L = 1$, which separates the wrinkled flamelets from the corrugated flamelets, and the line denoted by $Ka_\delta = 1$, which separates thin reaction zones from broken reaction zones.

The line $Re = 1$ separates all turbulent flame regimes characterized by $Re > 1$ from the regime of laminar flames ($Re < 1$), which is situated in the lower-left corner of the diagram. As stated in the introduction, we will consider turbulent combustion in the limit of large Reynolds numbers, which corresponds to a region sufficiently removed from the line $Re = 1$ toward the upper right in Figure 2.8. We will not consider the wrinkled flamelet regime, because it is not of much practical interest. In that regime, where $v' < s_L$, the turnover velocity v' of even the large eddies is not large enough to compete with the advancement of the flame front with the laminar burning velocity s_L. Laminar flame propagation therefore dominates over flame front corrugations by turbulence. We will also not consider the broken reaction zones regime in any detail for reasons to be discussed at the end of this section.

Among the remaining two regimes, the corrugated flamelets regime is characterized by the inequalities $Re > 1$ and $Ka < 1$. In view of (2.26) the latter inequality indicates that $\ell_F < \eta$, which means that the entire reactive-diffusive flame structure is embedded within eddies of the size of the Kolmogorov scale, where the flow is quasi-laminar. Therefore the flame structure is not perturbed by turbulent fluctuations and remains quasi-steady.

The boundary of the corrugated flamelets regime to the thin reaction zones regime is given by $Ka = 1$, which, according to (2.26), is equivalent to the condition that the flame thickness is equal to the Kolmogorov length scale. This is called the Klimov–Williams criterion. From (2.26) it also follows that for $Ka = 1$ the flame time is equal to the Kolmogorov time and the burning velocity is equal to the Kolmogorov velocity.

The thin reaction zones regime is characterized by $Re > 1$, $Ka_\delta < 1$, and $Ka > 1$, the last inequality indicating that the smallest eddies of size η can enter into the reactive-diffusive flame structure since $\eta < \ell_F$. These small eddies are still larger than the inner layer thickness ℓ_δ and can therefore not penetrate into that layer. The nondimensional thickness δ of the inner layer in a premixed flame is typically one tenth, such that ℓ_δ is one tenth of the preheat zone thickness which is of the same order of magnitude as the flame thickness ℓ_F. Using (2.28) we see that the line $Ka_\delta = 1$ corresponds with $\delta = 0.1$ to $Ka = 100$. This value is used in Figure 2.8 for the upper limit of the thin reaction zones regime. It seems roughly to agree with the flamelet boundary obtained in numerical studies by Poinsot et al. (1991), where two-dimensional interactions between a laminar premixed flame front and a vortex pair were analyzed. These simulations correspond to $Ka = 180$ for cases without heat loss and $Ka = 25$ with small heat loss. The authors argued that since quenching by vortices occurs only for larger Karlovitz numbers, the region below the limiting value of the Karlovitz number should correspond to the flamelet regime.

We will now enter into a more detailed discussion of the two flamelet regimes. In the regime of corrugated flamelets there is a *kinematic* interaction between turbulent eddies and the advancing laminar flame. Here we have with $Ka < 1$

$$v' \geq s_L \geq v_\eta. \qquad (2.30)$$

To determine the size of the eddy that interacts locally with the flame front, we set the turnover velocity v_n in (1.57) equal to the burning velocity s_L. This

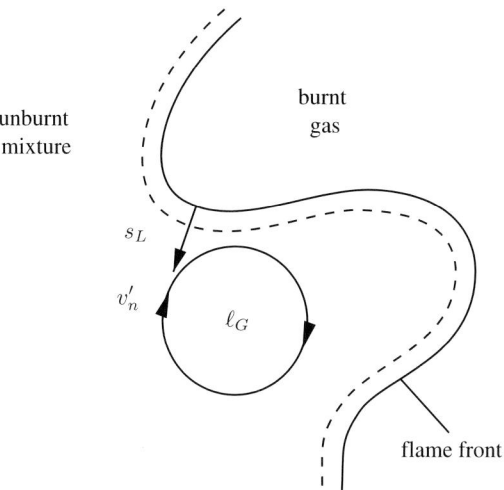

Figure 2.9. Kinematic interaction between a propagating flame front and an eddy of the size $\ell_n = \ell_G$. The dashed line marks the thickness of the preheat zone.

determines the corresponding length ℓ_n as the Gibson scale (cf. Peters, 1986)

$$\ell_G = \frac{s_L^3}{\varepsilon}. \tag{2.31}$$

Only eddies of size ℓ_G, which have a turnover velocity $v_n = s_L$, can interact with the flame front. This is illustrated in Figure 2.9. Since the turnover velocity of the large eddies exceeds the laminar burning velocity, these eddies will push the flame front around, causing a substantial corrugation. Smaller eddies of size $\ell_n < \ell_G$ having a turnover velocity smaller than s_L will not even be able to wrinkle the flame front. Replacing ε by v'^3/ℓ one may also write (2.31) in the form

$$\frac{\ell_G}{\ell} = \left(\frac{s_L}{v'}\right)^3. \tag{2.32}$$

A graphical derivation of the Gibson scale ℓ_G within the inertial range is shown in Figure 2.10. Here, following Kolmogorov scaling in the inertial range given by (1.57), the logarithm of the velocity v_n is plotted over the logarithm of the length scale ℓ_n. We assume v' and ℓ and thereby ε, and also ν and thereby v_η and η, to be fixed. If one enters on the vertical axis setting the burning velocity s_L equal to v_n, one obtains ℓ_G as the corresponding length scale on the horizontal axis. Also shown is the laminar flame thickness ℓ_F, which is smaller than η in the corrugated flamelets regime. This diagram illustrates the limiting values of ℓ_G: If the burning velocity is equal to v', ℓ_G is equal to the integral length scale ℓ.

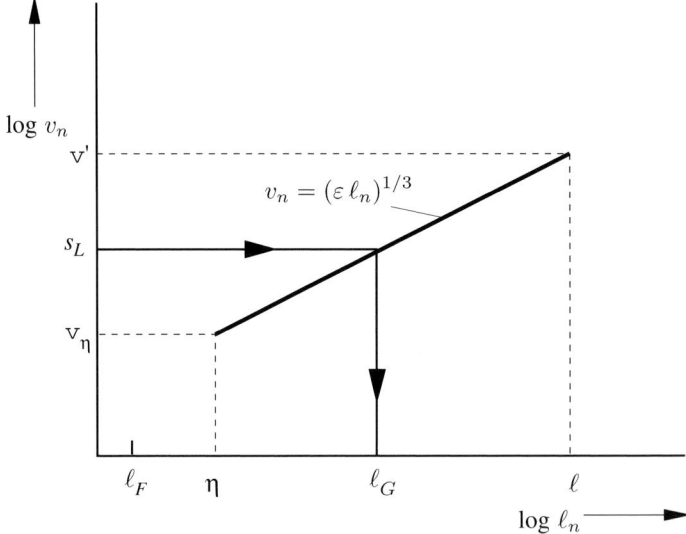

Figure 2.10. Graphical illustration of the Gibson scale ℓ_G within the inertial range for the corrugated flamelets regime.

This case corresponds to the borderline between corrugated and wrinkled flamelets in Figure 2.8. Conversely, if s_L is equal to the Kolmogorov velocity v_η, ℓ_G is equal to η, which corresponds to the line $Ka = 1$ in Figure 2.8.

It has been shown by Peters (1992) that the Gibson scale ℓ_G is the lower cutoff scale of the scalar spectrum function in the corrugated flamelets regime. At that cutoff there is only a weak change of slope in the scalar spectrum function. This is the reason why the Gibson scale is difficult to measure (cf. Menon and Kerstein, 1992, discussed in Section 2.10 below). The stronger diffusive cutoff occurs at the Obukhov–Corrsin scale defined by

$$\ell_C = \left(\frac{D^3}{\varepsilon}\right)^{1/4}. \tag{2.33}$$

Since we have assumed $D = \nu$ this scale is equal to the Kolmogorov scale η.

The next flamelet regime in Figure 2.8 is the regime of thin reaction zones. As noted earlier, since $\eta < \ell_F$ in this regime, small eddies can enter into the preheat zone and increase scalar mixing, but they cannot penetrate into the inner layer since $\eta > \ell_\delta$. The burning velocity is smaller than the Kolmogorov velocity, which would lead to a Gibson scale that is smaller than η. Therefore the Gibson scale has no meaning in this regime.

A time scale, however, can be used in the thin reaction zones regime to define a characteristic length scale using Kolmogorov scaling in the inertial

2.3 Regimes in premixed turbulent combustion

range. That time scale should represent the response of the thin reaction zone and the surrounding diffusive layer to unsteady perturbations. The time scale t_δ defined at the end of Section 1.6 in Chapter 1 is not appropriate because it acts only within the thin reaction zone. A more appropriate time is the quench time t_q, which is the inverse of the strain rate needed to extinguish a premixed flame. Peters (1991) has shown that this time scale is of the same order of magnitude as the flame time t_F. If one combines t_q with the diffusivity D, the resulting diffusion thickness ℓ_D

$$\ell_D = \sqrt{D\, t_q} \tag{2.34}$$

is of the order of the flame thickness ℓ_F. By setting $t_n = t_q$ in (1.59), one obtains the length scale

$$\ell_m = \left(\varepsilon t_q^3\right)^{1/2}. \tag{2.35}$$

This length scale was introduced by Peters (1991), where it was falsely interpreted as a quench scale. A more appropriate interpretation is that of a mixing length scale, which has been advocated, based on the concept of thin reaction zones, by Peters (1999). It is the size of an eddy within the inertial range that has a turnover time equal to the time needed to diffuse scalars over a distance equal to the diffusion thickness ℓ_D. During its turnover time an eddy of size ℓ_m will interact with the advancing reaction front and will be able to transport preheated fluid from a region of thickness ℓ_D in front of the reaction zone over a distance corresponding to its own size. This is schematically shown in Figure 2.11. Much

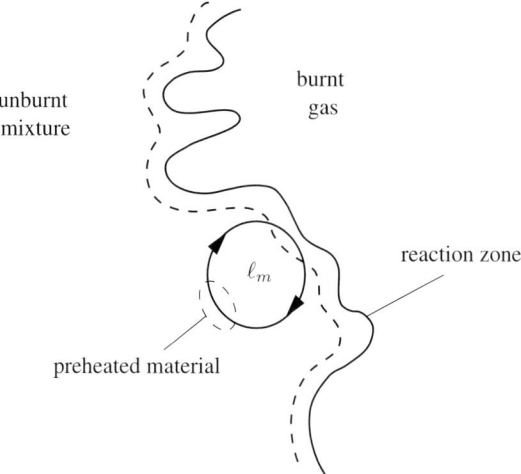

Figure 2.11. Transport of preheated gas from a region of thickness ℓ_D by an eddy of size $\ell_n = \ell_m$ during half a turnover time $t_n = t_q$.

smaller eddies will also do this but since their size is smaller, their action will be masked by eddies of size ℓ_m. Larger eddies have a longer turnover time and would therefore be able to transport thicker structures than those of thickness ℓ_D. They will therefore corrugate the broadened flame structure at scales larger than ℓ_m. The physical interpretation of ℓ_m is therefore that of the maximum distance that preheated fluid can be transported ahead of the flame. As a mixing length scale ℓ_m had already been identified by Zimont (1979).

In contrast to the Gibson length scale the mixing length scale can be observed experimentally. Changes of the instantaneous flame structure with increasing Karlovitz numbers have been measured by Buschmann et al. (1996) who used 2D-Rayleigh thermometry combined with 2D laser-induced fluorescence on a turbulent premixed Bunsen flame. They varied the Karlovitz number between 0.03 and 13.6 and observed at $Ka > 5$ thermal thicknesses that largely exceed the size of the smallest eddies in the flow.

The derivation of ℓ_m also is illustrated in Figure 2.12, which shows (1.59) in a log–log plot of t_n over ℓ_n. If one enters the time axis at $t_q = t_n$, on the length scale axis the mixing length scale ℓ_m is obtained. If t_q is equal to the Kolmogorov time t_η, Figure 2.12 shows that ℓ_m is equal to the Kolmogorov scale η at the border between the thin reaction zones regime and the corrugated flamelets regime. Similarly, from Figure 2.12, if the flame time t_q is equal to the integral

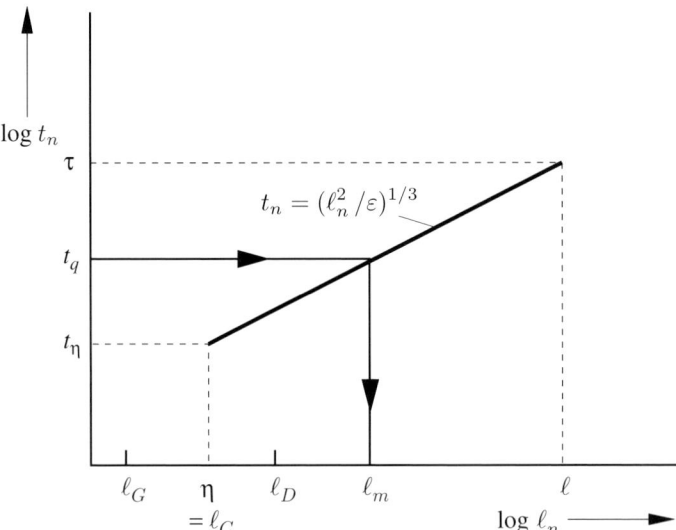

Figure 2.12. Graphical illustration of the mixing scale ℓ_m within the inertial range for the thin reaction zones regime.

time $\tau = k/\varepsilon \approx \ell/v'$, ℓ_m is equal to the integral length scale. This corresponds to $Da = 1$, which Borghi (1985) and Peters (1986) interpreted as the borderline between two regimes in turbulent combustion. However, it merely sets a limit for the mixing scale ℓ_m, which cannot increase beyond the integral scale ℓ.

The diffusion thickness ℓ_D, lying between η and ℓ_m, is also marked in Figure 2.12 as is the Obukhov–Corrsin scale ℓ_C, which is the lower cutoff scale of the scalar spectrum in the thin reaction zones regime. Since we have assumed $\nu = D$, the Obukhov–Corrsin scale ℓ_C is equal to the Kolmogorov length scale η.

As a final remark related to the corrugated flamelets regime and the thin reaction zones regime, it is important to realize that turbulence in high Reynolds number turbulence is intermittent and ε has a statistical distribution. This refinement of Kolmogorov's theory has led to the notion of intermittency or "spottiness" of the activity of turbulence in a flow field (cf. Monin and Yaglom, 1975, Chapter 25). This may have important consequences on the physical appearance of turbulent flames at sufficiently large Reynolds numbers. One may expect that the flame front shows manifestations of strong local mixing by small eddies as in the thin reaction zones regime as well as rather smooth regions where corrugated flamelets appear. The two regimes discussed above may therefore both be apparent in the same experimentally observed turbulent flame.

Beyond the line $Ka_\delta = 1$ there is a regime called the broken reaction zones regime where Kolmogorov eddies are smaller than the inner layer thickness ℓ_δ. These eddies may therefore enter into the inner layer and perturb it with the consequence that chemistry breaks down locally owing to enhanced heat loss to the preheat zone followed by temperature decrease and the loss of radicals. When this happens the flame will extinguish and fuel and oxidizer will interdiffuse and mix at lower temperatures where combustion reactions have ceased.

In a series of papers Mansour et al. (1992), Chen et al. (1996), Chen and Mansour (1997), and Mansour et al. (1998) have investigated highly stretched premixed flames on a Bunsen burner, which were surrounded by a large pilot. Among the flames F1, F2, and F3 that were investigated, the flame F1 with an exit velocity of 65 m/s was close to total flame extinction, which occured on this burner at 75 m/s. A photograph of the flame is shown in Chen et al. (1996). Mansour (1999) has reviewed the recent results obtained from laser diagnostics applied to turbulent premixed and partially premixed flames.

Mansour et al. (1998) have shown that the flame F1 lies on the borderline to the broken flamelets regime in Figure 2.8 having a Karlovitz number of 91. In simultaneous temperature and CH measurements shown in Figure 2.13 they found a thin reaction zone, as deduced from the CH profile and steep temperature

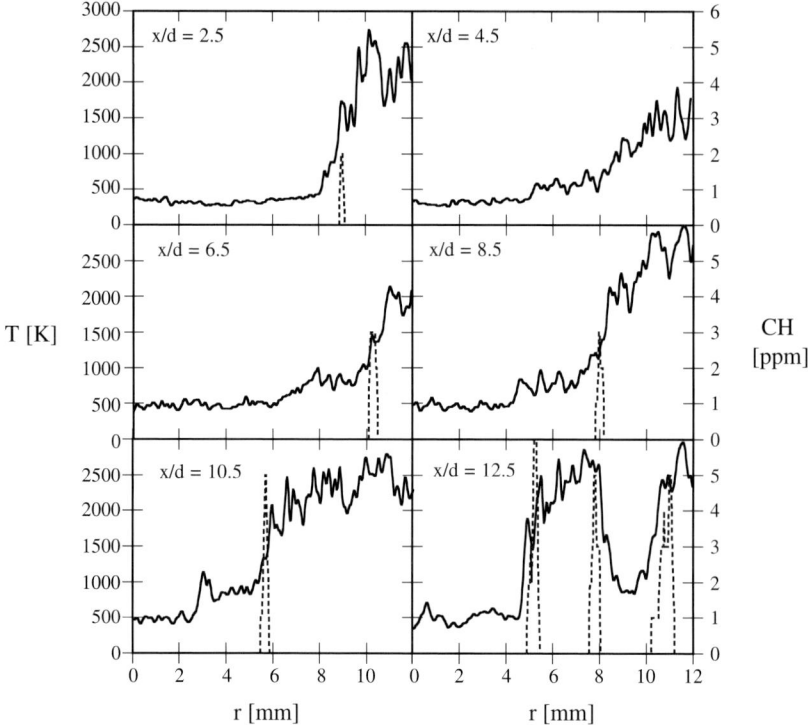

Figure 2.13. Line cuts of one-dimensional temperature (——) and CH concentration (----) profiles in the images presented in Mansour et al. (1998). (Reprinted with permission by The Combustion Institute.)

gradients in the vicinity of that zone. There also was evidence of occasional extinction of the reaction zone. This corresponds to instantaneous shots where the CH profile was absent as in the plot on the upper right in Figure 2.13. Such extinction events do not occur in the flame F3, which has a Karlovitz number of 23 and is located in the middle of the thin reaction zones regime. It can be expected that local extinction events would appear more frequently, if the exit velocity is increased and the flame enters into the broken reaction zones regime. This will occur at an exit velocity close to 75 m/s so frequently that the entire flame extinguishes. Therefore one may conclude that in the broken reaction zones regime a premixed flame is unable to survive.

The plots in Figure 2.13 also show strong perturbations of the temperature profile on the unburnt side of the reaction zone. This is most evident in the plot on the lower left where the temperature reaches more than 1,100 K but falls back to 800 K again. This seems to be due to small eddies that enter into the preheat zone and confirms the concept of the thin reaction zones regime.

2.4 The Bray–Moss–Libby Model and the Coherent Flame Model

The classical flamelet concept for premixed turbulent combustion is the Bray–Moss–Libby (BML) model. It introduces the progress variable c as a scalar quantity, where c is viewed either as a normalized temperature or as a normalized product mass fraction:

$$c = \frac{T - T_u}{T_b - T_u} \quad \text{or} \quad c = \frac{Y_P}{Y_{P,b}}. \tag{2.36}$$

The flamelet concept in the BML model is formulated by assuming that the pdf of c in a turbulent flow field is given by two delta functions as shown in Figure 2.14, at $c = 0$ and at $c = 1$:

$$P(c, \boldsymbol{x}, t) = \alpha(\boldsymbol{x}, t)\delta(c) + \beta(\boldsymbol{x}, t)\delta(1 - c), \tag{2.37}$$

where α and β are the probabilities of finding unburnt or burnt mixture, respectively, at location $\boldsymbol{x}$ and time t. Integration of (2.37) from $c = 0$ to $c = 1$ yields the condition

$$\alpha(\boldsymbol{x}, t) + \beta(\boldsymbol{x}, t) = 1. \tag{2.38}$$

The assumption of the pdf given by (2.37) is equivalent to assuming that the combustion system is either in chemical equilibrium or in the unburnt state and that intermediate states are highly improbable. This assumption was already discussed in Section 1.1 in the context of the S-shaped curve. As a flamelet model the BML model assumes scale separation between combustion and turbulence since the transition from the unburnt to the equilibrium state takes place in infinitely thin layers.

The BML model is very powerful in describing the consequences of this flamelet assumption on unclosed terms in the equation for the Favre averaged mean progress variable $\tilde{c}$. Assuming constant mean molecular weight the ratio

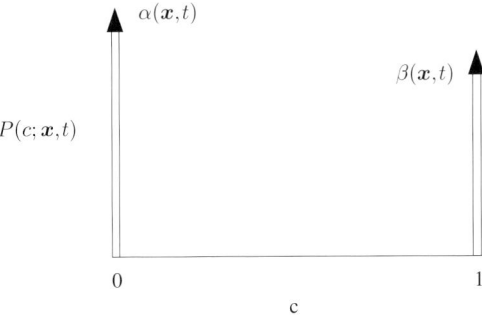

Figure 2.14. The presumed pdf of the progress variable in the BML model.

of the density to that in the unburnt mixture is inversely proportional to the temperature ratio. Then, using (2.36) and the thermal equation of state with constant pressure one may also express the density as a function of the progress variable:

$$\frac{\rho}{\rho_u} = \frac{T_u}{T} = \frac{1-\gamma}{1-\gamma(1-c)}, \tag{2.39}$$

where (2.17) was used. Then the ratio of the mean density $\bar{\rho}$ to the value ρ_u in the unburnt mixture can be calculated from the pdf by taking only the entries at $c = 0$ and $c = 1$ into account:

$$\frac{\bar{\rho}(x,t)}{\rho_u} = \int_0^1 \frac{\rho}{\rho_u} P(c;x,t)\,dc = \alpha(x,t) + \beta(x,t)(1-\gamma) = 1 - \beta(x,t)\gamma, \tag{2.40}$$

where (2.38) has been used. The Favre averaged progress variable $\tilde{c}$ may be obtained by using its definition as

$$\tilde{c}(x,t) \equiv \frac{\overline{\rho c}}{\bar{\rho}} = \frac{\rho_u}{\bar{\rho}} \int_0^1 \frac{(1-\gamma)c}{1-\gamma(1-c)} P(c;x,t)\,dc = \frac{\rho_u}{\bar{\rho}} \beta(x,t)(1-\gamma). \tag{2.41}$$

Combining (2.40) and (2.41) one may express the mean progress variable

$$\bar{c} = \int_0^1 c P(c;x,t)\,dc = \beta(x,t) \tag{2.42}$$

as a function of $\tilde{c}(x,t)$:

$$\bar{c}(x,t) = \frac{\tilde{c}}{1-\gamma(1-\tilde{c})}. \tag{2.43}$$

This also yields a simple expression for the density ratio in terms of $\tilde{c}$:

$$\frac{\bar{\rho}(x,t)}{\rho_u} = \frac{1-\gamma}{1-\gamma(1-\tilde{c})}, \tag{2.44}$$

which shows an interesting analogy to (2.39).

An extension of the methodology introduced by (2.37) is to consider the joint pdf of the progress variable and any velocity component, the flame normal velocity u in the x direction, for instance. This may then be written as

$$P(u,c;x,t) = \alpha(x,t)\delta(c)P(u_u;x,t) + \beta(x,t)\delta(1-c)P(u_b;x,t). \tag{2.45}$$

Here $P(u_u;x,t)$ and $P(u_b;x,t)$ are the conditional pdfs of the velocities in the unburnt and burnt mixture, respectively. Multiplying (2.45) by ρu and integrating over u and c one obtains with the use of (1.27) and (2.39) the Favre mean velocity as

$$\tilde{u}(x,t) = (1-\tilde{c})\bar{u}_u(x,t) + \tilde{c}\bar{u}_b(x,t), \tag{2.46}$$

2.4 The Bray–Moss–Libby model and the Coherent Flame model

where $\bar{u}_u$ and $\bar{u}_b$ are the conditional mean velocities in the unburnt and the burnt mixture, respectively. A similar procedure can be applied to the Favre correlation $\widetilde{u''c''} = \overline{\rho(u-\tilde{u})(c-\tilde{c})}/\bar{\rho}$. One then obtains

$$\widetilde{u''c''} = \tilde{c}(1-\tilde{c})(\bar{u}_b - \bar{u}_u). \tag{2.47}$$

In a steady planar turbulent flame, the mean velocity increases toward the burnt gas in the same way as the density decreases, since the mass flow rate remains constant. Therefore one also may expect that in most parts of the flame brush, the difference of the local conditional mean velocities $(\bar{u}_b - \bar{u}_u)$ is greater than zero and therefore $\widetilde{u''c''} > 0$. This conflicts with the gradient transport assumption

$$-\widetilde{u''c''} = D_t \frac{\partial \tilde{c}}{\partial x}, \tag{2.48}$$

which would require that $\widetilde{u''c''} < 0$ since $\partial \tilde{c}/\partial x > 0$ in the flame brush. This phenomenon has been called counter gradient diffusion (Libby and Bray, 1981). It was confirmed by many numerical and experimental studies. Reviews are given by Libby and Williams (1994) and Bray (1995). Recently, transition from countergradient to gradient diffusion was also found by Kalt et al. (1998) who measured the conditional velocities $\bar{u}_b$ and $\bar{u}_u$ in (2.47). Direct numerical simulations by Wenzel and Peters (2000) based on the G-equation with an expansion term included (cf. Section 2.12) are able to distinguish between the two contributions to scalar transport: the one due to turbulent mixing and the other due to gas expansion. The first term results in gradient diffusion and the second term in countergradient diffusion.

This and all the previous analyses clearly show that countergradient diffusion of the progress variable is due to the gas expansion at the flame front. Since this scales with the laminar burning velocity, it is less important if the turbulence intensity v' is larger than s_L. A simple model for $\widetilde{u''c''}$ was derived by Veynante et al. (1997) from DNS data as

$$\widetilde{u''c''} = \tilde{c}(1-\tilde{c})\left(\frac{\gamma}{1-\gamma}s_L - 2\alpha v'\right), \tag{2.49}$$

where the modeling coefficient α is of order unity but should depend on the length scale ratio ℓ/ℓ_F (cf. Bray, 1996). Effects of pressure gradients can also be included in this equation. This was shown by Veynante and Poinsot (1997), again analyzing DNS data.

The equation for the Favre mean value of the progress variable $\tilde{c}$ is written as

$$\bar{\rho}\frac{\partial \tilde{c}}{\partial t} + \bar{\rho}\tilde{\boldsymbol{v}}\cdot\nabla\tilde{c} + \nabla\cdot(\bar{\rho}\widetilde{\boldsymbol{v}''c''}) = \bar{\omega}_c, \tag{2.50}$$

where the molecular diffusion term has been neglected. In addition to the closure problem associated with the scalar transport term, the mean chemical source term $\bar{\omega}_c$ is unclosed in this equation. If the premixed flamelet structure at the interface between the unburnt and the burnt mixture is very thin, the reaction rate takes large values at that interface and vanishes outside. Therefore the mean chemical source term results from an averaging process over a series of very spiky events, which may be approximated by delta functions in the limit of an infinitely thin flamelet structure. Such an average involves the crossing frequency of flamelets at point x in a steady turbulent flame (Bray et al., 1984a,b). A model for the mean chemical reaction rate is given by

$$\bar{\omega}_c = \rho_u s_L^0 I_0 g \frac{\bar{c}(1-\bar{c})}{\hat{L}_y}. \qquad (2.51)$$

Here s_L^0 is the laminar burning velocity of the unstretched flame and I_0 is a stretch factor, which is quantified, for instance, in Bray and Peters (1994). In (2.51) g is a coefficient that depends on the pdf of the passage times. For an exponential distribution g takes the value two and for a gamma-two distribution the value unity (Bray and Libby, 1986). The mean progress variable $\bar{c}$ in (2.51) can be related to $\tilde{c}$ by (2.43). Finally, $\hat{L}_y$ is the crossing length scale, which needs to be modeled. A more detailed description of the basic results of the BML model may be found in Libby and Williams (1994) and Bray and Libby (1994).

An alternative expression to (2.51) is

$$\bar{\omega}_c = \rho_u s_L^0 I_0 \Sigma, \qquad (2.52)$$

where Σ is the flame surface density. Models based on a transport equation for Σ have been discussed in the context of the Coherent Flame Model in Section 1.11. A particular model, due to Trouvé and Poinsot (1994), has been presented as (1.155). Many of the models are based on direct numerical simulation, for instance Poinsot et al. (1990, 1991). A review is given by Poinsot et al. (1996). Poinsot et al. (1993) have developed a model for flame–wall interaction, while effects of flame instability due to gas expansion were included in a model by Paul and Bray (1996) in the Σ-equation. Similarities between the Σ-equation and the pdf transport equation are discussed in detail by Vervisch et al. (1995).

Formulations of the Coherent Flame Model have been reviewed by Duclos et al. (1993). More recent evaluations are by Choi and Huh (1998) and Prasad and Gore (1999) for unsteady flame propagation and a turbulent Bunsen flame, respectively. Wu and Bray (1997) apply the Coherent Flame Model to premixed combustion in a counterflow geometry.

The flame surface density Σ was determined from experimental data in both a spark-ignition engine and in a Bunsen-type burner by Deschamps et al. (1992) and Deschamps et al. (1996). Measurements of the flame surface density in a

2.5 The Level Set Approach for the Corrugated Flamelets Regime

stagnation point flow and a V-shaped geometry were related to the flame surface area ratio by Shepherd (1996). From flame front visualizations by high-speed cinematography different terms of the Σ-equation were analyzed by Veynante et al. (1994a). The terms related to flame curvature and flame propagation were found to act as source terms on the unburnt gas side and as consumption terms on the burnt gas side. According to the authors, this points at a lack of generality of the closure assumptions used for the last term in (1.155).

2.5 The Level Set Approach for the Corrugated Flamelets Regime

An alternative model for premixed turbulent combustion, based on the nonreacting scalar G rather than on the progress variable, has been developed in recent years. It avoids complications associated with counter-gradient diffusion and, since G is nonreacting, there is no need for a source term closure. An equation for G can be derived by considering an iso-scalar surface

$$G(\boldsymbol{x}, t) = G_0. \tag{2.53}$$

As shown in Figure 2.15 this surface divides the flow field into two regions where $G > G_0$ is the region of burnt gas and $G < G_0$ is that of the unburnt mixture. The choice of G_0 is arbitrary but fixed for a particular combustion event. This is called the level set approach (cf. Sethian, 1996).

In the introduction of this chapter two examples illustrating the burning velocity have been discussed: the steady state Bunsen flame and unsteady spherical flame propagation in a combustion vessel. In both cases there is a kinematic

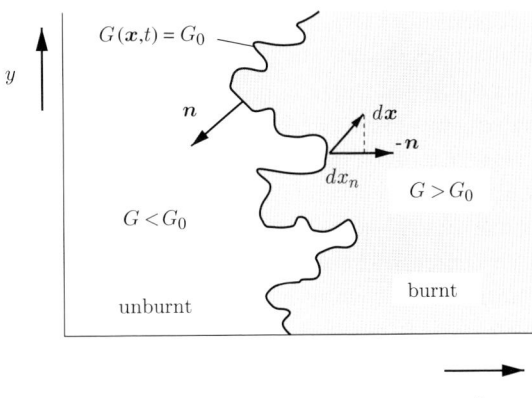

Figure 2.15. A schematic representation of the flame front as an iso-scalar surface $G(\boldsymbol{x}, t) = G_0$.

balance involving the flow velocity and the burning velocity normal to the flame. While in the case of the steady state Bunsen flame these two velocities are equal to each other but in opposite direction, in the case of unsteady spherical flame propagation, their sum defines the propagation velocity of the flame front. The resulting kinematic balance among the three terms in (2.14) can be generalized by introducing the vector normal to the front in the direction of the unburnt gas, as shown in Figure 2.15, by

$$\boldsymbol{n} = -\frac{\nabla G}{|\nabla G|}. \qquad (2.54)$$

In a general three-dimensional flow field the propagation velocity $d\boldsymbol{x}_f/dt$ of the front equals the sum of the flow velocity $\boldsymbol{v}_f$ at the front and the burning velocity in the normal direction:

$$\frac{d\boldsymbol{x}_f}{dt} = \boldsymbol{v}_f + \boldsymbol{n}\, s_L. \qquad (2.55)$$

A field equation can now be derived by differentiating (2.53) with respect to t,

$$\frac{\partial G}{\partial t} + \nabla G \cdot \frac{d\boldsymbol{x}_f}{dt} = 0, \qquad (2.56)$$

and by introducing (2.55) and $\nabla G = -\boldsymbol{n}|\nabla G|$ to obtain the field equation

$$\frac{\partial G}{\partial t} + \boldsymbol{v}_f \cdot \nabla G = s_L |\nabla G|. \qquad (2.57)$$

This equation was introduced by Williams (1985b) and is known as the G-equation in the combustion literature. It is applicable to thin flame structures that propagate with a well-defined burning velocity. It therefore is well suited for the description of premixed turbulent combustion in the corrugated flamelets regime, where it is assumed that the laminar flame thickness is smaller than the smallest turbulent length scale, the Kolmogorov scale. Therefore, the entire flame structure is embedded within a locally quasi-laminar flow field and the laminar burning velocity remains well defined. Equation (2.57) contains a local term and a convective term on the l.h.s, a propagation term with the burning velocity s_L on the r.h.s, but no diffusion term. The quantity G is a scalar, defined at the flame surface only, while the surrounding G-field is not uniquely defined. This is because the kinematic balance (2.55) describes the dynamics of a two-dimensional surface whereas the G-equation (2.57) is an equation in three-dimensional space. In this respect $G(\boldsymbol{x}, t)$ differs fundamentally from the mixture fraction $Z(\boldsymbol{x}, t)$ used in nonpremixed combustion, which is a conserved scalar that is well defined in the entire flow field.

The distance x_n from the flame surface in normal direction, however, can be uniquely defined by introducing its differential increase toward the burnt gas

2.5 The level set approach for the corrugated flamelets regime

side by

$$dx_n = -\mathbf{n} \cdot d\mathbf{x} = \frac{\nabla G}{|\nabla G|} \cdot d\mathbf{x}. \tag{2.58}$$

Here $d\mathbf{x}$ is a differential vector pointing from the front to its surroundings, as shown in Figure 2.15. If we consider a frozen G-field, a differential increase of the G-level is given by

$$dG = \nabla G \cdot d\mathbf{x}. \tag{2.59}$$

Introducing this into (2.58) one sees that the differential increase dx_n is related to dG by

$$dx_n = \frac{dG}{|\nabla G|}. \tag{2.60}$$

This relation will become of interest when flamelet equations will be introduced in Section 2.13. These equations describe the profiles of the reactive scalars in the vicinity of the flame surface. Since G is not uniquely defined outside of $G(\mathbf{x}, t) = G_0$ we will formulate the flamelet equations in terms of x_n, which resolves the distance normal to the instantaneous flame surface.

In the following the absolute value of the gradient of G at $G(\mathbf{x}, t) = G_0$ will be denoted by

$$\sigma = |\nabla G|. \tag{2.61}$$

For illustration purpose we choose as ansatz for the G-field

$$G(\mathbf{x}, t) - G_0 = x + F(y, z, t). \tag{2.62}$$

Thus the flame front displacement $F(y, z, t)$ is assumed to be a single-valued function of y and z as shown for the two-dimensional case in Figure 2.16. This assumption does not allow for multiple crossings of the flame surface. Note that x is the coordinate normal to the mean flame surface. Specifying the G-field by (2.62) it is defined everywhere in the flow field. In Figure 2.16 the scalar distance x_n resolves the vicinity normal to the flame surface, while according to (2.62) G is measured in the x direction. It is also seen that the angle β between the flame normal direction $-\mathbf{n}$ and the x axis is equal to the angle between the tangential direction $\mathbf{t}$ and the y axis.

In this context it is interesting to note that early asymptotic derivations of a flame front equation, essentially aimed at analyzing flame instabilities (cf. Sivashinsky, 1977b, Pelce and Clavin, 1982, and Matalon and Matkowsky, 1982) used an equation for the flame displacement $F(y, z, t)$ rather than for $G(\mathbf{x}, t)$, thereby excluding multiple crossings. The isotropic formulation of the G-equation (2.57) has the advantage that it leads to a field equation in three

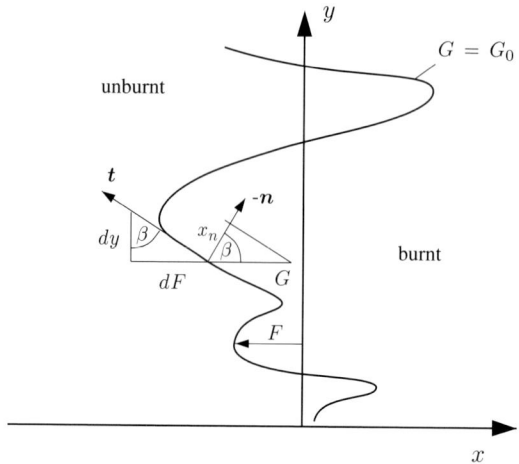

Figure 2.16. Graphical interpretation of the G-field. The movement of the instantaneous flame front position $G = G_0$ is related to spatial fluctuations F.

dimensions and therefore is formally consistent with other field equations in chemically reacting flows.

Below we will present the example of a Bunsen flame on a slot burner where two different solutions of the G-equation are obtained by choosing a different ansatz for $G(x, t)$. However, whatever ansatz is chosen, and whatever the G-field is outside of the flame surface, the location of the surface $G(x, t) = G_0$ itself is independent of that choice.

It is easily seen that a transformation of G, such as $G_1 = aG + b$, will also satisfy (2.57). Furthermore, any *monotonic* function $f(G)$ with $df/dG > 0$ will satisfy (2.57), as noted by Oberlack et al. (2000b). This can be seen by inserting $f(G)$ into (2.57) and canceling the derivative df/dG that will appear in all the terms.

In the corrugated flamelets regime the reactive-diffusive flame structure is assumed to be thin compared to all length scales of the flow. Therefore it may be approximated by a jump of temperature, reactants and products. For such a very thin flame structure the iso-scalar surface $G(x, t) = G_0$ is often defined to lie in the unburnt mixture immediately ahead of the flame structure. Other definitions will be used for flame structures of finite thickness and in a formulation in Section 2.12 where gas expansion effects are taken into account.

Since (2.57) was derived from (2.55), the velocity v_f and the burning velocity s_L are values defined at the surface $G(x, t) = G_0$. In numerical studies values for these quantities must be assigned in the entire flow field. The flow velocity

v_f can simply be replaced by the local flow velocity v, a notation that we will adopt in the following.

The burning velocity s_L appearing in (2.55) and (2.57) may be modified to account for the effect of flame stretch. Performing two-scale asymptotic analyses of corrugated premixed flames, Pelce and Clavin (1982) and Matalon and Matkowsky (1982) derived first-order correction terms for small curvature and strain. The expression for the modified burning velocity s_L becomes

$$s_L = s_L^0 - s_L^0 \mathcal{L} \kappa - \mathcal{L} S. \tag{2.63}$$

Here s_L^0 is the burning velocity of the unstretched planar flame, κ is the curvature, and S is the strain rate. The flame curvature κ is defined in terms of the G-field as

$$\kappa = \nabla \cdot \boldsymbol{n} = \nabla \cdot \left(-\frac{\nabla G}{|\nabla G|} \right) = -\frac{\nabla^2 G - \boldsymbol{n} \cdot \nabla(\boldsymbol{n} \cdot \nabla G)}{|\nabla G|}, \tag{2.64}$$

where $\nabla(|\nabla G|) = -\nabla(\boldsymbol{n} \cdot \nabla G)$ has been used. It is positive if the flame is convex with respect to the unburnt mixture. The strain rate imposed on the flame by velocity gradients is defined as

$$S = -\boldsymbol{n} \cdot \nabla v \cdot \boldsymbol{n}. \tag{2.65}$$

The Markstein length $\mathcal{L}$ appearing in (2.63) is of the same order of magnitude and proportional to the laminar flame thickness ℓ_F; their ratio $\mathcal{L}/\ell_F$ is called the Markstein number. For the case of a one-step reaction with a large activation energy, constant transport properties, and a constant heat capacity c_p, the Markstein length with respect to the unburnt mixture reads, for example,

$$\frac{\mathcal{L}_u}{\ell_F} = \frac{1}{\gamma} \ln \frac{1}{1-\gamma} + \frac{Ze(Le-1)}{2} \frac{(1-\gamma)}{\gamma} \int_0^{\gamma/(1-\gamma)} \frac{\ln(1+x)}{x} dx. \tag{2.66}$$

This expression was derived by Clavin and Williams (1982) and Matalon and Matkowsky (1982). Here $Ze = E(T_b - T_u)/RT_b^2$ is the Zeldovich number, where E is the activation energy and R the universal gas constant, and Le is the Lewis number of the deficient reactant. A different expression can be derived if both s_L and $\mathcal{L}$, are defined with respect to the burnt gas (cf. Clavin, 1985). Tien and Matalon (1991) show how these quantities vary within the flame structure, if a one-step reaction with a large activation energy is assumed. Whenever the location of the flame surface within the flame structure is not defined, we will use the notations s_L and $\mathcal{L}$ for simplicity.

Rogg and Peters (1990) have shown that an expression similar to (2.66) is valid for stoichiometric methane flames, if the four-step mechanism (1.92) is used. Then the activation energy can be replaced by an apparent activation energy and the Lewis numbers is that of the fuel. Markstein numbers have been measured by Searby and Quinard (1990), Dowdy et al. (1990), Tseng et al. (1993), Brown et al. (1996), Karpov et al. (1997), and Aung et al. (1998). Numerical calculations are due to Bradley et al. (1996). Approximations of burning velocities and Markstein numbers have been reported by Müller et al. (1997).

There is a large body of literature on the effect of curvature and strain on quasi-steady flames. Part of the early work is found in the books by Buckmaster and Ludford (1982) and Williams (1985a). Asymptotic studies assuming a one-step reaction with a large activation energy have pointed at the importance of nonunity Lewis numbers and at the possibility of extinction (cf. Clavin, 1985). Numerical solutions for steady strained methane flames with elementary chemistry have been reported by Rogg (1988) and Cant et al. (1994). Mishra et al. (1994) have extracted stretch effects from numerical simulations of unsteady outward propagating spherical flames using full chemistry. Similarly, Najm and Wyckoff (1997) have studied the unsteady interaction of a methane flame with a vortex pair, while Driscoll et al. (1994) and Mueller et al. (1996) have analyzed this interaction experimentally. Two-dimensional direct numerical simulations of methane flames in a stochastically imposed turbulence field were performed by Echekki and Chen (1996) and Chen and Im (1998) to extract statistics of flame stretch. A comparison between experimental and numerical distributions of curvature in turbulent premixed flames was made by Ashurst and Shepherd (1997).

Strain due to flow divergence can be interpreted as streamline curvature. Since strain and curvature have similar effects on the burning velocity they may be summarized as flame stretch (cf. Matalon, 1983). The concept of stretch was generalized to account for finite flame thickness (cf. de Goey and Ten Thije Boonkhamp, 1997, de Goey et al., 1997, and Echekki, 1997). In these papers a quasi-one-dimensional analysis of the governing equations was performed to identify different contributions to flame stretch. Experimental studies of stretched flames were performed by Egolfopoulos et al. (1990a,b), Erard et al. (1996), and Deshaies and Cambray (1990), and many others.

Modifications of the laminar burning velocity by curvature effects can lead to so-called diffusional-thermal instabilities. Their theoretical basis has been described by Sivashinsky (1977a). If, for instance, the Lewis number is sufficiently smaller than unity such that the Markstein length becomes negative, (2.63) shows that those parts of a corrugated flame front having a positive curvature κ will propagate faster than those with a negative curvature. Since the curvature is defined as positive when the flame front is convex with respect to the unburnt mixture, the more advanced parts of the front will accelerate relative to the negatively curved parts behind. This will enhance the original corrugations. On the contrary, flames with a Lewis number larger than unity are stabilized by this diffusional-thermal effect. Lewis numbers smaller than unity are typical for fuel-lean hydrogen and most fuel-rich hydrocarbon flames. Instabilities of such flames have been found experimentally, for instance, by Kwon et al. (1992). The Lewis number is approximately unity for methane flames and larger than unity for fuel-rich hydrogen and all fuel-lean hydrocarbon flames other than methane. Therefore, since the first term on the r.h.s. of (2.66) is always positive, the Markstein length is positive for most practical applications of premixed hydrocarbon combustion, occuring typically under stoichiometric or fuel-lean conditions. Whenever the Markstein length is negative, as in lean hydrogen–air mixtures, diffusional-thermal instabilities tend to increase the flame surface area. This is believed to be an important factor in gas cloud explosions of hydrogen–air mixtures. Although turbulence tends to dominate such local effects diffusional-thermal instabilities and instabilities induced by gas expansion (cf. Section 2.12 below) sum up and lead to strong flame accelerations.

If (2.63) is introduced into the G-equation (2.57) it may be written as

$$\frac{\partial G}{\partial t} + \boldsymbol{v} \cdot \nabla G = s_L^0 |\nabla G| - \mathcal{D}_\mathcal{L} \kappa |\nabla G| - \mathcal{L} S |\nabla G|. \tag{2.67}$$

Here

$$\mathcal{D}_\mathcal{L} = s_L^0 \mathcal{L} \tag{2.68}$$

is defined as the Markstein diffusivity.

The curvature term adds a second-order derivative to the G-equation. This avoids the formation of cusps that would result from (2.57) for a constant value s_L^0. If $\mathcal{L} > 0$, the mathematical nature of (2.67) is that of a Hamilton–Jacobi equation with a parabolic second-order differential operator coming from the curvature term. While the solution of the G-equation (2.57) with a constant s_L^0 is solely determined by specifying the initial conditions, the parabolic character of (2.67) requires that the boundary conditions for each iso-surface G must be specified. In particular, the flame is anchored by setting $G(\boldsymbol{x}, t) = G_0$ at the boundaries.

If $\mathcal{L}$ and thereby the Markstein diffusivity $\mathcal{D}_\mathcal{L}$ are negative, the G-equation (2.67) is ill-posed leading to instabilities. This is because the model (2.63) includes only first-order correction terms. Sivashinsky (1977b) has derived a fourth-order derivative that appears in a simplified version of the G-equation. In addition, he considered instabilities induced by gas expansion by including a nonlocal term in his equation. Effects of gas expansion will be discussed in the following example and also in Section 2.12.

2. Premixed turbulent combustion

In the following, to illustrate some properties of the level set approach, we want to consider three examples of laminar flames and determine the flame front position by solving the G-equation.

Example 1: The Laminar Bunsen Flame on a Slot Burner. As a first example we consider a laminar flame on a slot burner. Slot burners with turbulent flames are often used in household heating facilities. Assuming an infinite length of the slot it may be viewed as the two-dimensional version of the axisymmetric Bunsen burner. For simplicity we consider (2.57) with a burning velocity that depends on curvature only, defined in the unburnt mixture immediately ahead of the flame structure. If the vertical coordinate is denoted by x, and the horizontal coordinate by y, and if the vertical velocity u is constant and the horizontal velocity is zero in the unburnt mixture (cf. Figure 2.17a), the steady two-dimensional G-equation takes the form

$$u \frac{\partial G}{\partial x} = s_{L,u}^0 \left(\left(\frac{\partial G}{\partial x} \right)^2 + \left(\frac{\partial G}{\partial y} \right)^2 \right)^{1/2} (1 - \mathcal{L}_u \kappa). \tag{2.69}$$

We will only consider the G-field in the unburnt mixture and use the ansatz

$$G = x + F(y) + G_0. \tag{2.70}$$

Inserting this into (2.69) one obtains

$$u = s_{L,u} \left(1 + \left(\frac{\partial F}{\partial y} \right)^2 \right)^{1/2} (1 - \mathcal{L}_u \kappa). \tag{2.71}$$

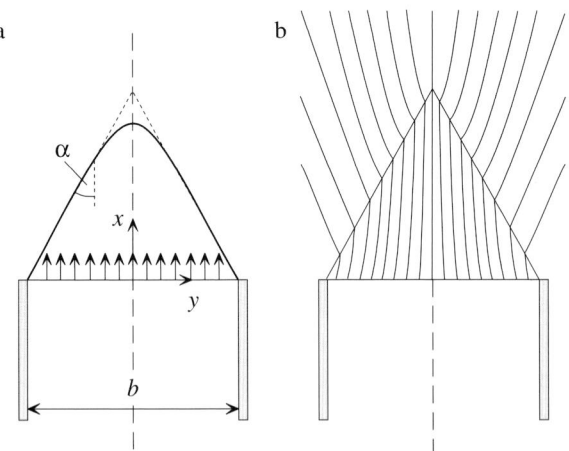

Figure 2.17. A planar premixed laminar flame on a slot burner.

2.5 The level set approach for the corrugated flamelets regime

For the special case $\mathcal{L}_u = 0$ integration of (2.71) leads to

$$F = \frac{\left(u^2 - s_{L,u}^{0\,2}\right)^{1/2}}{s_{L,u}^0}(|y| - \text{const}). \tag{2.72}$$

Steady flames on a Bunsen burner can only be obtained when $u > s_L$. With the burner rim located at $x = 0$, $y = \pm b/2$ where $F = 0$, the constant in (2.72) is evaluated as $b/2$. This leads to the solution of the G-equation as

$$G = \frac{\left(u^2 - s_{L,u}^{0\,2}\right)^{1/2}}{s_{L,u}^0}\left(|y| - \frac{b}{2}\right) + x + G_0. \tag{2.73}$$

The flame surface $x = x_f$ is given by setting $G = G_0$ as

$$x_f = \frac{\left(u^2 - s_{L,u}^{0\,2}\right)^{1/2}}{s_{L,u}^0}\left(\frac{b}{2} - |y_f|\right). \tag{2.74}$$

The flame tip lies on the axis of symmetry $y_f = 0$:

$$x_0 = \frac{b}{2}\frac{\left(u^2 - s_{L,u}^{0\,2}\right)^{1/2}}{s_{L,u}^0}. \tag{2.75}$$

The flame angle α is then given by

$$\tan\alpha = \frac{b/2}{x_0} = \frac{s_{L,u}^0}{\left(u^2 - s_{L,u}^{0\,2}\right)^{1/2}}. \tag{2.76}$$

With $\tan^2\alpha = \sin^2\alpha/(1 - \sin^2\alpha)$ it follows that

$$\sin\alpha = \frac{s_{L,u}^0}{u}, \tag{2.77}$$

which is equivalent to (2.13).

Equation (2.73) represents a cusp at the flame tip on the centerline, since $\mathcal{L}_u$ was set equal to zero. To obtain a rounded flame tip, curvature effects must be accounted for. Inserting (2.70) into (2.64) gives the flame curvature

$$\kappa = -\frac{\partial^2 F/\partial y^2}{[1 + (\partial F/\partial y)^2]^{3/2}}, \tag{2.78}$$

so that instead of (2.71) one has to solve the following equation:

$$\frac{u}{s_{L,u}^0}\left[1 + \left(\frac{\partial F}{\partial y}\right)^2\right] - \left[1 + \left(\frac{\partial F}{\partial y}\right)^2\right]^{3/2} = \mathcal{L}_u \frac{\partial^2 F}{\partial y^2}. \tag{2.79}$$

An exact, but lengthy, implicit solution of this second-order ordinary differential equation can be obtained, which will not be reported here. The equation shows that at the flame tip, where $\partial F/\partial y = 0$, the second derivative of F is positive, which, according to (2.78), corresponds to a negative curvature. The rounded flame contour is schematically shown in Figure 2.17a, together with the cusp-shaped solution obtained for $\mathcal{L}_u = 0$. An experimental analysis showing the tip curvature but also the streamlines in Figure 2.17b was performed by Echekki and Mungal (1990).

We now want to examine an alternative ansatz to solve (2.69) for the case $\mathcal{L}_u = 0$. Instead of using (2.70) one may set $|\nabla G| = 1$ to solve (2.69). This procedure is often used in numerical simulations, where after integrating (2.57) over a certain time interval Δt, the hyperbolic differential equation

$$|\nabla G| = 1 \tag{2.80}$$

is solved outside the flame surface with boundary conditions at that surface. This procedure is called reinitialization and will be presented with the numerical example at the end of this chapter. If one introduces (2.80) into (2.69) one obtains the two equations

$$u\frac{\partial G}{\partial x} = s_{L,u}, \qquad \left(\frac{s_{L,u}}{u}\right)^2 + \left(\frac{\partial G}{\partial y}\right)^2 = 1. \tag{2.81}$$

The G-field is then obtained as

$$G = \frac{(u^2 - s_{L,u}^{0\,2})^{1/2}}{u}\left(|y| - \frac{b}{2}\right) + \frac{s_{L,u}^0}{u}x + G_0, \tag{2.82}$$

which differs from (2.73) by the constant factor $s_{L,u}^0/u$. However, setting $G = G_0$ it yields the same results (2.75)–(2.77). This shows that, while the field outside of the surface $G(\mathbf{x}, t) = G_0$ is not uniquely defined, the location of that surface is independent of the ansatz that was introduced.

We want to use the example of a two-dimensional Bunsen flame to illustrate the effect of gas expansion. For that purpose we consider a model where the velocity field generated by gas expansion is superimposed on the constant vertical velocity field considered before. For the thin flame approximation at zero Mach number combustion, Majda and Sethian (1985) have decomposed the velocity $\mathbf{v}$ into a solenoidal (i.e., divergence-free) rotational part $\mathbf{v}_{\text{rot}}$ and an irrotational part $\mathbf{v}_\phi$:

$$\mathbf{v} = \mathbf{v}_{\text{rot}} + \mathbf{v}_\phi. \tag{2.83}$$

Here the irrotational part is defined by

$$\mathbf{v}_\phi = \nabla \phi, \tag{2.84}$$

2.5 The level set approach for the corrugated flamelets regime

where ϕ is a velocity potential. It is known that a curved flame front will produce vorticity in the flow behind the flame, if gas expansion is present. This would show up in the rotational part of (2.83).

Majda and Sethian (1985) have shown that the continuity equation combined with the energy equation leads to a Poisson equation for ϕ with a singular source at the flame front. We want to derive that equation for the case of a thin flame at constant pressure where the densities in the unburnt gas and in the burnt gas are constant. With that assumption the velocity field on either side of the flame front is divergence free, resulting in the potential equation

$$\nabla^2 \phi = 0, \tag{2.85}$$

which is valid everywhere except at the flame surface.

Generalizing (2.16) to a front propagating in a three-dimensional coordinate system, one can write the balance of the mass flow rate through the surface in the thin flame approximation as

$$\rho_u \boldsymbol{n} \cdot \left(\boldsymbol{v}_u - \frac{d\boldsymbol{x}_f}{dt} \right) = \rho_b \boldsymbol{n} \cdot \left(\boldsymbol{v}_b - \frac{d\boldsymbol{x}_f}{dt} \right). \tag{2.86}$$

Introducing (2.55) applied with respect to the unburnt mixture one obtains the velocity jump normal to the front as

$$\boldsymbol{n} \cdot (\boldsymbol{v}_u - \boldsymbol{v}_b) = \frac{\gamma}{1-\gamma} s_{L,u}, \tag{2.87}$$

while there is no jump of the tangential velocities. The flow field in the vicinity of the flame front therefore is given by

$$\boldsymbol{v} = \boldsymbol{v}_u + H(x_n)(\boldsymbol{v}_b - \boldsymbol{v}_u), \tag{2.88}$$

where H is the Heaviside step function and the normal coordinate x_n introduced in (2.58) has been used. Taking the divergence of $\boldsymbol{v}$ one obtains with Equations (2.83), (2.84), and (2.87)

$$\nabla^2 \phi = \delta(x_n) \boldsymbol{n} \cdot (\boldsymbol{v}_u - \boldsymbol{v}_b) = \delta(G(\boldsymbol{x}, t) - G_0) \frac{\gamma}{1-\gamma} s_{L,u}. \tag{2.89}$$

This is a Poisson equation with a singular volume source confined to the flame surface.

Keller (1993) has derived a closed-form solution of (2.89) for a fixed flame surface given by (2.74) in terms of a Green's function solution. The flow field was obtained by adding the constant flow velocity u to that obtained from the solution of the Poisson equation. The resulting streamlines are shown in Figure 2.17b. It shows the deflection of the streamlines at the flame surface that was discussed in the context of Figure 2.4. Note that the superposition of a potential flow field with an independently defined solenoidal flow field is a model that neither satisfies the Navier–Stokes equations nor the boundary conditions at the burner exit. We will use this model in Section 2.12.

Example 2: The Propagating Spherical Flame. As a second example we want to explore the influence of curvature on the burning velocity for the case of a spherical propagating flame. Since the flow velocity is zero in the burnt gas, it is advantageous to formulate the G-equation with respect to the burnt gas as in (2.21). The burning velocity in (2.67) is then $s_{L,b}^0$ and the Markstein length is that with respect to the burnt gas, $\mathcal{L}_b$, which differs from that given by (2.66) (cf. Clavin, 1985). Here we assume $\mathcal{L}_b > 0$ to avoid complications associated with thermo-diffusive instabilities. In a spherical coordinate system (2.67) becomes

$$\frac{\partial G}{\partial t} = s_{L,b}^0 \left(\left|\frac{\partial G}{\partial r}\right| + \frac{2\mathcal{L}_b}{r} \frac{\partial G}{\partial r} \right), \tag{2.90}$$

where the second term in the parentheses represents the curvature term in spherical coordinates. We introduce the ansatz

$$G = r_f(t) - r, \tag{2.91}$$

where $r_f(t)$ is the radial flame position, to obtain at the flame front $r = r_f$

$$\frac{\partial r_f}{\partial t} = s_{L,b}^0 \left(1 - \frac{2\mathcal{L}_b}{r_f} \right). \tag{2.92}$$

This equation may also be found in Clavin (1985). It reduces to (2.21) if $\mathcal{L}_b$ is set equal to zero. It may be integrated to obtain

$$s_{L,b}^0 t = r_f - r_{f,0} + 2\mathcal{L}_b \ln\left(\frac{r_f - 2\mathcal{L}_b}{r_{f,0} - 2\mathcal{L}_b}\right), \tag{2.93}$$

where the initial radius at $t = 0$ is denoted by $r_{f,0}$. This expression has no meaningful solutions for $r_{f,0} < 2\mathcal{L}_b$, indicating that there needs to be a minimum initial flame kernel for flame propagation to take off. It should be recalled that (2.63) is only valid if the product $\mathcal{L}\kappa$ is small compared to unity. For $r_{f,0} > 2\mathcal{L}_b$ curvature corrections are important at early times only.

Example 3: Two Strained Premixed Flames in an Oscillating Counterflow. For illustration purpose we want to consider two laminar premixed flames in a back-to-back configuration (cf. Rogg, 1988) in an oscillating diverging flow field. Such a flow field can be generated by a counterflow burner, which was shown in Figure 2.6 for turbulent premixed flames. For simplicity we neglect

2.5 The level set approach for the corrugated flamelets regime

density changes on the flow field but impose an oscillating flow field, which we assume to be given by

$$u(x,t) = a(t)x, \qquad v = -a(t)y. \tag{2.94}$$

Here $a(t)$ is the strain rate. (For the governing equations with nonconstant density refer to (3.31)–(3.33) in Chapter 3.)

With the simplifying assumption of two infinitely wide ducts the two oscillating flames are planar and the curvature κ vanishes. Therefore G depends only on time t and the y coordinate. Assuming a constant value of s_L we can write the G-equation as

$$\frac{\partial G}{\partial t} - a(t)y \frac{\partial G}{\partial y} = s_L \left| \frac{\partial G}{\partial y} \right|. \tag{2.95}$$

From the requirement $G < G_0$ in the unburnt mixture and $G > G_0$ in the burnt gas we expect $\partial G/\partial y < 0$ for the upper flame and $\partial G/\partial y > 0$ for the lower flame. The gradient σ will be a function of time. Defining $\sigma(t) = |\partial G/\partial y| > 0$ and $G_0 = 0$, R. Klein (private communication) has suggested introducing the ansatz

$$G(y,t) = \mp \sigma(t) y + F(t) + G_0, \tag{2.96}$$

where the minus sign stands for the upper flame and the plus sign for the lower flame. Inserting (2.96) into (2.95) one obtains

$$\frac{\partial F}{\partial t} \mp y \left[\frac{\partial \sigma}{\partial t} - a(t)\sigma \right] = s_L^0 \sigma. \tag{2.97}$$

Comparing powers of y shows that the term in square brackets must be zero, leading to

$$\frac{\partial \ln \sigma}{\partial t} = a(t). \tag{2.98}$$

From (2.98) the gradient σ is obtained as

$$\sigma(t) = \exp \int_0^t a(t)\,dt, \tag{2.99}$$

where $\sigma(0) = 1$ was used as an initial condition. The flame displacement $F(t)$ is then calculated from the remaining two terms in (2.97) as

$$F(t) = s_L^0 \int_0^t \sigma(t)\,dt \tag{2.100}$$

with the initial displacement set as $F(0) = 0$. The flame positions of the two flames are then calculated from (2.96) by setting $G(y,t) = G_0$:

$$y_F(t) = \pm s_L^0 \frac{\int_0^t \sigma(t)\,dt}{\sigma(t)}. \tag{2.101}$$

For a constant value of the strain rate (2.101) leads to the flame positions of the two flames:

$$y_F = \pm \frac{s_L^0}{a}. \quad (2.102)$$

This solution could have been obtained directly from the steady state version of (2.95). For an oscillating strain rate $a(t)$, history effects enter through the time integrations in (2.99) and (2.100). This was studied for the case of a sinusoidal function by Huang et al. (1998) who chose the ansatz

$$G = -y + D(t) \quad (2.103)$$

instead of (2.96), which leads to

$$\frac{\partial D}{\partial t} + a(t)D = s_L^0 \quad (2.104)$$

instead of (2.97). They also included the effect of gas expansion at the flame front. As a consequence, (2.95) is only valid in the unburnt gas and G depends on the appropriate solution of the Euler equations in the burnt gas. This coupling was shown to lead to a nonlinear response including blowoff in the high frequency limit.

2.6 The Level Set Approach for the Thin Reaction Zones Regime

Equation (2.67) is suitable for thin flame structures in the corrugated flamelets regime, where the entire flame structure is quasi-steady and the laminar burning velocity is well defined, but not for the thin reaction zones regime. We now want to derive a level set formulation for the case where the flame structure cannot be assumed quasi-steady because Kolmogorov eddies enter into the preheat zone and cause unsteady perturbations. The resulting equation will be valid in the thin reaction zones regime.

Since the inner layer shown in Figure 1.6 is responsible for keeping the reaction process alive, we identify the thin reaction zone as the inner layer. Its location will be determined by the iso-scalar surface of the temperature setting $T(\mathbf{x}, t) = T_0$, where T_0 is the inner layer temperature. We now consider the temperature equation

$$\rho \frac{\partial T}{\partial t} + \rho \mathbf{v} \cdot \nabla T = \nabla \cdot (\rho D \nabla T) + \omega_T, \quad (2.105)$$

where D is the thermal diffusivity and ω_T the chemical source term. Similar to (2.56) for the scalar G the iso-temperature surface $T(\mathbf{x}, t) = T_0$ satisfies the

2.6 The level set approach for the thin reaction zones regime

condition

$$\frac{\partial T}{\partial t} + \nabla T \cdot \frac{d\mathbf{x}}{dt}\bigg|_{T=T_0} = 0. \tag{2.106}$$

Gibson (1968) has derived an expression for the displacement speed s_d for an iso-surface of nonreacting diffusive scalars. Extending this result to the reactive scalar T leads to

$$\frac{d\mathbf{x}}{dt}\bigg|_{T=T_0} = \mathbf{v}_0 + \mathbf{n}\, s_d, \tag{2.107}$$

where the displacement speed s_d is given by

$$s_d = \left[\frac{\nabla \cdot (\rho D \nabla T) + \omega_T}{\rho |\nabla T|}\right]_0. \tag{2.108}$$

Here the index 0 defines conditions at the thin reaction zone. The normal vector on the iso-temperature surface is defined as

$$\mathbf{n} = -\frac{\nabla T}{|\nabla T|}\bigg|_{T=T_0}. \tag{2.109}$$

We want to formulate a G-equation that describes the location of the thin reaction zones such that the iso-surface $T(\mathbf{x}, t) = T_0$ coincides with the iso-surface defined by $G(\mathbf{x}, t) = G_0$. Then the normal vector defined by (2.109) is equal to that defined by (2.54) and also points toward the unburnt mixture. Using (2.54) and (2.56) together with (2.108) leads to

$$\frac{\partial G}{\partial t} + \mathbf{v} \cdot \nabla G = \left[\frac{\nabla \cdot (\rho D \nabla T) + \omega_T}{\rho |\nabla T|}\right]|\nabla G|, \tag{2.110}$$

where the index 0 is omitted here and in the following for simplicity of notation.

Echekki and Chen (1999) and Peters et al. (1998) show that the diffusive term appearing in the square brackets in this equation may be split into one term accounting for curvature and another for diffusion normal to the iso-surface:

$$\nabla \cdot (\rho D \nabla T) = -\rho D |\nabla T| \nabla \cdot \mathbf{n} + \mathbf{n} \cdot \nabla (\rho D \mathbf{n} \cdot \nabla T). \tag{2.111}$$

This is consistent with the definition of the curvature in (2.64) if the iso-surface $G(\mathbf{x}, t) = G_0$ is replaced by the iso-surface $T(\mathbf{x}, t) = T_0$ and if ρD is assumed constant. Introducing (2.111) into (2.110) one obtains

$$\frac{\partial G}{\partial t} + \mathbf{v} \cdot \nabla G = (s_n + s_r)|\nabla G| - D\kappa|\nabla G|. \tag{2.112}$$

Here $\kappa = \nabla \cdot \boldsymbol{n}$ is to be expressed by (2.64) in terms of the G-field. The quantities s_n and s_r are contributions due to normal diffusion and reaction to the displacement speed of the thin reaction zone and are defined as

$$s_n = \frac{\boldsymbol{n} \cdot \nabla(\rho D \boldsymbol{n} \cdot \nabla T)}{\rho |\nabla T|}, \tag{2.113}$$

$$s_r = \frac{\omega_T}{\rho |\nabla T|}. \tag{2.114}$$

In a steady, unstretched, planar laminar flame the sum of s_n and s_r would be equal to the burning velocity s_L^0. In the thin reaction zones regime, however, the unsteady mixing and diffusion of chemical species and the temperature in the regions ahead of the thin reaction zone will influence the local displacement speed. Then the sum of s_n and s_r, denoted by

$$s_{L,s} = s_n + s_r, \tag{2.115}$$

is not equal to s_L^0 but is a fluctuating quantity that couples the G-equation to the solution of the balance equations of the reactive scalars. There is reason to expect, however, that $s_{L,s}$ is of the same order of magnitude as the laminar burning velocity. The evaluation by Peters et al. (1998) of DNS data confirms this estimate. In that paper it was also found that the mean values of s_n and s_r slightly depend on curvature. This leads to a modification of the diffusion coefficient that partly takes Markstein effects into account. We will ignore these modifications here and consider the following level set equation for flame structures of finite thickness:

$$\frac{\partial G}{\partial t} + \boldsymbol{v} \cdot \nabla G = s_{L,s} |\nabla G| - D\kappa |\nabla G|. \tag{2.116}$$

This equation is defined at the thin reaction zone and $\boldsymbol{v}$, $s_{L,s}$, and D are values at that position. Equation (2.116) is very similar to (2.67), which was derived for thin flame structures in the corrugated flamelets regime. An important difference, apart from that between s_L^0 and $s_{L,s}$, is the difference between $\mathcal{D}_\mathcal{L}$ and D and the disappearance of the strain term. The latter is implicitly contained in the burning velocity $s_{L,s}$.

In an analytical study of the response of one-dimensional constant density flames to time-dependent strain and curvature, Joulin (1994) has shown that in the limit of high frequency perturbations the effect of strain disappears entirely and Lewis-number effects also disappear in the curvature term such that $\mathcal{D}_\mathcal{L}$ approaches D. This analysis was based on one-step large activation energy asymptotics with the assumption of a single thin reaction zone. It suggests that (2.116) could also have been derived from (2.67) for the limit of high frequency perturbations of the flame structure. This strongly supports it as a level set equation for flame structures of finite thickness and shows that unsteadiness of that structure is an important feature in the thin reaction zones regime.

Since the derivation of (2.116) was based on the balance Equation (2.105) for the temperature, the diffusion coefficient is the thermal diffusivity. However, a similar derivation could have been based on any other reactive scalar defining the position of the inner layer. Then the diffusivity of that particular scalar would appear in (2.116). To obtain the same result we therefore must assume equal diffusivities for all reactive scalars. This is in agreement with the analysis in Section 2.14 below, where it will be shown that equal diffusivities are a good choice for the flamelet equations in the thin reaction zones regime. Since the temperature plays a particular role in combustion owing to the strong temperature sensitivity of chemistry, the use of the thermal diffusivity D is appropriate.

The important difference between the level set formulation (2.116) and the equation for the reactive scalar (2.105), from which it has been derived, is the appearance of a burning velocity, which replaces normal diffusion and reaction at the flame surface. It should be noted that both level set equations, (2.67) and (2.116), are only defined at the flame surface, whereas (2.105) is valid in the entire field.

2.7 A Common Level Set Equation for Both Regimes

It has been anticipated that the two different formulations (2.67) and (2.116) of the G-equation apply to different regimes in premixed turbulent combustion, namely to the corrugated flamelets regime and the thin reaction zones regime of Figure 2.8, respectively. To show this we will analyze the order of magnitude of the different terms in (2.116). This can be done by normalizing the independent variables and the curvature in this equation with respect to Kolmogorov length, time, and velocity scales:

$$t^* = t/t_\eta, \qquad x^* = x/\eta, \qquad v^* = v/v_\eta,$$
$$\kappa^* = \eta\kappa, \qquad \nabla^* = \eta\nabla. \qquad (2.117)$$

Using $\eta^2/t_\eta = \nu$ one obtains

$$\frac{\partial G}{\partial t^*} + v^* \cdot \nabla^* G = \frac{s_{L,s}}{v_\eta}|\nabla^* G| - \frac{D}{\nu}\kappa^*|\nabla^* G|. \qquad (2.118)$$

Since Kolmogorov eddies can perturb the flow field as well as the G-field, all derivatives, the curvature, and the velocity v^* are typically of order unity. In flames D/ν is also of order unity. However, since $s_{L,s}$ is of the same order of magnitude as s_L, (2.26) shows that the ratio $s_{L,s}/v_\eta$ is proportional to $Ka^{-1/2}$. Since $Ka > 1$ in the thin reaction zones regime it follows that

$$s_{L,s} < v_\eta \tag{2.119}$$

in that regime. The propagation term therefore is small and the curvature term will be dominant. Relative small mean values of $s_{L,s}$ may, for instance, result from instantaneously negative values of the burning velocity. Even though wrinkling of the reaction zone by small eddies lead to large local curvatures, it is the mixing within the preheat zone that is rate determining for the advancement of the mean flame front.

It can be shown by a similar analysis that in the corrugated flamelets regime where $Ka < 1$ and therefore

$$s_L^0 > v_\eta, \tag{2.120}$$

the propagation term $s_L^0 |\nabla G|$ is dominant in (2.67) and the curvature and strain terms are of higher order.

We want to base the following analysis on an equation that contains only the leading order terms in both regimes. Therefore we take the propagation term with a constant laminar burning velocity s_L^0 from the corrugated flamelets regime and the curvature term multiplied with the diffusivity D from the thin reaction zones regime. The strain term $\mathcal{L}S$ in the G-equation (2.67) will be neglected in both regimes. Since the Markstein length $\mathcal{L}$ is of the order of the flame thickness, this term is unimportant in the corrugated flamelets regime, where $\mathcal{L}$ is smaller than the Kolmogorov scale. A term called scalar-strain covariance resulting from this term is effective in the diffusive subrange of the scalar spectrum only (cf. Peters, 1992). It therefore does not interact with the inertial range of the spectrum and is unimportant for leading order scaling arguments required for turbulent closure. In the thin reaction zone regime there is no quasi-steady laminar flame structure and a Markstein length cannot be defined.

The leading order equation valid in both regimes then reads

$$\rho \frac{\partial G}{\partial t} + \rho v \cdot \nabla G = (\rho s_L^0)\sigma - (\rho D)\kappa\sigma. \tag{2.121}$$

For consistency with other field equations that will be used as a starting point for turbulence modeling, we have multiplied all terms in this equation by ρ. This will allow us to apply Favre averaging to all equations. Furthermore, we have set (ρs_L^0) constant and denoted this by brackets. This accounts for the mass flow rate (ρs_L^0) through a planar steady flame being constant as shown by (2.4). The parentheses of (ρD) also denote that this product was assumed constant in deriving (2.112). There it was defined at T_0, and since $\rho D = \lambda/c_p$, it is equal to $(\lambda/c_p)_0$ used in the definition of the flame thickness in (2.5). With that definition the last term in (2.121) can also be expressed as $(\rho s_L^0)\ell_F\kappa\sigma$. Again, (2.121) is defined at the flame surface $G(\boldsymbol{x}, t) = G_0$ only.

2.8 Modeling Premixed Turbulent Combustion Based on the Level Set Approach

If the G-equation is to be used as a basis for turbulence modeling, it is convenient to ignore at first its nonuniqueness outside the surface $G(\boldsymbol{x}, t) = G_0$. Then the G-equation would have properties similar to other field equations used in fluid dynamics and scalar mixing. This would allow us to define, at point $\boldsymbol{x}$ and time t in the flow field, a probability density function $P(G; \boldsymbol{x}, t)$ for the scalar G. From $P(G; \boldsymbol{x}, t)$ the first two moments of G, the mean and the variance, can be calculated as

$$\bar{G}(\boldsymbol{x}, t) = \int_{-\infty}^{+\infty} G P(G; \boldsymbol{x}, t) \, dG, \tag{2.122}$$

$$\overline{G'^2}(\boldsymbol{x}, t) = \int_{-\infty}^{+\infty} (G - \bar{G})^2 P(G; \boldsymbol{x}, t) \, dG. \tag{2.123}$$

If modeled equations for these two moments are formulated and solved, one could, for instance, use the presumed shape pdf approach to calculate $P(G; \boldsymbol{x}, t)$ by presuming a two-parameter shape function. However, since G is only defined at the flame front, $P(G; \boldsymbol{x}, t)$ and its moments carry the nonuniqueness of its definition outside $G(\boldsymbol{x}, t) = G_0$.

There is, nevertheless, a quantity that is well defined and of physical relevance, which may be derived from $P(G; \boldsymbol{x}, t)$. This is the probability density of finding the flame surface $G(\boldsymbol{x}, t) = G_0$ at $\boldsymbol{x}$ and t given by

$$P(G_0, \boldsymbol{x}, t) = \int_{-\infty}^{+\infty} \delta(G - G_0) P(G; \boldsymbol{x}, t) \, dG = P(\boldsymbol{x}, t). \tag{2.124}$$

This quantity can be measured, for instance, by counting the number of flame crossings in a small volume ΔV located at $\boldsymbol{x}$ over a small time difference Δt.

In Figures 2.18 and 2.19 two experimental examples of this pdf are shown. The pdf $P(G')$ in Figure 2.18 was obtained by Wirth et al. (1993) by evaluating photographs of the flame front in the transparent spark-ignition engine described in Section 2.1. Smoke particles, which burn out immediately in the flame front, were added to the unburnt mixture, allowing the front to be visualized by a laser sheet as the borderline of the region where Mie scattering of particles could be detected. Experimental details may be found in Wirth and Peters (1992) and

110   2. Premixed turbulent combustion

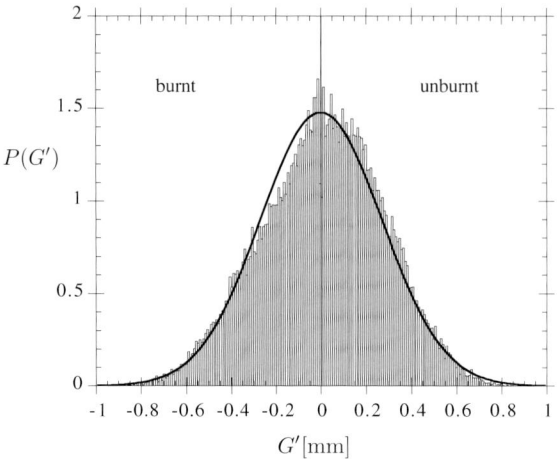

Figure 2.18. Probability density function of flame front fluctuations in an internal combustion engine. —— Gaussian distribution. Measurements by Wirth et al. (1993). (Reprinted with permission from SAE paper 932646 © 1993 Society of Automotive Engineers Inc.)

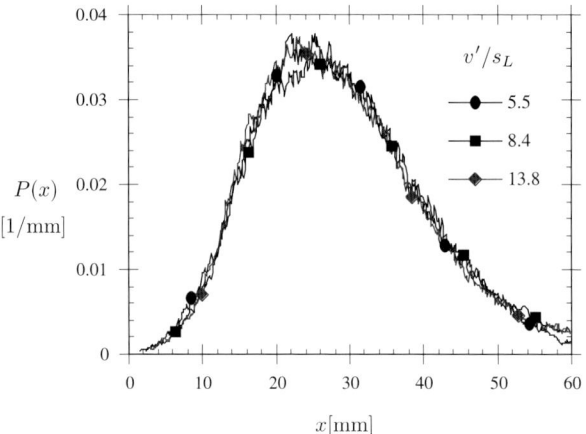

Figure 2.19. The probability density of finding the instantaneous flame front at the axial position x in a turbulent flame stabilized on a weak swirl burner (cf. Figure 2.7). Measurements by Plessing et al. (2000). (Reprinted with permission by The Combustion Institute.)

Wirth et al. (1993). The pdf $P(G')$ represents the pdf of fluctuations around the mean flame contour of several instantaneous images. The relation between G' and the coordinate x normal to the flame brush will be given in (2.131) below.

A comparison of the measured pdf in Figure 2.18 with a Gaussian distribution shows it to be slightly skewed to the unburnt gas side. This is due to

2.8 Modeling premixed turbulent combustion

the nonsymmetric influence of the laminar burning velocity on the shape of the flame front: There are rounded leading edges toward the unburnt mixture but sharp and narrow troughs toward the burnt gas.

This nonsymmetry is also found in the experimental pdfs shown in Figure 2.19. Plessing et al. (1999, 2000) have measured the probability density of finding the flame surface in steady turbulent premixed flames on a weak swirl burner. The flames were stabilized nearly horizontally on the burner thus representing one-dimensional steady turbulent flames. The pdfs were obtained by averaging over 300 temperature images obtained from Rayleigh scattering. The three profiles of $P(x)$, shown in Figure 2.19 for three velocity ratios v'/s_L, nearly coincide and are slightly skewed toward the unburnt gas side.

Without loss of generality, we now want to consider, for illustration purpose, a one-dimensional steady turbulent flame propagating in the x direction. We will analyze its structure by introducing the flame-normal coordinate x, such that all turbulent quantities are a function of this coordinate only. Then the pdf of finding the flame surface at a particular location x within the flame brush simplifies to $P(G_0; x)$, which we write as $P(x)$. We normalize $P(x)$ by

$$\int_{-\infty}^{+\infty} P(x)\,dx = 1 \tag{2.125}$$

and define the mean flame position x_f as

$$x_f = \int_{-\infty}^{+\infty} x P(x)\,dx. \tag{2.126}$$

The turbulent flame brush thickness $\ell_{F,t}$ can also be defined using $P(x)$. With the definition of the variance

$$\overline{(x - x_f)^2} = \int_{-\infty}^{+\infty} (x - x_f)^2 P(x)\,dx \tag{2.127}$$

a plausible definition for $\ell_{F,t}$ is

$$\ell_{F,t} = (\overline{(x - x_f)^2})^{1/2}. \tag{2.128}$$

We note that from $P(x)$ two important properties of a premixed turbulent flame, namely the mean flame position and the flame brush thickness, can be calculated. Now we ask the question: How can these quantities be related to the mean $\bar{G}(x)$ and the variance $\overline{G'^2}(x)$ of the G-field? For the present example we choose a linear dependence of $\bar{G}$ on x:

$$\bar{G}(x) - G_0 = x - x_f. \tag{2.129}$$

The mean flame front position $x = x_f$ is thereby defined by $\bar{G}(x = x_f) = G_0$. Using (2.129), we can relate the spatial fluctuation $(x - x_f)$ of the flame front location to the scalar fluctuation $G' = G - \bar{G}$ as

$$G' = G - G_0 - (x - x_f). \tag{2.130}$$

Since G is only defined at the flame front, one may set $G = G_0$ in (2.130) to obtain the simple equivalence

$$G' = -(x - x_f). \tag{2.131}$$

This shows that spatial fluctuations of the flame front position correspond to fluctuation of the scalar G, conditioned at $G(x, t) = G_0$. Since the variance defined by (2.127) is a property of the entire flame brush, it is by definition independent of the position x within the flame brush. This must also hold for the conditional variance derived from (2.131)

$$\overline{(G'^2)}_0 = \overline{(x - x_f)^2}. \tag{2.132}$$

This requirement distinguishes the conditional variance $\overline{(G'^2)}_0$ from the unconditional variance $\overline{G'^2}(x, t)$ introduced in (2.123). Since it is easier to derive an equation for the unconditional variance, Equation (2.140) below, the additional requirement imposed on the conditional variance may be used to calculate the latter from the former. There are different ways to satisfy this requirement: In numerical simulations where $\overline{G'^2}(x, t)$ is calculated in the entire flow field the requirement can be satisfied by using a reinitialization technique to assign the value of $\overline{G'^2}$ calculated at the mean flame position $\bar{G}(x, t) = G_0$ to the adjacent values in normal direction. Alternatively, if gradients of $\overline{G'^2}$ in the normal direction are small, one may simply use the value of $\overline{G'^2}$ calculated at $\bar{G}(x, t) = G_0$ as a first approximation for $\overline{(G'^2)}_0$. While $\overline{(G'^2)}_0$ is independent of the flame normal coordinate, it may vary, however, in the tangential direction along the mean flame front and with time.

Equation (2.131) leads to an useful interpretation of the scalar G as a fluctuating quantity in turbulent combustion: It represents the scalar distance between the mean and the instantaneous flame front measured in the direction normal to the mean turbulent flame. This also holds for multiple crossings since multiple flame front locations also enter into the pdf $P(x)$ shown in Figure 2.19. This interpretation shall guide the subsequent modeling of equations for the mean and the variance of G.

This analysis may be generalized to any shape of a turbulent flame front $\bar{G}(x, t) = G_0$. Once the $\bar{G}$-field is calculated and the reinitialization condition

2.8 Modeling premixed turbulent combustion

$|\nabla \bar{G}| = 1$ is applied outside of the mean flame surface, the coordinate normal to iso-surfaces of $\bar{G}(x, t)$ is defined by

$$x = \frac{\bar{G}(x, t) - G_0}{|\nabla \bar{G}|} + x_f. \tag{2.133}$$

Integrations across the flame brush then must be performed along those directions.

In recent years, the G-equation has been used in a number of studies to investigate quantities relevant to premixed turbulent combustion. An early review was given by Ashurst (1994). Kerstein et al. (1988) have performed direct numerical simulations of (2.57) in a cubic box, assigning a stationary turbulent flow field and constant density. The constant density assumption has the advantage that the flow field is not altered by gas expansion effects. The gradient $\partial \bar{G}/\partial x$ in the direction of mean flame propagation was fixed at unity and cyclic boundary conditions in the two other directions were imposed. In this formulation all instantaneous G-levels can be interpreted as representing different flame fronts. Therefore G_0 was considered as a variable and averages over all G-levels were taken to show that for large times the mean gradient $\bar{\sigma}$ can be interpreted as the flame surface area ratio. A similar approach has been used by Ulitsky and Collins (1997) to determine the effect of large coherent structures on the turbulent burning velocity.

Peters (1992) considered turbulent modeling of the G-equation in the corrugated flamelets regime and derived Reynolds-averaged equations for the mean and the variance of G. Constant density was assumed and G and the velocity v were each split into a mean and a fluctuation. The main sink term in the variance equation resulted from the propagation term $s_L^0 |\nabla G| = s_L^0 \sigma$ in (2.67) and was defined as

$$\bar{\omega} = -2 s_L^0 \overline{G' \sigma'}. \tag{2.134}$$

The quantity $\bar{\omega}$ was called kinematic restoration in order to emphasize the effect of local laminar flame propagation in restoring the G-field and thereby the flame surface. Corrugations produced by turbulence, which would exponentially increase the area of a nondiffusive iso-scalar surface with time (cf. Batchelor, 1952), are restored by this kinematic effect. Closure of the kinematic restoration term was achieved by deriving the scalar spectrum function of two-point correlations of G in the limit of large Reynolds numbers. That analysis resulted in a closure assumption relating $\bar{\omega}$ to the variance $\overline{G'^2}$ and the integral time scale k/ε:

$$\bar{\omega} = c_\omega \frac{\varepsilon}{k} \overline{G'^2}, \tag{2.135}$$

where $c_\omega = 1.62$ is a modeling constant. This expression shows that kinematic restoration plays a similar role in reducing fluctuations of the flame front as scalar dissipation does in reducing fluctuations of diffusive scalars.

It was also shown by Peters (1992) that kinematic restoration is active at the Gibson scale ℓ_G, since the cutoff of the inertial range in the scalar spectrum function occurs at that scale. A dissipation term involving a positive Markstein diffusivity $D_\mathcal{L}$ was shown to be effective at the Obukhov–Corrsin scale ℓ_C and a term called scalar-strain covariance was shown to be most effective at the Markstein length $\mathcal{L}$. In the corrugated flamelets regime the Gibson scale ℓ_G is larger than both ℓ_C and $\mathcal{L}$. Therefore these additional terms are higher order corrections, which, in view of the order of magnitude assumptions used in turbulence modeling, should be neglected.

A similar analysis was performed by Peters (1999) for the thin reaction zones regime. In that regime the diffusion term in (2.116) is dominant as shown by the order of magnitude analysis of (2.118). This leads to a dissipation term replacing kinematic restoration as the leading order sink term in the variance equation. It is defined as

$$\tilde{\chi} = 2D\overline{(\nabla G')^2}. \qquad (2.136)$$

Closure of that term is obtained in a similar way as for nonreacting diffusive scalars and leads to

$$\tilde{\chi} = c_\chi \frac{\varepsilon}{k} \overline{G'^2}. \qquad (2.137)$$

We will use the two closure relations (2.135) and (2.136) as the basis for the modeling of the turbulent burning velocity in the two different regimes in Section 2.10.

As was the conditional variance $\overline{(G'^2)}_0$, the kinematic restoration $\bar{\omega}$ and the scalar dissipation $\tilde{\chi}$ are conditional quantities defined for the entire flame brush and therefore must be independent of the normal coordinate x within the flame brush. Using the conditional variance in (2.135) and (2.137), respectively, they are to be calculated at the mean flame front. For simplicity of notation we will not distinguish between conditional and unconditional quantities in the following. From the balance equations that will be derived in the next section for unconditional quantities only those calculated at the mean flame front are of physical significance. It is assumed that they represent the respective conditional quantities.

2.9 Equations for the Mean and the Variance of G

To obtain a formulation that is consistent with the well-established use of Favre averages in turbulent combustion, we split G and the velocity vector v into

2.9 Equations for the mean and the variance of G

Favre means and fluctuations:
$$G = \tilde{G} + G'', \qquad v = \tilde{v} + v''. \tag{2.138}$$

Here $\tilde{G}$ and $\tilde{v}$ are at first viewed as unconditional averages. Since in a turbulent flame G was interpreted as the scalar distance between the instantaneous and the mean flame front, evaluated at $G(\mathbf{x}, t) = G_0$, the Favre mean $\tilde{G} = \overline{\rho G}/\bar{\rho}$ represents the Favre average of that distance. If $G(\mathbf{x}, t) = G_0$ is defined to lie in the unburnt mixture immediately ahead of the thin flame structure, as is often assumed for the corrugated flamelets regime, the density at $G(\mathbf{x}, t) = G_0$ is constant and equal to ρ_u. Similarly, if it is an iso-temperature surface, as is assumed for the thin reaction zones regime, changes of the density along that surface are expected to be small. In both cases the Favre average $\tilde{G}$ is approximately equal to the conventional mean value $\bar{G}$. Using Favre averages rather than conventional averages, which might have appeared more appropriate for a nonconserved quantity like the scalar G, therefore has no practical consequences.

The Favre mean velocity $\tilde{v}$ is an unconditional average, but in the end of the analysis of the analysis its conditional counterpart is needed. The conditional mean velocity can be measured by taking averages over the entire flame brush, conditioned at the location of instantaneous flame front. It will be modeled in a similar way as the conditional variance by assuming that it is equal to the unconditional value at the mean flame front position.

Introducing (2.138) into (2.121) leads to the following equation for the Favre mean value of G:

$$\bar{\rho}\frac{\partial \tilde{G}}{\partial t} + \bar{\rho}\tilde{v} \cdot \nabla \tilde{G} + \nabla \cdot (\overline{\rho v'' G''}) = \left(\overline{\rho s_L^0}\right)\bar{\sigma} - \overline{(\rho D)\kappa\sigma}. \tag{2.139}$$

The last term in this equation is proportional to the molecular diffusivity D. In the large Reynolds number limit it is expected to be small compared to the other terms in the equation. Since turbulent modeling is consistent in that limit only, we will neglect this term in the following. We will find its turbulent equivalent, however, below.

An equation for the variance $\widetilde{G''^2}$ may be derived by subtracting (2.139) from (2.121) to obtain an equation for G''. After multiplying this by $2G''$ and averaging one obtains

$$\bar{\rho}\frac{\partial \widetilde{G''^2}}{\partial t} + \bar{\rho}\tilde{v}\cdot\nabla\widetilde{G''^2} + \nabla\cdot(\overline{\rho v'' G''^2}) = -2\overline{\rho v'' G''}\cdot\nabla\tilde{G} - \bar{\rho}\tilde{\omega} - \bar{\rho}\tilde{\chi} - \overline{(\rho D)\mathcal{K}\sigma}. \tag{2.140}$$

Details of the derivation may be found in Peters (1999). As noted before, the condition $\tilde{G}(\mathbf{x}, t) = G_0$ now defines the location of the mean flame front, while the Favre variance $\widetilde{G''^2}$ represents the turbulent flame brush thickness.

Let us at first consider the sink terms in (2.140). The Favre kinematic restoration $\tilde{\omega}$ corresponds to $\bar{\omega}$ defined before and is important in the corrugated flamelets regime only. It is defined as

$$\tilde{\omega} = -2(\rho s_L^0)\overline{G''\sigma}/\bar{\rho} \tag{2.141}$$

and is to be modeled similarly to (2.135). The Favre scalar dissipation is defined as

$$\tilde{\chi} = 2(\rho D)\widetilde{(\nabla G'')^2}/\bar{\rho} \tag{2.142}$$

and is to be modeled in the thin reaction zone regime similarly to (2.137).

The last term $(\rho D)\widetilde{\mathcal{K}\sigma}$ in (2.140) represents a curvature term (cf. Peters, 1999). Since it is proportional to the molecular diffusivity, it is small compared to the other terms in (2.140) in the limit of large Reynolds numbers and will not be considered further.

Since closure of the sink terms $\tilde{\omega}$ and $\tilde{\chi}$ is different in the two regimes, a combined expression must be sought if (2.140) is to be used for a general model of premixed turbulent combustion. The order of magnitude analysis performed on (2.118) shows that the dominant term in the corrugated flamelets regime is kinematic restoration whereas in the thin reaction zones regime it is scalar dissipation. In Peters (1999) the sum of both sink terms $\tilde{\omega}$ and $\tilde{\chi}$, valid for both regimes, was modeled as

$$\tilde{\omega} + \tilde{\chi} = c_s \frac{\tilde{\varepsilon}}{\tilde{k}} \widetilde{G'^2} , \tag{2.143}$$

where c_s is a modeling constant. Peters (1999) suggests a value of 2.0 for this constant and this will be used in the following. If gas expansion effects, which were not considered by Peters (1999), are taken into account, a smaller value appears to be appropriate, as will be discussed in Section 2.12.

Let us now consider the modeling of the correlation $\widetilde{v''G''}$ appearing in both (2.139) and (2.140). The last term $\nabla \cdot (\bar{\rho}\widetilde{v''G''})$ on the l.h.s of (2.139) is a turbulent transport term. A classical gradient transport approximation cannot be used for this term, because this would lead to an elliptic equation for $\tilde{G}$, which is inconsistent with the mathematical character of the G-equation. To obtain for (2.139) the same mathematical form as that of (2.121) the transport term must be modeled as a curvature term. In fact, a transformation similar to (2.111) shows that the second-order elliptic operator that would result from a gradient flux approximation can be split into a normal diffusion and a curvature term:

$$-\nabla \cdot (\bar{\rho}\widetilde{v''G''}) = \nabla \cdot (\bar{\rho} D_t \nabla \tilde{G}) = \tilde{n} \cdot \nabla(\bar{\rho} D_t \tilde{n} \cdot \nabla \tilde{G}) - \bar{\rho} D_t \tilde{\kappa} |\nabla \tilde{G}|. \tag{2.144}$$

2.9 Equations for the mean and the variance of G

Here D_t is the turbulent diffusivity and $\tilde{\boldsymbol{n}}$ and $\tilde{\kappa}$ are defined as in (2.54) and (2.64), respectively, but with $\tilde{G}$ instead of G. Since diffusion normal to G-isolines is already contained in the burning velocity and therefore does not appear in the instantaneous G-equation, it cannot appear in the equation for $\tilde{G}$ for the same reason. We therefore must remove the normal diffusion term in (2.144) to obtain

$$-\nabla \cdot (\widetilde{\bar{\rho} v'' G''}) = -\bar{\rho} D_t \tilde{\kappa} |\nabla \tilde{G}|. \qquad (2.145)$$

This curvature term avoids the formation of cusps of the mean flame front. It is worth noting that by imposing $|\nabla \tilde{G}| = 1$ by reinitialization of the $\tilde{G}$-field outside $\tilde{G}(\boldsymbol{x}, t) = G_0$, normal diffusion automatically vanishes and the formal removal of the normal diffusion term would not have been necessary.

For the turbulent production term in (2.140) classical gradient transport modeling is appropriate since second-order derivatives are not involved, and we have

$$-\widetilde{v'' G''} \cdot \nabla \tilde{G} = D_t (\nabla \tilde{G})^2. \qquad (2.146)$$

For the same reason as above the turbulent transport term in the variance Equation (2.140) must be modeled in a way to avoid turbulent diffusion normal to the mean flame front. It has been noted in Section 2.8 that the conditional variance should not depend on the coordinate normal to the flame front. However, since transport in the tangential direction is permitted, we replace the turbulent transport term in (2.140) by a gradient transport approximation in tangential direction only:

$$-\nabla(\widetilde{\bar{\rho} v'' G''^2}) = \nabla_\| \cdot (\bar{\rho} D_t \nabla_\| \widetilde{G''^2}). \qquad (2.147)$$

The tangential diffusion operator may be calculated by subtracting the normal diffusion $\tilde{\boldsymbol{n}} \cdot \nabla(\bar{\rho} D_t \tilde{\boldsymbol{n}} \cdot \nabla \widetilde{G''^2})$ from the diffusive operator of $\widetilde{G''^2}$.

The equation for $\tilde{G}$, (2.139), models the propagation of the mean turbulent premixed flame front. As in laminar combustion, where the mass flow rate through a steady one-dimensional flame determines the laminar burning velocity s_L^0, as defined in (2.4), we will consider the case of a steady planar turbulent flame, in order to determine the turbulent burning velocity s_T^0, assuming that it is a quantity that depends on local mean quantities only. Whether this assumption is justified will be discussed at the end of Section 2.12 in detail. If it is, the mass flow rate through the flame is constant and equal to $(\bar{\rho} s_T^0)$. One then obtains from (2.139) the equation

$$(\bar{\rho} s_T^0) \frac{\partial \tilde{G}}{\partial x} = (\rho s_L^0) \bar{\sigma}. \qquad (2.148)$$

Here, as before, x is the coordinate normal to the mean turbulent flame surface pointing toward the burnt gas. Similarly to (2.60) one has for the mean turbulent flame

$$dx = \frac{d\tilde{G}}{|\nabla \tilde{G}|}. \tag{2.149}$$

The gradient $d\tilde{G}/dx$ in (2.148) may therefore be replaced by $|\nabla \tilde{G}|$, which leads to

$$(\bar{\rho} s_T^0)|\nabla \tilde{G}| = (\rho s_L^0) \bar{\sigma}. \tag{2.150}$$

This equation relates the turbulent burning velocity s_T^0 to the mean gradient $\bar{\sigma}$. Both are conditional quantities to be evaluated at the mean flame front. Using this and the closure assumptions derived above, Equations (2.139) and (2.140) for $\tilde{G}$ and $\widetilde{G''^2}$ become

$$\bar{\rho}\frac{\partial \tilde{G}}{\partial t} + \bar{\rho}\tilde{v} \cdot \nabla \tilde{G} = (\bar{\rho} s_T^0)|\nabla \tilde{G}| - \bar{\rho} D_t \tilde{\kappa} |\nabla \tilde{G}|, \tag{2.151}$$

$$\bar{\rho}\frac{\partial \widetilde{G''^2}}{\partial t} + \bar{\rho}\tilde{v} \cdot \nabla \widetilde{G''^2} = \nabla_\| \cdot (\bar{\rho} D_t \nabla_\| \widetilde{G''^2}) + 2\bar{\rho} D_t (\nabla \tilde{G})^2 - c_s \bar{\rho} \frac{\tilde{\varepsilon}}{\tilde{k}} \widetilde{G''^2}. \tag{2.152}$$

It is easily seen that (2.151) has the same form as (2.121) and therefore shares its mathematical properties. It also is valid at $\tilde{G}(x, t) = G_0$ only, while the solution outside of that surface depends on the ansatz for $\tilde{G}(x, t)$ that is introduced. The same argument holds for (2.152) since the conditional variance is a property defined at the flame front. The solution of that equation will provide the conditional value $(\widetilde{G''^2})_0$ at the mean flame surface $\tilde{G}(x, t) = G_0$. Following (2.128) and (2.132), we see that its square root is a measure of the flame brush thickness $\ell_{F,t}$, which for an arbitrary value of $|\nabla \tilde{G}|$ at the front, will be defined as

$$\ell_{F,t} = \frac{(\widetilde{G''^2}(x,t))^{1/2}}{|\nabla \tilde{G}|}\bigg|_{\tilde{G}=G_0}. \tag{2.153}$$

To solve (2.151), a model for the turbulent burning velocity s_T^0 must be provided. A first step would be to use empirical correlations from the literature. Alternatively, since s_T^0 is related to $\bar{\sigma}$ by (2.150), a modeled balance equation for the mean gradient $\bar{\sigma}$ will be derived. It will turn out that $\bar{\sigma}$ represents the flame surface area ratio, which is a conditional mean quantity to be calculated from its balance equation at $\tilde{G}(x, t) = G_0$.

An Example Solution for the Turbulent Flame Brush Thickness

For illustration purpose we want to solve the variance equation for a one-dimensional unsteady planar flame using $|\nabla \tilde{G}| = 1$. We pose the problem such that at time $t = 0$ a one-dimensional steady laminar flame with flame thickness ℓ_F is already present and that the laminar flow is suddenly replaced by a fully developed turbulent flow field. We assume that the turbulence quantities D_t, $\tilde{k}$, and $\tilde{\varepsilon}$ are constant, independent of time. Since the flame is planar and, furthermore, since the variance must not depend on the coordinate normal to the mean flame, if it is supposed to represent the conditional variance, all gradients of $\widetilde{G''^2}$ must vanish. Therefore, the convective and diffusive terms in (2.152) disappear entirely.

For modeling purposes we will use the empirical relations given in Table 2.1 in Section 2.11. They follow from (1.38) and (1.50) and relate $\tilde{k}$, $\tilde{\varepsilon}$, and D_t to v', ℓ, and τ. Nondimensionalizing the time in (2.152) by the integral time scale $\tau = \tilde{k}/\tilde{\varepsilon}$ transforms the variance equation into an equation for the turbulent flame brush thickness,

$$\frac{\partial \ell_{F,t}^2}{\partial (t/\tau)} = 2 a_3 a_4 \, \ell^2 - c_s \ell_{F,t}^2, \tag{2.154}$$

which has the solution

$$\ell_{F,t}^2 = b_2^2 \, \ell^2 [1 - \exp(-c_s t/\tau)] + \ell_F^2 \exp(-c_s t/\tau), \tag{2.155}$$

where $b_2 = (2a_3 a_4 / c_s)^{1/2} = 1.78$ for $c_s = 2.0$. Here ℓ_F was used as an initial value. In the limit $\ell_F / \ell \to 0$ one obtains

$$\ell_{F,t} = b_2 \, \ell [1 - \exp(-c_s t/\tau)]^{1/2}. \tag{2.156}$$

The unsteady development of the flame brush thickness in this limit is shown in Figure 2.20. For large times it becomes proportional to the integral length scale ℓ.

2.10 The Turbulent Burning Velocity

It was noted in the beginning of this chapter that one of the most important unresolved problems in premixed turbulent combustion is determining the turbulent burning velocity. This statement implies that the turbulent burning velocity is a well-defined quantity that only depends on local mean quantities. The mean turbulent flame front is expected to propagate with that burning velocity relative to the flow field. Gas expansion effects induced at the mean front will change the surrounding flow field and may generate Darrieus–Landau instabilities in a way similar to those generated by a laminar flame front

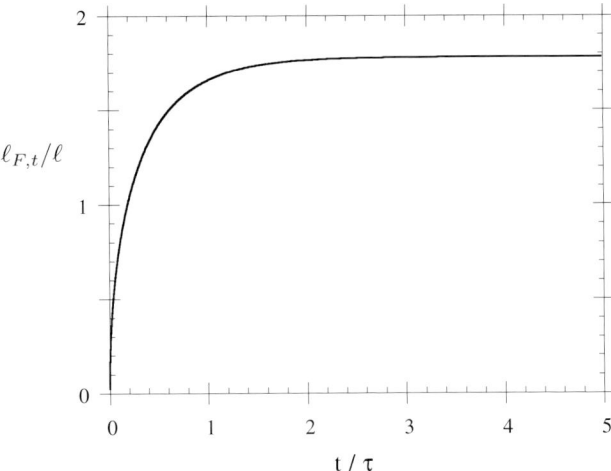

Figure 2.20. Time evolution of the turbulent flame brush thickness $\ell_{F,t}$ normalized by the integral length scale ℓ. The time t is normalized by the integral time scale $\tau = k/\varepsilon$.

(cf. Clavin, 1985). A preliminary analysis of these phenomena will be presented in Section 2.12, suggesting that turbulence suppresses these instabilities at scales that are smaller than the integral length scales. They may, however, develop at larger scales.

Damköhler (1940) was the first to present theoretical expressions for the turbulent burning velocity. He identified two different regimes of premixed turbulent combustion, which he called large scale and small scale turbulence. We will identify these two regimes with the corrugated flamelets regime and the thin reaction zones regime, respectively.

Damköhler equated the mass flux $\dot{m}$ through the instantaneous turbulent flame surface area A_T with the mass flux through the cross-sectional area A, using the laminar burning velocity for the mass flux through the instantaneous surface and the turbulent burning velocity s_T for the mass flux through the cross-sectional area A as

$$\dot{m} = \rho_u s_L A_T = \bar{\rho}_u s_T A. \tag{2.157}$$

This is schematically shown in Figure 2.21. In (2.157) the burning velocities s_L and s_T are defined with respect to the conditions in the unburnt mixture and the density ρ_u is assumed constant. From that equation it follows that the burning velocity ratio s_T/s_L is equal to the flame surface area ratio A_T/A:

$$\frac{s_T}{s_L} = \frac{A_T}{A}. \tag{2.158}$$

2.10 The turbulent burning velocity

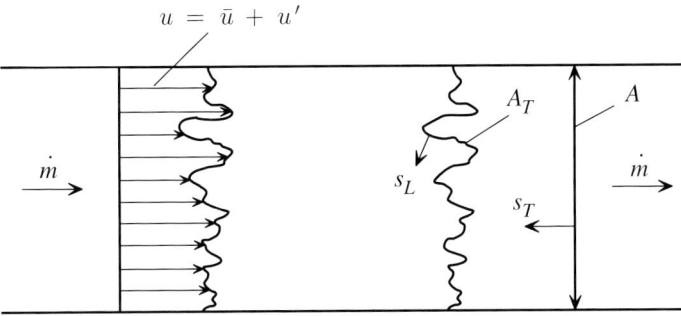

Figure 2.21. An idealized steady premixed flame in a duct.

Since only continuity is involved, averaging of the flame surface area can be performed at any length scale Δ within the inertial range. If Δ is interpreted as a filter width one obtains a filtered flame surface area $\hat{A}_T$. Equation (2.157) then implies that the product of the filtered burning velocity $\hat{s}_T$ and a filtered area $\hat{A}$ is also equal to $s_L A_T$ and to $s_T A$:

$$s_L A_T = \hat{s}_T \hat{A}_T = s_T A. \tag{2.159}$$

This shows that the product $\hat{s}_T \hat{A}$ is inertial range invariant, similar to dissipation in the inertial range of turbulence. This similarity is intriguing, since it implies that the product $s_T A$, like ε, is to be determined at the integral scales and thereby is fixed for a given state of turbulence, while the instantaneous flame surface A_T increases as s_L decreases in a similar way as the velocity gradient in (1.41) increases as ν decreases. As a consequence, by analogy to the large Reynolds number limit used in turbulent modeling, the additional limit of large values of v'/s_L is the backbone of premixed turbulent combustion modeling.

For large scale turbulence, Damköhler (1940) assumed that the interaction between a wrinkled flame front and the turbulent flow field is purely kinematic. Using the geometrical analogy with a Bunsen flame, he related the area increase of the wrinkled flame surface area to the velocity fluctuation divided by the laminar burning velocity,

$$\frac{A_T}{A} \sim \frac{v'}{s_L}. \tag{2.160}$$

Combining (2.158) and (2.160) leads to

$$s_T \sim v' \tag{2.161}$$

in the limit of large v'/s_L, which is a kinematic scaling. We now want to show that this is consistent with the modeling assumption for the G-equation in the corrugated flamelets regime.

The closure model (2.135) shows that in the limit of large values of v'/s_L kinematic restoration is independent of s_L^0 and can be modeled in terms of quantities defined at the integral scales. Using this as the only sink term in (2.140) shows that the variance equation becomes independent of s_L^0. This suggests that the mean propagation term $s_L^0 \bar{\sigma}$ in (2.139) should also be independent of s_L^0. Dimensional arguments (cf. Peters, 1992) then lead to the scaling relations

$$s_L^0 \bar{\sigma} \sim \left(\frac{\varepsilon}{k}\bar{\omega}\right)^{1/2} \sim \frac{\varepsilon}{k}(\widetilde{G''^2})^{1/2}. \tag{2.162}$$

For a steady planar turbulent flame the quantity $(\widetilde{G''^2})^{1/2}/|\nabla \tilde{G}|$ is proportional to the integral length scale ℓ as shown in (2.156). Using in addition $\varepsilon/k \sim v'/\ell$ such that (2.162) becomes $s_L^0 \bar{\sigma} \sim v'|\nabla \tilde{G}|$, and introducing this into (2.150) where the burning velocities are defined with respect to the unburnt mixture, one obtains Damköhler's result (2.161), valid in the corrugated flamelets regime in the limit of large values of v'/s_L.

For small scale turbulence, which we will identify with the thin reaction zones regime, Damköhler (1940) argued that turbulence only modifies the transport between the reaction zone and the unburnt gas. In analogy to the scaling relation for the laminar burning velocity

$$s_L \sim \left(\frac{D}{t_c}\right)^{1/2}, \tag{2.163}$$

where t_c is the chemical time scale and D the molecular diffusivity, he proposes that the turbulent burning velocity can simply be obtained by replacing the laminar diffusivity D by the turbulent diffusivity D_t,

$$s_T \sim \left(\frac{D_t}{t_c}\right)^{1/2}, \tag{2.164}$$

while the chemical time scale remains the same. Thus it is implicitly assumed that the chemical time scale is not affected by turbulence.[†] Combining (2.163) and (2.164) gives the ratio of the turbulent to the laminar burning velocity:

$$\frac{s_T}{s_L} \sim \left(\frac{D_t}{D}\right)^{1/2}. \tag{2.165}$$

Since the turbulent diffusivity D_t is proportional to the product $v'\ell$, and the laminar diffusivity is proportional to the product of the laminar burning velocity

---

[†] This assumption breaks down when Kolmogorov eddies penetrate into the thin reaction zone. This implies that there is an upper limit for the thin reaction zones regime, which was identified as the condition $Ka_\delta = 1$ in Figure 2.8.

2.10 The turbulent burning velocity

s_L and the flame thickness ℓ_F, one may write (2.165) as

$$\frac{s_T}{s_L} \sim \left(\frac{v'}{s_L}\frac{\ell}{\ell_F}\right)^{1/2}, \qquad (2.166)$$

showing that for small scale turbulence the burning velocity ratio not only depends on the velocity ratio v'/s_L but also on the length scale ratio ℓ/ℓ_F.

Performing a similar analysis as before, but for the thin reaction zone regime, suggests that the scaling of $\bar{\sigma}$ must involve the molecular diffusivity D rather than the laminar burning velocity s_L. This points at scalar dissipation rather than at kinematic restoration as the quantity on which dimensional analysis involving $\bar{\sigma}$ must be based. For scaling purposes we write (2.136) and (2.137) as

$$\bar{\chi} \sim D\bar{\sigma}^2 \sim \frac{\varepsilon}{k}\overline{G'^2}. \qquad (2.167)$$

With $\varepsilon/k \sim D_t/\ell^2$ and since $\overline{G'^2}/|\nabla\bar{G}|^2$ is proportional to ℓ^2 for steady planar turbulent flames one obtains in the limit of large Reynolds numbers

$$D\bar{\sigma}^2 \sim D_t|\nabla\bar{G}|^2. \qquad (2.168)$$

If this scaling relation is introduced into (2.150), where the burning velocities are defined with respect to the unburnt mixture, one obtains Damköhler's expression (2.165) for small scale turbulence.

There were many attempts to modify Damköhler's analysis and to derive expressions that would reproduce the large amount of experimental data on turbulent burning velocities. By introducing an adjustable exponent n, where $0.5 < n < 1.0$, (2.161) and (2.166) may be combined to obtain expressions of the form

$$\frac{s_T}{s_L} = 1 + C\left(\frac{v'}{s_L}\right)^n. \qquad (2.169)$$

This includes the limit $v' \to 0$ for laminar flame propagation where $s_T = s_L$. The constant C is expected to depend on the length scale ratio ℓ/ℓ_F. By comparison with experiments the exponent n is often found to be in the vicinity of 0.7 (cf. Williams, 1985a, p. 429ff). Attempts to justify a single exponent on the basis of dimensional analysis, however, fall short even of Damköhler's pioneering work who had recognized the existence of two different regimes in premixed turbulent combustion. It is interesting to note that these two regimes have recently been rediscovered by Marti et al. (1997) using a numerical model based on a Langevin equation.

There is a large amount of data on turbulent burning velocities in the literature. Correlations of this material, mostly presented in terms of the burning

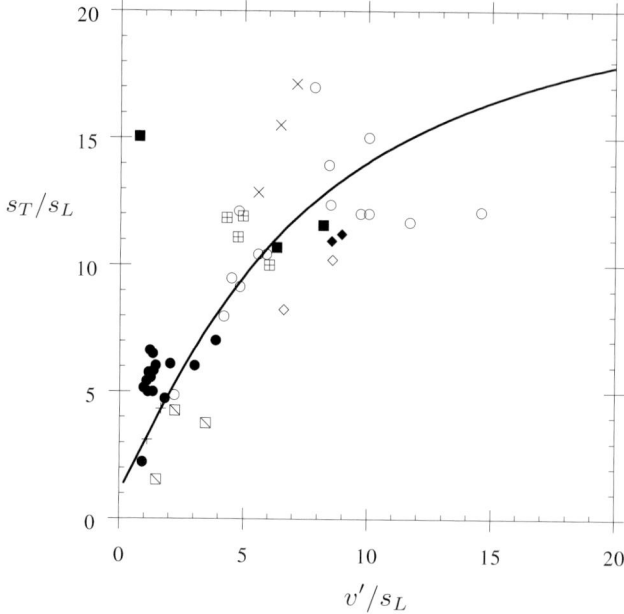

Figure 2.22. Comparison of the burning velocity ratio calculated from (2.195) (solid line), using $Re = 625$, with data from Abdel-Gayed and Bradley (1981) for Reynolds number ranging between 500 and 750. The origin of the individual data points may be found in that reference. From Peters (1999).

velocity ratio s_T/s_L plotted as a function of v'/s_L, called the burning velocity diagram, date back to the 1950s and 1960s. An example taken from Peters (1999) is shown in Figure 2.22. When experimental data from different authors are collected in such a diagram, they usually differ considerably. In the review articles by Bray (1990) and Bradley (1992) the many physical parameters that affect the turbulent burning velocity are discussed.

In a very careful review Abdel-Gayed and Bradley (1981) collected and interpreted all the material that was available at that time. More recent correlations are due to Abdel-Gayed et al. (1987), Gülder (1990b), Bradley et al. (1992), and Zimont and Lipatnikov (1995). Another correlation that focuses on curvature effects alone has been proposed by Karpov et al. (1996). A critical discussion of the various correlations may also be found in Liñán and Williams (1993). More recent data from turbulent Bunsen flames were obtained by Kobayashi et al. (1996, 1997, 1998) in a high pressure environment and by Aldredge et al. (1998) in Taylor–Couette flow. This flow configuration had already been used by Ronney et al. (1995) to study freely propagating chemical fronts in

2.10 The turbulent burning velocity

aqueous solutions (cf. also Shy et al., 1996). Another interesting scaling is that by Khoklov et al. (1996) who found that for buoyancy-driven turbulent premixed flames the turbulent burning velocity scales with the buoyancy induced turbulence intensity $v' \sim (gL)^{1/2}$. Here g is the acceleration due to gravity and L measures the size of the system under consideration. It is suggested that this scaling could be used to calibrate the turbulence model to be used in supernova explosions.

There are also a few theoretical papers that derive expressions for the turbulent burning velocity from first principles. These papers mostly ignore diffusive effects and therefore are valid only in the corrugated flamelets regime. Yakhot (1988) used a renormalization group method leading to a spectral representation of the flow field and the G-equation. He obtains a burning velocity formula that includes a logarithmic term and does not contain adjustable parameters. An extension to include turbulent scales smaller than the flame thickness was formulated by Ronney and Yakhot (1992). Sivashinsky (1988) used an alternative renormalization theory, which leads to a power law expression in the limit $v'/s_L \to \infty$. The renormalization procedure by Pocheau (1992) also predicts a power law dependence. Recently, Klimenko (1998) has presented a scaling law for the turbulent burning velocity based on the cascade hypothesis, which he validates with experimental data by Bradley et al. (1992). He also discusses the relation to fractal dimension and inner cutoff scales. As the theories by Yakhot (1988), Sivashinsky (1988), and Pocheau (1992) are not restricted to the limit $v'/s_L \to \infty$, they could potentially be used to explore deviations from that limit.

The apparently fractal geometry of the flame surface and the fractal dimension that can be extracted from it also has led to predictions of the turbulent burning velocity. Gouldin (1987) has derived a relationship between the flame surface area ratio A_T/A and the ratio of the outer and inner cutoff of the fractal range:

$$\frac{A_T}{A} = \left(\frac{\varepsilon_0}{\varepsilon_i}\right)^{D_f - 2}. \tag{2.170}$$

Here D_f is the fractal dimension. While there is general agreement that the outer cutoff scale ε_0 should be the integral length scale, there are different suggestions by different authors concerning the inner cutoff scale ε_i. Whereas Peters (1986) and Kerstein (1988a) propose, based on theoretical grounds, that ε_i should be the Gibson scale, most experimental studies reviewed by Gülder (1990a) and Gülder (1999) favor the Kolmogorov scale η or a multiple thereof.

The reason for this discrepancy probably lies in the fact that, as long as the flames lie in the corrugated flamelets regime, the changes in the slope of the fractal representation is too weak for the Gibson scale to be accurately determined, and the stronger change at the Kolmogorov scale appears to dominate. Linear Eddy Model calculations by Menon and Kerstein (1992) using the G-equation support these observations. Although turbulent Reynolds numbers up to 1,000 were used in that investigation, the results could not distinguish between the Gibson scale and the Kolmogorov scale as a fractal cutoff. This is attributed to a still insufficient inertial range in the simulations "reflecting an analogous limitation of typical experimental configurations."

As far as the fractal dimension D_f is concerned, the reported values in the literature also vary considerably. Kerstein (1988a) has suggested the value $D_f = 7/3$, which, in combination with the Gibson scale as the inner cutoff, is in agreement with Damköhler's result $s_T \sim v'$ in the corrugated flamelets regime. This is easily seen by inserting (2.170) into (2.158) using (2.32). However, if the Kolmogorov scale is used as inner cutoff, one obtains $s_T/s_L \sim Re^{1/4}$ as Gouldin (1987) has pointed out. This power law dependence seems to have been observed by Kobayashi et al. (1998) in high pressure flames. Gülder (1999) shows in his recent review that most of the measured values for the fractal dimension are smaller than $D_f = 7/3$. He concludes that the available fractal parameters are not capable of correctly predicting the turbulent burning velocity.

It is interesting to note that the closure model (2.135) can also be used to define a Taylor length λ_G of the G-equation in the corrugated flamelets regime. By analogy to the definition of the fluid-dynamical Taylor length scale λ defined by (1.53), λ_G can be related to the mean gradient $\bar{\sigma}$ as

$$\bar{\sigma} \sim G'/\lambda_G. \tag{2.171}$$

Here G' is the r.m.s. of G-fluctuations:

$$G' = (\overline{G'^2})^{1/2}. \tag{2.172}$$

From the definition of $\bar{\omega}$ in (2.134) and the closure (2.135) it follows by dimensional scaling that

$$\bar{\omega} \sim s_L^0 \frac{G'^2}{\lambda_G} \sim \frac{v'}{\ell} G'^2. \tag{2.173}$$

This shows that the ratio of the Taylor length scale λ_G to the integral length scale is proportional to s_L^0/v':

$$\frac{\lambda_G}{\ell} \sim \frac{s_L^0}{v'}. \tag{2.174}$$

Bray and Peters (1994) have pointed at the equivalence of λ_G with the flamelet crossing length scale $\hat{L}_y$. Using (2.160) together with (2.174) one obtains the relations

$$\frac{A_T}{A} \sim \frac{\ell}{\lambda_G} \sim \frac{\ell}{\hat{L}_y}, \qquad (2.175)$$

showing that flame surface area increases with decreasing crossing length scales $\hat{L}_y$. This indicates that an increase of the flame surface area ratio is mainly due to a closer packing of the flame surface by the decrease in the lateral distance $\hat{L}_y$, while the flame brush thickness remains essentially constant proportional to the integral length scale.

In the thin reaction zones regime the analogy with the definition of the fluid-dynamical Taylor length scale is more obvious. From (2.167) and (2.171) a Taylor length of the G-equation can be defined in this regime by

$$\bar{\chi} \sim D \frac{G'^2}{\lambda_G^2}. \qquad (2.176)$$

A geometrical interpretation of the Taylor scale λ_G for the thin reaction zone regime may be found in Peters (1999).

2.11 A Model Equation for the Flame Surface Area Ratio

At the end of Section 2.9 it was stated that the mean gradient $\bar{\sigma}$ represents the flame surface area ratio. By comparing (2.150) with (2.158) this becomes evident. In the two-dimensional illustration in Figure 2.23 the instantaneous flame surface area A_T is identified with the length of the line $G = G_0$. The blowup in that figure shows that a differential section dS of that line and the corresponding differential section dy of the cross sectional area A are related to each other by

$$\frac{dS}{dy} = \frac{1}{|\cos \beta|}. \qquad (2.177)$$

However, in two dimensions the gradient σ is given by

$$\sigma = \left(1 + \left(\frac{\partial F}{\partial y}\right)^2\right)^{1/2}. \qquad (2.178)$$

It can be seen from Figure 2.23 that $\partial F/\partial y = \tan \beta$, which relates σ to the angle β as

$$\sigma = \frac{1}{|\cos \beta|} \qquad (2.179)$$

and therefore, combining this with (2.177), the differential flame surface area ratio is equal to the gradient σ:

$$\frac{dS}{dy} = \sigma. \qquad (2.180)$$

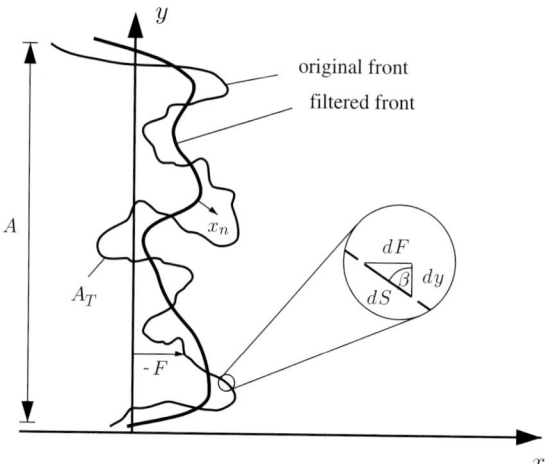

Figure 2.23. Illustration of the filtering of flame front corrugations showing that (2.180) remains valid even if multiple crossings occur.

We therefore expect to be able to calculate the mean flame surface area ratio from a model equation for the mean gradient $\bar{\sigma}$. There remains, however, the question whether this is also valid for multiple crossings of the flame surface with respect to the x axis. To resolve this conceptual difficulty one may define a filtered flame surface by eliminating large wavenumber contributions in a Fourier representation of the original surface, so that in a projection of the original surface on the filtered surface no multiple crossings occur. This is also shown in Figure 2.23. The normal coordinate x_n on the filtered surface then corresponds to x in Figure 2.16, showing that the ansatz assuming a single-valued function of x is again applicable. A successive filtering procedure can then be applied, so that the flame surface area ratio is related to the gradient σ at each level of filtering. Within a given section dy, (2.180) is then replaced by

$$\frac{dS^\nu}{dS^{\nu+1}} = \frac{\sigma^\nu}{\sigma^{\nu+1}}, \qquad (2.181)$$

where ν is an iteration index of successive filtering. The quantities dS^0 and σ^0 correspond to the instantaneous differential flame surface area dS and the respective gradient σ, and dS^1 and σ^1 to those of the first filtering level. At the next iteration one has $dS^1/dS^2 = \sigma^1/\sigma^2$ and so on. At the last filtering level for $\nu \to \infty$ the flame surface becomes parallel to the y coordinate, so that $dS^\infty = dy$ and $\sigma^\infty = 1$. Canceling all intermediate iterations we obtain again (2.180).

2.11 A model equation for the flame surface area ratio

This analysis assumes that the original flame surface is unique and continuous. There may be situations where pockets are formed, as shown in the 2D-simulation by Chen and Im (1998). In a subsequent paper Kollmann and Chen (1998) have shown, however, that singularities in the source terms of the σ-equation, to be presented next, cancel out exactly even during pocket formation.

We now want to derive a modeled equation for the flame surface area ratio $\bar{\sigma}$ in order to determine the turbulent burning velocity. An earlier attempt in this direction is due to Rutland et al. (1990). An equation for σ can be derived from (2.121). For illustration purpose we assume constant density and constant values of s_L^0 and D. Applying the ∇ operator to both sides of the resulting equation and multiplying this with $-\boldsymbol{n} = \nabla G/\sigma$ one obtains

$$\frac{\partial \sigma}{\partial t} + \boldsymbol{v} \cdot \nabla \sigma = -\boldsymbol{n} \cdot \nabla \boldsymbol{v} \cdot \boldsymbol{n} \sigma + s_L^0(\kappa \sigma + \nabla^2 G) + D\boldsymbol{n} \cdot \nabla(\kappa \sigma). \tag{2.182}$$

The terms on the l.h.s. of this equation describe the local rate of change and convection of σ. The first term on the r.h.s accounts for straining by the flow field, which amounts to a production of flame surface area. The next term containing the laminar burning velocity has a similar effect as kinematic restoration has in the variance equation. The last term is proportional to D and its effect is similar to that of scalar dissipation in the variance equation.

To derive a model equation for the mean value $\bar{\sigma}$ we could, in principle, take the appropriate averages of (2.182). There is, however, no standard two-point closure procedure for a balance equation of a gradient, as there is none for deriving the ε-equation from velocity gradients appearing in the definition for the viscous dissipation. Therefore another approach was adopted in Peters (1999): The scaling relations among $\bar{\sigma}$, $\tilde{\varepsilon}$, $\tilde{k}$, and $\widetilde{G''^2}$, namely (2.162) and (2.167), were used separately in both regimes to derive equations for $\bar{\sigma}$ from a combination of the $\tilde{k}$-, $\tilde{\varepsilon}$-, and $\widetilde{G''^2}$-equations. The resulting equations contain the local rate of change and convection of $\bar{\sigma}$, a production term by mean gradients, and another due to turbulence. However, each of them contains a different sink term: In the corrugated flamelets regime the sink term is proportional to $s_L^0 \bar{\sigma}^2$ and in the thin reaction zones it is proportional to $D\bar{\sigma}^3$. Finally, to obtain a common equation for $\bar{\sigma}$ valid in both regimes, the two sink terms are assumed to be additive as are the two terms in (2.182), which also are proportional to s_L^0 and D, respectively.[†]

[†] This does not take the analysis by Kollmann and Chen (1998) into account, who showed that during pocket formation, contributions to the kinematic restoration as well as to the scalar dissipation term in (2.182) become singular. They are of opposite sign and only their sum cancels. This shows that their means cannot be modeled independently in cases where pocket formation occurs frequently.

Since the scaling relations (2.162) and (2.167) were derived for the limit of large values of v'/s_L and large Reynolds numbers, respectively, they account only for the increase of the flame surface area ratio caused by turbulence, beyond the laminar value $\bar{\sigma} = |\nabla \tilde{G}|$ for $v' \to 0$. We will therefore simply add the laminar contribution and write $\bar{\sigma}$ as

$$\bar{\sigma} = |\nabla \tilde{G}| + \bar{\sigma}_t, \tag{2.183}$$

where $\bar{\sigma}_t$ now is the turbulent contribution to the flame surface area ratio $\bar{\sigma}$.

The resulting model equation for the unconditional quantity $\bar{\sigma}_t$ from Peters (1999), which covers both regimes, is written as

$$\bar{\rho}\frac{\partial \bar{\sigma}_t}{\partial t} + \bar{\rho}\tilde{v} \cdot \nabla \bar{\sigma}_t = \nabla_\| \cdot (\bar{\rho} D_t \nabla_\| \sigma_t) + c_0 \bar{\rho} \frac{(-\widetilde{v''v''}) : \nabla \tilde{v}}{\tilde{k}} \bar{\sigma}_t$$
$$+ c_1 \bar{\rho} \frac{D_t (\nabla \tilde{G})^2}{\widetilde{G''^2}} \bar{\sigma}_t - c_2 \bar{\rho} \frac{s_L^{0} \bar{\sigma}_t^2}{(\widetilde{G''^2})^{1/2}} - c_3 \bar{\rho} \frac{D \bar{\sigma}_t^3}{\widetilde{G''^2}}. \tag{2.184}$$

The terms on the l.h.s. represent the local rate of change and convection. Turbulent transport is modeled here by gradient transport in the tangential direction only, since similar arguments as in the variance Equation (2.152) apply with respect to the need to avoid turbulent diffusion in the direction normal to the turbulent flame surface. This is the first term on the r.h.s. of (2.184). The second term models production of the flame surface area ratio due to mean velocity gradients. The constant $c_0 = c_{\varepsilon_1} - 1 = 0.44$ originates from the $\tilde{\varepsilon}$-equation (1.40). The last three terms in (2.184) represent turbulent production, kinematic restoration, and scalar dissipation of the flame surface area ratio, respectively, and correspond to the three terms on the r.h.s of (2.182).

Equation (2.184) has some features in common with the Σ-equation (1.155) of the Coherent Flame Model in the sense that both contain a production term that is linear in the respective variable and a kinematic restoration term that is quadratic. There is, however, no equivalent of the dissipation term in the Σ-equation. There also is no straight-forward way to convert one equation into the other using the exact relation between $\bar{\sigma}$ and Σ to be derived at the end of this section.

We now want to determine values for the modeling constants c_1, c_2, and c_3 in (2.184). The average of the production term in (2.182) is equal to $-\overline{S\sigma}$. Wenzel (2000) has performed DNS of the constant density G-equation in an isotropic homogeneous field of turbulence (cf. also Wenzel, 1997). He has shown that the strain rate at the flame surface is statistically independent of σ and that the mean strain on the flame surface is always negative. The latter reflects the alignment of scalar gradients with the most compressive (negative) strain rate,

2.11 A model equation for the flame surface area ratio

as shown by Ashurst et al. (1987a) in analyzing DNS data of isotropic and homogeneous shear turbulence. When $-\overline{S\sigma}$ divided by $s_L^0 \bar{\sigma}$ is plotted against v'/s_L^0 one obtains a linear dependence. This leads to the closure model

$$-\overline{S\sigma} \sim \frac{v'}{\ell}\bar{\sigma}. \tag{2.185}$$

By taking the average of (2.182) and comparing the production terms in (2.182) and (2.184) Wenzel (2000) determined the modeling constant c_1 in (2.184) as $c_1 = 4.63$.

To determine the remaining constants c_2 and c_3 we consider again the steady planar flame. In the planar case the convective term on the l.h.s. and the turbulent transport term on the r.h.s. of (2.184) vanish and since the flame is steady, so does the unsteady term. The production term due to velocity gradients, being in general much smaller than production by turbulence, may also be neglected. Using the definition (2.153) for the flame brush thickness $\ell_{F,t}$, the balance of turbulent production, kinematic restoration, and scalar dissipation in (2.184) leads to the algebraic equation

$$c_1 \frac{D_t}{\ell_{F,t}^2} - c_2 \frac{s_L^0}{\ell_{F,t}} \frac{\bar{\sigma}_t}{|\nabla \tilde{G}|} - c_3 \frac{D}{\ell_{F,t}^2} \frac{\bar{\sigma}_t^2}{|\nabla \tilde{G}|^2} = 0. \tag{2.186}$$

In the limit of a steady state planar flame the flame brush thickness $\ell_{F,t}$ is proportional to the integral length scale ℓ. We may therefore use $\ell_{F,t} = b_2 \ell$ obtained from (2.156) in that limit and write (2.186) as

$$c_1 \frac{D_t}{\ell^2} - c_2 b_2 \frac{s_L^0}{\ell} \frac{\bar{\sigma}_t}{|\nabla \tilde{G}|} - c_3 \frac{D}{\ell^2} \frac{\bar{\sigma}_t^2}{|\nabla \tilde{G}|^2} = 0. \tag{2.187}$$

This equation covers two limits: In the corrugated flamelets regime the first two terms balance, while in the thin reaction zones regime there is a balance of the first and the last term. Using $D_t = a_4 v' \ell$ from Table 2.1 it follows for the corrugated flamelets regime that

$$c_2 b_2 s_L^0 \bar{\sigma}_t = a_4 c_1 v' |\nabla \tilde{G}|. \tag{2.188}$$

Experimental data (cf. Abdel-Gayed and Bradley, 1981) for fully developed turbulent flames in that regime show that for $Re \to \infty$ and $v'/s_L \to \infty$ the turbulent burning velocity is $s_T^0 = b_1 v'$ where $b_1 = 2.0$. If the burning velocities s_T^0 and s_L^0 in (2.150) are evaluated at a constant density it follows in that limit that $s_L^0 \bar{\sigma}_t = b_1 v' |\nabla \tilde{G}|$ and therefore one obtains by comparison with (2.188)

$$b_1 b_2 c_2 = a_4 c_1, \tag{2.189}$$

which leads to $c_2 = 1.01$ using the constants from Table 2.1.

Table 2.1. *Constants used in the modeling of premixed and partially premixed turbulent combustion*

Constant	Equation	Suggested value	Origin
a_1	$\tilde{\varepsilon} = a_1 v'^3/\ell$	0.37	Bray (1990)
a_2	$\tilde{k} = a_2 v'^2$	1.5	definition (1.150)
a_3	$\tau = a_3 \ell/v'$	4.05	$\tau = \tilde{k}/\tilde{\varepsilon}$
a_4	$D_t = a_4 v' \ell$	0.78	$D_t = v_t/0.7$
b_1	$s_T = b_1 v'$	2.0	experimental data
b_2	$\ell_{F,t} = b_2 \ell$	1.78	$(2a_3 a_4/c_s)^{1/2}$
b_3	$s_T^0/s_L^0 = b_3 (D_t/D)^{1/2}$	1.0	Damköhler (1940)
c_0	$c_0 = c_{\varepsilon 1} - 1$	0.44	standard value
c_1	(2.183)	4.63	DNS
c_2	(2.183)	1.01	$a_4 c_1/(b_1 b_2)$
c_3	(2.183)	4.63	$c_1 = c_3$
c_s	(2.143)	2.0	spectral closure

Similarly, for the thin reaction zones regime we obtain from (2.186) the balance

$$c_3 D \bar{\sigma}_t^2 = c_1 D_t |\nabla \tilde{G}|^2. \qquad (2.190)$$

This must be compared with (2.165) written as

$$\frac{s_T^0}{s_L^0} = b_3 \left(\frac{D_t}{D}\right)^{1/2}. \qquad (2.191)$$

Damköhler (1940) believed that the constant of proportionality b_3 should be unity. Recently, Wenzel (1997) has performed DNS simulations similar to those of Kerstein et al. (1988) based on (2.121) to calculate $\bar{\sigma}$ in the thin reaction zones regime. He finds $b_3 = 1.07$, which is very close to Damköhler's suggestion. Therefore we will use $b_3 = 1.0$ which leads with $b_3^2 = c_1/c_3$ to

$$c_3 = c_1. \qquad (2.192)$$

For consistency, (2.6) will be used to define the diffusivity D as $(\lambda/c_p)_0/\rho_u$. Then, with $\ell_F = D/s_L^0$ and the relations in Table 2.1, (2.186) leads to the quadratic equation

$$\frac{\bar{\sigma}_t^2}{|\nabla \tilde{G}|^2} + \frac{a_4 b_3^2}{b_1} \frac{\ell}{\ell_F} \frac{\bar{\sigma}_t}{|\nabla \tilde{G}|} - a_4 b_3^2 \frac{v' \ell}{s_L^0 \ell_F} = 0. \qquad (2.193)$$

Using (2.183) and (2.150) evaluated at a constant density, the difference Δs between the turbulent and the laminar burning velocity is

$$\Delta s = s_T^0 - s_L^0 = s_L^0 \frac{\bar{\sigma}_t}{|\nabla \tilde{G}|}. \qquad (2.194)$$

2.11 A model equation for the flame surface area ratio

Taking only the positive root in the solution of (2.193) gives an algebraic expression for Δs:

$$\frac{\Delta s}{s_L^0} = -\frac{a_4 b_3^2}{2 b_1} \frac{\ell}{\ell_F} + \left[\left(\frac{a_4 b_3^2}{2 b_1} \frac{\ell}{\ell_F} \right)^2 + a_4 b_3^2 \frac{v'\ell}{s_L^0 \ell_F} \right]^{1/2}. \quad (2.195)$$

This expression covers in the limit $\ell/\ell_F \to \infty$ the correlation $\Delta s = b_1 v'$ of the corrugated flamelets regime and in the limit $\ell/\ell_F \to 0$ the correlation (2.191) of the thin reaction zones regime.

The modeling constants used in the final equations for $\tilde{G}$, $\widetilde{G''^2}$, and $\bar{\sigma}_t$ are summarized in Table 2.1. Note that b_1 is the only constant that has been adjusted using experimental turbulent burning velocity data while the constant b_3 was suggested by Damköhler (1940). The constant c_1 was obtained from DNS and all other constants are related to constants in standard turbulence models.

If (2.195) is compared with experimental data as in the burning velocity diagram in Figure 2.22, the turbulent Reynolds number $Re = v'\ell/s_L^0 \ell_F$ appears as a parameter. From the viewpoint of turbulence modeling this seems disturbing, since in free shear flows any turbulent quantity should be independent of the Reynolds number in the large Reynolds number limit. The apparent Reynolds number dependence of (2.195) turns out to be an artifact resulting from the normalization of Δs by s_L^0, which is a molecular quantity whose influence should disappear in the limit of large Reynolds numbers and large values of v'/s_L. If the burning velocity difference Δs is normalized by v' rather than by s_L^0, (2.195) may be expressed as a function of the turbulent Damköhler number $Da = s_L^0 \ell/v'\ell_F$ instead, and one obtains the form

$$\frac{\Delta s}{v'} = -\frac{a_4 b_3^2}{2 b_1} Da + \left[\left(\frac{a_4 b_3^2}{2 b_1} Da \right)^2 + a_4 b_3^2 Da \right]^{1/2}. \quad (2.196)$$

This is Reynolds number independent and only a function of a single parameter, the turbulent Damköhler number. In the limit of large scale turbulence ($\ell/\ell_F \to \infty$ or $Da \to \infty$) it becomes Damköhler number independent. In the small scale turbulence limit ($\ell/\ell_F \to 0$ or $Da \to 0$), it is proportional to the square root of the Damköhler number. This was noted in Chapter 1, Section 1.1.

A Damköhler number scaling has also been suggested by Gülder (1990b) who has proposed

$$\frac{\Delta s}{v'} = 0.62 \, Da^{1/4} \quad (2.197)$$

as an empirical fit to a large amount of burning velocity data. A similar correlation with the same Damköhler number dependence, but using a constant of 0.51 instead of 0.62, was proposed by Zimont and Lipatnikov (1995).

Bradley et al. (1992), pointing at flame stretch as a determinant of the turbulent burning velocity, propose using the product of the Karlovitz stretch factor K and the Lewis number as the appropriate scaling parameter:

$$\frac{s_T^0}{v'} = 0.88(K\,Le)^{-0.3}, \tag{2.198}$$

where the Karlovitz stretch factor is related to the Damköhler number by

$$K = 0.157 \frac{v'}{s_L^0} Da^{-1/2}. \tag{2.199}$$

This leads to the expression

$$\frac{\Delta s}{v'} = 1.53 \left(\frac{s_L^0}{v'}\right)^{0.3} Da^{0.15} Le^{-0.3} - \frac{s_L^0}{v'}. \tag{2.200}$$

The correlations (2.196), (2.197), and (2.200) are compared in Figure 2.24 with each other and with data from the experimental data collection used by Bradley et al. (1992), which was kindly provided by M. Lawes. The data points show a large scatter because the experimental conditions were not always well defined. Since unsteady effects have been neglected in deriving (2.195), only data based on steady state experiments were retained from this collection.

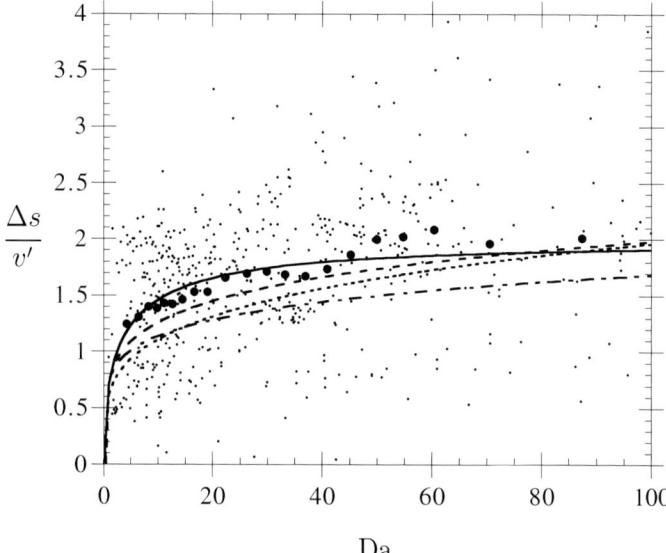

Figure 2.24. The burning velocity difference $\Delta s = s_T^0 - s_L^0$, normalized by v', as a function of the turbulent Damköhler number. ——— (2.196), ······ (2.197), - - - - (2.200) with $v'/s_L^0 = 2$, — · — · — (2.200) with $v'/s_L^0 = 5$.

2.11 A model equation for the flame surface area ratio 135

These 598 data points and their averages within fixed ranges of the turbulent Damköhler number are shown in Figure 2.24 as small and large dots, respectively. To make such a comparison possible, the Lewis number was assumed equal to unity in (2.200) and two values of v'/s_L^0 were chosen. As a common feature of all three correlations one may note that $\Delta s/v'$ strongly increases in the range of turbulent Damköhler numbers up to ten, but levels off for larger turbulent Damköhler numbers. The correlation (2.196) is the only one that predicts Damköhler number independence in the large Damköhler number limit.

The model for the turbulent burning velocity derived here is based on (2.121) in which the mass diffusivity D (rather than the Markstein diffusivity) appears. As a consequence, flame stretch and hence Lewis number effects do not enter into the model. Lewis number effects are often found to influence the turbulent burning velocity (cf. Abdel-Gayed et al., 1984). This is supported by two-dimensional numerical simulations by Ashurst et al. (1987b) and Haworth and Poinsot (1992), and by three-dimensional simulations by Rutland and Trouvé (1993), all being based on simplified chemistry. There are additional experimental data on Lewis number effects in turbulent flames at moderate intensities by Lee et al. (1993, 1995).

Lewis number effects could, in principle, be included by modifying the laminar burning velocity as in (2.63) by adding first-order correction terms. This would be similar to introducing a stretch factor as in (2.51). There the stretch factor I_0 was supposed to account for strain and curvature and thereby Lewis number effects in the corrugated flamelets regime. Kostiuk and Bray (1994) have calculated the mean effects of stretch by assuming Gaussian distributions for strain and curvature and a linear dependence of the laminar burning velocity on the Markstein number. They obtained a stretch factor I_0 that was a weaker function of the Karlovitz number than those deduced by Bray (1990) and Bradley et al. (1992), who had compared (2.51) with experimental data. They concluded that there must be an additional dependence of the turbulent burning velocity on the Karlovitz number. In the model Equation (2.184) for the flame surface area ratio, this much stronger influence is accounted for by the last term, which originates from diffusive effects in the thin reaction zones regime.

In some experimental data (cf. for instance Abdel-Gayed et al., 1984) the turbulent burning velocity ratio s_T^0/s_L^0 plotted in a burning velocity diagram first increases with v'/s_L^0 but then shows a maximum and strongly decreases afterwards. This behavior has been attributed to local quenching of flamelets. Meneveau and Poinsot (1991) have developed a statistical model for the stretch rate based on quasi-steady local quenching conditions to deduce global quenching criteria. In the present analysis the decrease of turbulent burning velocity ratio is attributed to local extinction phenomena occuring at the transition from the thin reaction zones regime to the broken reaction zones regime in Figure 2.8.

We now want to relate the flame surface area ratio $\bar{\sigma}$ to the flame surface density Σ. Since σ is defined only at the flame surface $G(x, t) = G_0$, it is convenient to introduce the quantity

$$\Sigma' = \sigma \delta(G - G_0), \qquad (2.201)$$

which has been called the "fine-grained surface-to-volume ratio" by Pope (1988) or the "flame area per unit volume" by Candel and Poinsot (1990). The equations derived for Σ' in those two papers are equivalent to (2.182), except that the last term of (2.182) is missing.

Let us introduce the joint probability density of σ and G, where G accounts for fluctuations of the instantaneous flame front in the direction normal to the mean front and σ for fluctuations of its angle. Decomposing the joint pdf into the marginal pdf of G and the conditional pdf of σ using Bayes' theorem as

$$P(\sigma, G; x, t) = P(\sigma \mid G; x, t) \cdot P(G; x, t) \qquad (2.202)$$

the mean value of Σ', called the flame surface density Σ, becomes

$$\Sigma(x, t) = \int_{-\infty}^{+\infty} \int_0^{\infty} \delta(G - G_0) \sigma \, P(\sigma \mid G; x, t) P(G; x, t) \, d\sigma \, dG. \qquad (2.203)$$

Applying the delta-function in the integration over G and using (2.124) one obtains

$$\Sigma(x, t) = \langle \sigma \mid G = G_0 \rangle \, P(x, t), \qquad (2.204)$$

where $\langle \sigma \mid G = G_0 \rangle$ is the conditional mean gradient at $G(x, t) = G_0$,

$$\langle \sigma \mid G = G_0 \rangle = \int_0^{\infty} \sigma P(\sigma \mid G = G_0; x, t) \, d\sigma. \qquad (2.205)$$

Equation (2.204) shows that the flame surface density is the product of the conditional mean gradient multiplied by the probability density of finding the flame surface at point x and time t.

To obtain a relation between $\bar{\sigma}$ and Σ, we consider, as before, a one-dimensional, steady, turbulent, planar flame propagating in the x direction, such that $P(x, t)$ simplifies to $P(x)$ and the conditional mean gradient becomes a function of x only. Since all parts of the flame surface at different locations x within the flame brush contribute to the flame surface area ratio, we integrate the flame surface density over x to obtain the flame surface area ratio $\bar{\sigma}$:

$$\bar{\sigma} = \int_{-\infty}^{+\infty} \Sigma \, dx = \int_{-\infty}^{+\infty} \langle \sigma \mid G = G_0 \rangle P(x) \, dx. \qquad (2.206)$$

As an integral over the flame brush, $\bar{\sigma}$ is a global property of the turbulent flame front, which, by definition, does not vary in the normal direction. As the conditional variance $\widetilde{(G''^2)}_0$ it may vary, however, in the tangential direction along the flame front. This was taken into account in the modeling of the turbulent transport term in (2.184).

At this stage there arises the question of how the conditional mean gradient $\langle \sigma \mid G = G_0 \rangle$ depends on x. For finite values of v'/s_L^0 the front is likely to be rounded toward the unburnt mixture and develops troughs toward the burnt gas. The conditional mean gradient σ therefore is expected to increase with x in most parts of the flame brush. This dependence should vanish, however, in the limit of large values of v'/s_L^0 where the flame surface tends toward a nonpropagating material surface. Further work is needed to quantify the dependence of the conditional mean gradient on x for finite values of v'/s_L^0.

2.12 Effects of Gas Expansion on the Turbulent Burning Velocity

By considering the effect of gas expansion on the flow field of a Bunsen flame in the first example in Section 2.5, we have seen that the velocity induced by gas expansion scales with $\gamma/(1-\gamma)s_{L,u}$. If the turbulent velocity fluctuations are smaller than that velocity, we therefore may expect that the turbulent burning velocity is also affected by gas expansion. However, if the turbulent velocity fluctuations are much larger than the induced velocity, this effect is expected to be negligible.

We have seen in Section 2.4 that gas expansion is responsible for countergradient diffusion of the progress variable. In (2.49) the difference between the induced velocity $\gamma/(1-\gamma)s_{L,u}$ and the velocity fluctuation v' enters into a criterion for the transition from countergradient to gradient diffusion. Although that criterion differs from the criterion $Ka = 1$ for the transition from the corrugated flamelets regime to the thin reaction zones regime it may be expected that gas expansion is primarily of importance in the corrugated flamelets regime. The following analysis will therefore be restricted to that regime only.

In the first example at the end of Section 2.5 we presented a model in which the velocity field generated by gas expansion at the flame front was superimposed on the velocity field of the oncoming flow issuing from the burner. It was noted that the sum of these two velocity fields satisfies neither the Navier–Stokes equations nor the boundary conditions. The model also ignores vorticity generation at the flame front. Nevertheless, this model has been the basis of some intriguing numerical simulations by Ashurst (1996, 1997) who showed its capability of reproducing the Darrieus–Landau instability and flame acceleration by gas expansion. The model can be formulated in a way to contain gas expansion effects within a modified G-equation itself, rather than to take them into account as a modification of the flow field. While the latter formulation is more general, the modification of the G-equation has the advantage that the feedback of the expansion induced flow field on the flame motion can be studied more directly.

We will use this model here to gain an understanding of how gas expansion could influence the turbulent burning velocity. As in the first example in Section 2.5 we assume that a prescribed turbulent flow field v and the flow field induced by gas expansion can be superimposed. Gas expansion induces at the flame front velocities having components in normal directions that are of equal magnitude but opposite sign toward the unburnt and toward the burnt gas, respectively. The midpoint within the flame structure, where the induced velocity from the other parts of the flame front will be evaluated, is presumed to define the location of the flame front $G(\boldsymbol{x}, t) = G_0$. It corresponds to the location where half the density change has taken place. If the flame front propagates with a burning velocity $s_{L,u}$ into the unburnt mixture, the burning velocity at the midpoint is according to (2.22)

$$s_{L,b}(1 - \gamma/2). \tag{2.207}$$

Adding at the midpoint to this velocity the velocity $\boldsymbol{v}_f$ of the turbulent flow field and the induced normal velocity $\boldsymbol{n} \cdot v_{\phi,n}$ resulting from gas expansion at the other parts of the flame front, one obtains the propagation velocity

$$\frac{d\boldsymbol{x}_f}{dt} = \boldsymbol{v}_f + \boldsymbol{n} v_{\phi,n} + \boldsymbol{n} s_{L,b}(1 - \gamma/2). \tag{2.208}$$

This equation is due to Ashurst (1997). It replaces the kinematic relation (2.55) that had led to the G-equation for the corrugated flamelets regime. Here the corresponding G-equation becomes

$$\frac{\partial G}{\partial t} + \boldsymbol{v} \cdot \nabla G = s_{L,b}(1 - \gamma/2)|\nabla G| + v_{\phi,n}|\nabla G|. \tag{2.209}$$

The induced normal velocity $v_{\phi,n}$ may be expressed by a Green's function solution of the Poisson equation. Frankel (1990) has shown that in the two-dimensional case $v_{\phi,n}$ is

$$v_{\phi,n} = \gamma \frac{s_{L,b}}{2\pi} \int_S \frac{(\boldsymbol{x} - \boldsymbol{x}') \cdot \boldsymbol{n}}{(\boldsymbol{x} - \boldsymbol{x}')^2} dS, \tag{2.210}$$

where the integral is to be taken along the closed curve $S(\boldsymbol{x}')$. He also has shown that for weak gas expansion ($\gamma \to 0$) resulting in small amplitude perturbations $F(y, t)$ of a planar one-dimensional flame front, where F is defined by

$$G = x - F(y, t), \tag{2.211}$$

2.12 Effects of gas expansion on the turbulent burning velocity

the last term in (2.209) becomes

$$v_{\phi,n}|\nabla G| = \gamma \frac{s_{L,b}}{2}(1 - I_{1D}). \tag{2.212}$$

Here

$$I_{1D}(F, x, t) = \frac{1}{2\pi} \int_{-\infty}^{+\infty} \int_{-\infty}^{+\infty} |k| e^{ik(x-x')} F(x', t) \, dx' dk \tag{2.213}$$

is the one-dimensional form of Sivashinsky's integral (cf. Sivashinsky, 1977b). The integration in (2.213) is to be performed over all other parts of the flame surface located at x', and over all wavenumbers k.

Sivashinsky (1977b) has derived an equation for small perturbations of a planar two-dimensional flame front that includes two destabilizing effects: the diffusional-thermal instability that occurs if the Markstein length is negative and the hydrodynamic (or Darrieus–Landau) instability due to gas expansion. A reduced form of that equation that does not contain the fourth-order derivative is obtained by setting in (2.209) the normal component of the flow velocity equal to the burning velocity $s_{L,b}^0$ and setting the tangential velocities equal to zero. To show this we set

$$G(\mathbf{x}, t) = x - F(\boldsymbol{\eta}, t), \tag{2.214}$$

where $F(\boldsymbol{\eta}, t)$ is a two-dimensional perturbation depending on the two-dimensional vector $\boldsymbol{\eta} = (y, z)$, which stands for the two coordinates normal to the x direction. If the modification of the burning velocity due to curvature is taken into account as

$$s_{L,b} = s_{L,b}^0 (1 - \mathcal{L}_b \kappa) \tag{2.215}$$

and $|\nabla G|$ is expanded as

$$|\nabla G| = 1 + \frac{1}{2}(\nabla F)^2, \tag{2.216}$$

where ∇ is a two-dimensional derivative, the equation for $F(\boldsymbol{\eta}, t)$ becomes for small perturbations and small γ

$$F_t + \frac{s_{L,b}^0}{2}(\nabla F)^2 = s_{L,b}^0 \mathcal{L}_b \nabla^2 F + s_{L,b}^0 \frac{\gamma}{2} I_{2D}. \tag{2.217}$$

Here the two-dimensional version of Sivashinsky's integral I_{2D} is defined as

$$I_{2D}(F, \boldsymbol{\eta}, t) = \frac{1}{(2\pi)^2} \iiiint_{-\infty}^{+\infty} |\mathbf{k}| e^{i\mathbf{k}(\boldsymbol{\eta}-\boldsymbol{\eta}')} F(\boldsymbol{\eta}', t) \, d\boldsymbol{\eta}' d\mathbf{k}, \tag{2.218}$$

where $\mathbf{k}$ is the two-dimensional wave vector corresponding to $\boldsymbol{\eta}$.

In the case of negative Markstein numbers $\mathcal{L}_b$ the term containing the second-order derivative in (2.217) is destabilizing and the problem is not well defined, if the fourth-order derivative is not retained. Here we will assume $\mathcal{L}_b = 0$ and not consider diffusional-thermal instabilities. In addition, we will include the influence of a turbulent flow field v as in (2.209). We retain Sivashinsky's integral for mathematical convenience, although this conflicts with the assumption of small perturbations that had led to (2.218).

For a three-dimensional generalization of that integral we observe that two volume sources at equal but opposite distance in the x direction from a planar, slightly perturbed flame front located at $x = 0$ would induce at position $(x = 0, y, z)$ normal velocities of equal magnitude but opposite sign. Therefore, inclusion of the x' dimension and the corresponding wavenumber k_x in the expansion integral will not change I_{2D} in the limit of small amplitude perturbations. Therefore we introduce into (2.209) the integral

$$I_{3D}(G, \boldsymbol{x}, t) = \frac{1}{(2\pi)^3} \iiint\!\!\iiint_{-\infty}^{+\infty} |k| e^{ik(x-x')} G(\boldsymbol{x}', t) \, d\boldsymbol{x}' d\boldsymbol{k}. \quad (2.219)$$

By introducing (2.214) and integrating over the x' and k_x directions, it can be shown that (2.219) reduces to $-I_{2D}(F, \eta, t)$. A similar formulation has been used by Dandekar and Collins (1995) who have studied the effect of Lewis number effects and turbulence intensities on the basis of a modified version of the Sivashinsky equation. Using the three-dimensional analogue of (2.212) the resulting G-equation finally reads

$$\frac{\partial G}{\partial t} + \boldsymbol{v} \cdot \nabla G = s_{L,b}^0 \left(|\nabla G| + \frac{\gamma}{2} I_{3D}(G, \boldsymbol{x}, t) \right). \quad (2.220)$$

Since the analysis will be performed for the corrugated flamelets regime, effects of curvature and strain, being higher order terms, will not be included. Therefore $s_{L,b}$ has been set equal to $s_{L,b}^0$. We follow Peters (1992) and regard (2.220) as a three-dimensional field equation in a constant density flow field. As before we derive equations for the unconditional mean and its variance of G, keeping in mind that the conditional values correspond to those at the mean flame surface $\bar{G}(\boldsymbol{x}, t) = G_0$. Splitting G and v into a mean and a fluctuation,

$$G = \bar{G} + G', \qquad v = \bar{v} + v', \quad (2.221)$$

we obtain the equation for the mean value of G as

$$\frac{\partial \bar{G}}{\partial t} + \bar{\boldsymbol{v}} \cdot \nabla \bar{G} + \nabla \cdot \overline{v'G'} = s_{L,b}^0 \bar{\sigma} + \frac{\gamma}{2} s_{L,b}^0 I_{3D}(\bar{G}, \boldsymbol{x}, t). \quad (2.222)$$

Similar to the modeling of (2.139) the turbulent transport term $\nabla \cdot \overline{v'G'}$ is to be replaced by a curvature term. Furthermore, as in the derivation that led to (2.150) the term $s_{L,b}^0 \bar{\sigma}$ in (2.222) can be replaced by $s_{T,b}^0 |\nabla \bar{G}|$, where $s_{T,b}^0$ is the turbulent burning velocity with respect to the burnt gas.

2.12 Effects of gas expansion on the turbulent burning velocity

Equation (2.222) implies that the present model, where gas expansion is included within the G-equation itself (rather than inducing changes of the flow field), can be applied to predict the propagation of the mean flame surface, which now propagates with the turbulent burning velocity, but Darrieus–Landau instabilities act on the mean flame front position of a turbulent flame in a similar way as in the laminar case.

The equation for the variance becomes in the corrugated flamelets regime, with expansion effects included,

$$\frac{\partial \overline{G'^2}}{\partial t} + \bar{v} \cdot \nabla \overline{G'^2} + \nabla \cdot (\overline{v'G'^2}) = -2\overline{v'G'} \cdot \nabla \tilde{G} - \bar{\omega} + \gamma \bar{\pi}. \tag{2.223}$$

Similar to the modeling of the transport term in (2.140), only tangential turbulent diffusion is to be retained. As before, kinematic restoration is defined by (2.134), but with $s_{L,b}^0$ instead of $s_{L,u}^0$. The last term in (2.223), the covariance integral $\bar{\pi}$, defined as

$$\bar{\pi} = \frac{s_{L,b}^0}{(2\pi)^3} \iiint \int_{-\infty}^{+\infty} |k| e^{ik(x-x')} \overline{G'(x',t)G'(x,t)} \, dx' dk \tag{2.224}$$

describes nonlocal influences of gas expansion on the variance and thereby on the flame brush thickness. As is the kinematic restoration term $\bar{\omega}$ it is proportional to $s_{L,b}^0$. Whereas $\bar{\omega}$ reduces fluctuations, $\bar{\pi}$ counteracts this mechanism by further corrugating the flame surface via local gas expansion.

We now proceed in the same way as in Peters (1992) and consider for the case of isotropic homogeneous turbulence the two-point correlation

$$g^2(r,t) = \overline{G'(x,t)G'(x+r,t)} \tag{2.225}$$

Because of isotropy, the correlation g^2 depends only on the distance $r = |r|$. Using standard techniques in homogeneous turbulence an equation for g^2 can be derived in a similar way as in Peters (1992) to obtain

$$\frac{\partial g^2}{\partial t} + 2\nabla \cdot \overline{v'(x+r,t)G'(x,t)G(x+r,t)} + 2s_{L,b}^0 S_1 - \gamma s_{L,b}^0 S_4 = 0. \tag{2.226}$$

Here

$$S_1(r,t) = -\overline{G'(x+r,t)\sigma(x,t)} \tag{2.227}$$

describes the two-point correlation between G and σ and will lead to kinematic restoration. Differently from Peters (1992) Equation (2.227) now contains the additional term

$$S_4(\mathbf{r}, t) = \overline{G'(\mathbf{x}+\mathbf{r}, t) I_{3D}(G', \mathbf{x}, t)}$$
$$= \frac{1}{(2\pi)^3} \iiint\iiint_{-\infty}^{+\infty} |\boldsymbol{\ell}| e^{i\boldsymbol{\ell}(\mathbf{x}-\mathbf{x}')} \overline{G'(\mathbf{x}', t) G'(\mathbf{x}+\mathbf{r}, t)} \, d\mathbf{x}' d\boldsymbol{\ell}. \tag{2.228}$$

Here we have denoted the wavenumber vector by $\boldsymbol{\ell}$.

The Fourier transform $\hat{f}(\mathbf{k}, t)$ of a two-point correlation $f(\mathbf{r}, t)$ is defined as

$$\hat{f}(\mathbf{k}, t) = \frac{1}{(2\pi)^3} \iiint_{-\infty}^{+\infty} f(\mathbf{r}, t) e^{-i\mathbf{k}\mathbf{r}} \, d\mathbf{r} \tag{2.229}$$

and its inverse as

$$f(\mathbf{r}, t) = \iiint_{-\infty}^{+\infty} \hat{f}(\mathbf{k}, t) e^{i\mathbf{k}\mathbf{r}} \, d\mathbf{k}. \tag{2.230}$$

The spectrum function $\Gamma(k, t)$ of G fluctuations is related to the Fourier transform of g^2, namely $\widehat{g^2}(\mathbf{k}, t)$, by

$$\Gamma(k, t) = k^2 \oint \widehat{g^2}(\mathbf{k}, t) d\Omega = 4\pi k^2 \widehat{g^2}(\mathbf{k}, t), \tag{2.231}$$

where $k = |\mathbf{k}|$ is the wavenumber and integration is performed over the solid angle Ω. Fourier transformation over the divergence of the triple correlation in (2.226) leads to a spectral transport term $T(k, t)$ in wavenumber space.

Peters (1992) has applied the assumption of locality not only to the spectral transport term but also to the nonlinear term $S_1(\mathbf{r}, t)$ in (2.226). Dimensional analysis using the transformation properties of the G-equation then has led to a linear model for its Fourier transform $\hat{S}_1(k, t)$:

$$8\pi k^2 \hat{S}_1 = c_1 C_s^{-1} k \Gamma, \tag{2.232}$$

where C_s is the universal constant of the scalar spectrum and c_1 is a constant that was assumed to be independent of the wavenumber.

Since S_4 defined by (2.228) is linear in the two-point correlation, it can be expressed in terms of $\Gamma(k, t)$. This can be shown by introducing the vector $\mathbf{s}$ by

$$\mathbf{x} + \mathbf{r} = \mathbf{x}' + \mathbf{s}. \tag{2.233}$$

Thereby the two-point correlation appearing in (2.228) becomes $g^2(\mathbf{s}, t)$. Since $\mathbf{x}$ and $\mathbf{r}$ are fixed, $d\mathbf{x}' = -d\mathbf{s}$ and one may write $S_4(\mathbf{r}, t)$ as

$$S_4(\mathbf{r}, t) = -\iiint_{-\infty}^{+\infty} e^{-i\boldsymbol{\ell}\mathbf{r}} \left[\frac{|\boldsymbol{\ell}|}{(2\pi)^3} \iiint_{-\infty}^{+\infty} g^2(\mathbf{s}, t) e^{i\boldsymbol{\ell}\mathbf{s}} d\mathbf{s} \right] d\boldsymbol{\ell}. \tag{2.234}$$

2.12 Effects of gas expansion on the turbulent burning velocity

Identifying ℓ with $-k$ and comparing (2.230) with (2.234) we see that the term in square brackets in (2.234) represents the Fourier transform of $S_4(r, t)$, namely

$$\hat{S}_4(k, t) = \frac{|k|}{(2\pi)^3} \iiint_{-\infty}^{+\infty} g^2(s, t) e^{-iks} ds. \qquad (2.235)$$

Using (2.229) and (2.231) this may be written as

$$\hat{S}_4(k, t) = |k| \widehat{g^2}(k, t) = \frac{\Gamma(k, t)}{4\pi k}. \qquad (2.236)$$

Taking the Fourier transform of (2.226) and using (2.232) the equation for the spectrum function becomes finally

$$\frac{\partial \Gamma(k, t)}{\partial t} - T(k, t) + c_1 s_{L,b}^0 C_s^{-1} k \Gamma(k, t) - \gamma s_{L,b}^0 k \Gamma(k, t) = 0. \qquad (2.237)$$

In the limit $\mathcal{L} = 0$ the equation for $\Gamma(k, t)$ in Peters (1992) contains the first three terms of (2.237), while now an additional term modeling gas expansion effects appears. It is interesting to note that the gas expansion term has the same wavenumber dependence as the kinematic restoration term. Combining these two terms in (2.237) leads to the equation

$$\frac{\partial \Gamma(k, t)}{\partial t} - T(k, t) + c_1^* C_s^{-1} s_{L,b} k \Gamma(k, t) = 0, \qquad (2.238)$$

where

$$c_1^* = c_1 - \gamma C_s. \qquad (2.239)$$

This is the same equation as that in Peters (1992) for $\mathcal{L} = 0$, except that c_1^* instead of c_1 multiplies the kinematic restoration term. Integration of (2.237) over k from k_0 to $k = \infty$ leads to

$$\frac{\partial \overline{G'^2}}{\partial t} = -\bar{w}^*, \qquad (2.240)$$

where $\overline{G'^2}$ is defined as in Peters (1992) by

$$\overline{G'^2} = \int_{k_0}^{\infty} \Gamma \, dk \qquad (2.241)$$

and

$$\bar{w}^* = \bar{w} - \gamma \bar{\pi} = c_1^* C_s^{-1} s_{L,b}^0 \int_{k_0}^{\infty} k \Gamma \, dk \qquad (2.242)$$

is the modified kinematic restoration. In (2.241) and (2.242) $k_0 = \ell^{-1}$ denotes the wavenumber corresponding to the integral scale ℓ. As long as $c_1^* > 0$ the modified kinematic restoration reduces scalar fluctuations. Using the same arguments as in Peters (1992), one then obtains the closure

$$\bar{w}^* = c_w^* \frac{\varepsilon}{k} \overline{G'^2}. \tag{2.243}$$

In the limit of large Reynolds numbers and large values of v'/s_L, the modeling constant c_w^* should be the same as c_w. In many practical cases, however, where these limits are not satisfied, we expect c_w^* to be smaller. This would lead to smaller values of c_s and to larger values of b_2 in the model for the flame brush thickness than those given in Table 2.1.

The condition $c_1^* > 0$ is expected to be valid only in the universal subrange $k_0 < k < \infty$ of the spectrum where the model (2.232) is meaningful. Since kinematic restoration describes the interaction between turbulent fluctuations and local laminar flame propagation, it ceases to be dominant for $k < k_0$, corresponding to length scales larger than the integral scale. Therefore large scale instabilities will grow at those length scales. To account for this behavior the condition reduces to

$$c_1^* > 0 \quad \text{for} \quad k > k_0. \tag{2.244}$$

This has been tested by Wenzel (2000) using DNS for the isotropic G-equation in a cubic box. He finds that the coefficient c_1 depends on v'/s_L^0 and typically is larger than unity in the range $k > k_0$. (Details may be found in Peters et al. (2000).) These results are shown in Figure 2.25. Since C_s is approximately 1.2, which is the value prefered by Monin and Yaglom (1975), the product γC_s rarely exceeds unity. Although the DNS calculations have been performed in the limit $\gamma = 0$, they provide a first confirmation of the hypothesis that (2.244) might be valid. This result implies that within the flame brush thickness, which is of the order of the integral length scale ℓ, expansion effects are dominated by kinematic restoration. In physical terms this means that the energy-containing eddies at the integral scale are strong enough to suppress local gas expansion effects. Among the last two terms in (2.237) the kinematic restoration term, being the larger one, is then able to maintain the cascade process of scalar fluctuations.

If, however, the gas expansion term in (2.237) was larger than the kinematic restoration term below a certain wavenumber k^* within the inertial range such that $c_1^* > 0$ is only valid for $k > k^*$, corrugations of the flame front would be enhanced at all wavenumbers smaller than k^*. As a consequence, RANS modeling based on integral quantities could not be applied to premixed turbulent combustion, but LES models that resolve wavenumbers smaller than k^* could be used.

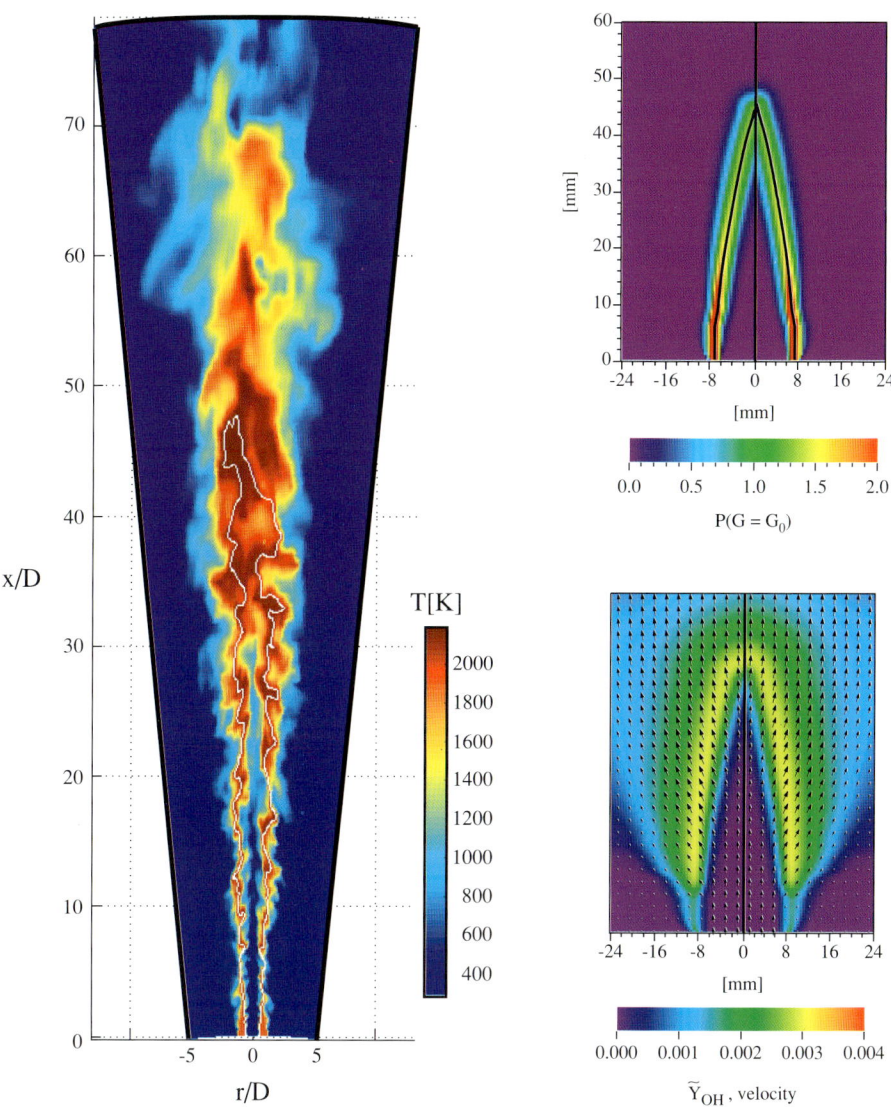

COLORPLATE I. (Left) Large eddy simulations of a turbulent jet diffusion flame by Pitsch and Steiner (2000). (Reprinted with permission by H. Pitsch and H. Steiner.) See text, p. 62.

COLORPLATE I. (Right) Mean scalar and velocity fields obtained by Herrmann (2000) for the Bunsen flame on a slot burner. These pictures are to be compared with those on the middle row in Colorplate II. Details are explained in the text. (Reprinted with permission by M. Herrmann.) See text, p. 165.

COLORPLATE II. (Top) Two-dimensional laser measurements of scalar and velocity fields in a Bunsen flame on a slot burner. These results were kindly provided by T. Plessing and A. Joedicke. The pictures are explained in the text. (Reprinted with permission by T. Plessing and A. Joedicke.) See text, p. 164.

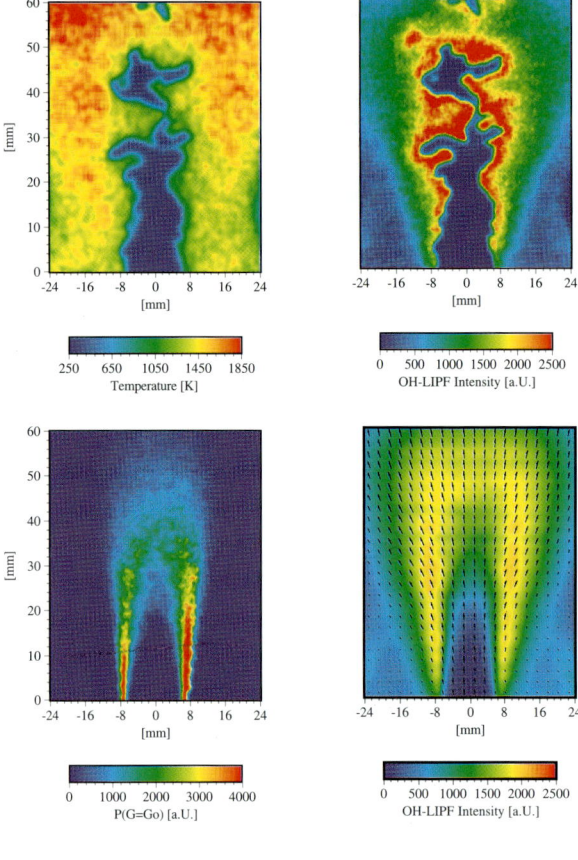

COLORPLATE II. (Bottom) The laminar triple flame investigated by Plessing et al. (1998). The picture on the l.h.s. shows a comparison between the calculated (left) and the measured (right) temperature field. The picture on the r.h.s. shows the three branches on both sides of the axisymmetric flame in terms of calculated heat release rates. (Reprinted by permission of Elsevier Science from "An experimental and numerical study of a laminar triple flame," by T. Plessing, P. Terhoeven, M. S. Mansour and N. Peters, Combustion & Flame, Vol. 115, pp. 335–353, Copyright 1998 by The Combustion Institute.) See text, p. 249.

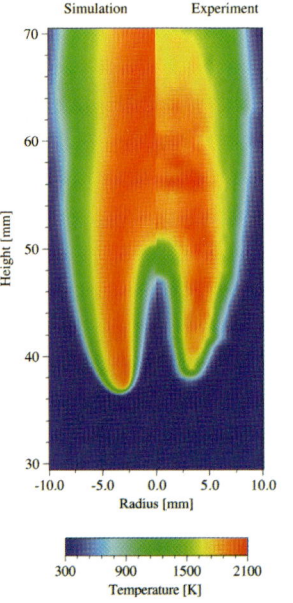 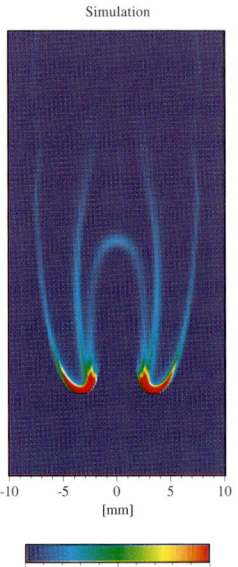

2.12 Effects of gas expansion on the turbulent burning velocity

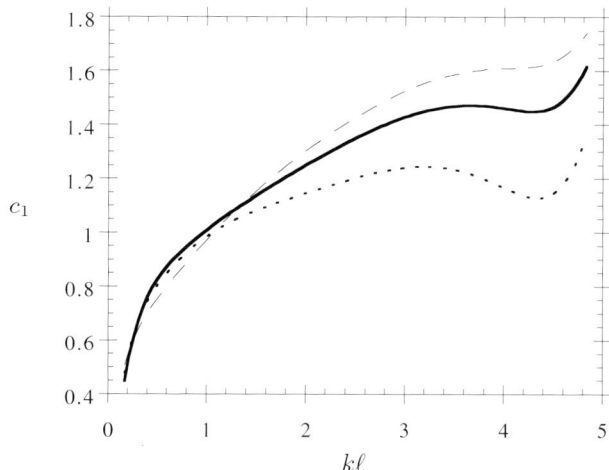

Figure 2.25. The coefficient c_1 in (2.232) obtained from DNS data. - - - - $v'/s_L = 1.0$, ——— $v'/s_L = 1.6$, – – – $v'/s_L = 2.0$.

Since the closure for the kinematic restoration is based on the cascade process and since the scaling of the turbulent burning velocity in the corrugated flamelets regime follows from this closure, (2.243) can be used with the assumption (2.244) in the same way as (2.135) was used previously. Then (2.162) and the subsequent scaling arguments leading to the turbulent burning velocity $s_T \sim v'$ are valid even in the presence of gas expansion. This is a scaling in terms of local mean quantities. Therefore it may be concluded that the turbulent burning velocity is a well-defined quantity that depends, as before, on local mean quantities only. It is worth noting that the constant b_1 in Table 2.1 was derived from experiments, which naturally include gas expansion effects. Therefore, one may expect that the modeling that led to the constant $b_1 = 2.0$ already contains the effect of gas expansion on the turbulent burning velocity.

In discussing Equation (2.222) for the mean value of $\bar{G}$ with gas expansion effects included, we have noted that nonlocal hydrodynamic instabilities due to gas expansion may influence the development of the mean flame front. If (2.244) is valid, these instabilities develop at scales larger than the integral length scale. For industrial applications of premixed turbulent combustion, where the integral length scale is typically one order of magnitude smaller than the smallest width of the combustion chamber, the development of such large scale instabilities is less likely to occur than, for instance, in large gas cloud explosions. By using the modeled Equation (2.151) or an equivalent LES formulation within a

2.13 Laminar Flamelet Equations for Premixed Combustion

The modeled equations for $\tilde{G}$, $\widetilde{G''^2}$, and $\bar{\sigma}_t$ constitute a closed set that allows us to calculate the position of the mean turbulent flame front, provided that the laminar burning velocity s_L^0 and the diffusivity D are known. If, in addition, the mean temperature field or those of other reactive scalars are needed, the laminar flame structure must be resolved. The easiest way to do this is to assume, as in the BML model, a jump of temperature and all reactive scalars from their values in the unburnt mixture to the corresponding equilibrium values in the burnt gas. This, however, does not take finite rate chemistry into account. For a more complete description we want to derive flamelet equations for premixed combustion. In the context of the level set approach these equations were first derived, using a similar coordinate transformation as shown in Figure 1.8 for nonpremixed combustion, by Peters (1993).

Here we will base the derivation on a two-scale asymptotic analysis following Keller and Peters (1994). The large length scale Λ is defined as

$$\Lambda = \frac{G'}{\bar{\sigma}}, \qquad (2.245)$$

where $\bar{\sigma}$ is the mean gradient and G' is the r.m.s. fluctuation of G. The length scale Λ represents the characteristic size of corrugations of the flame surface and is equal to the Taylor scale λ_G in the corrugated flamelets regime as can be seen from (2.171).

By introducing flamelet equations, we want to resolve the laminar flame structure in the vicinity of the flame surface. Since the propagation term is dominant in the corrugated flamelets regime, while the curvature term dominates in the thin reaction zones regime, flamelet equations will be different in the two regimes. We will first consider the corrugated flamelets regime where the flamelet structure extends over distances that are of the order of the flame thickness ℓ_F. We assume that the ratio of the flame thickness to the length scale Λ is small and introduce the small expansion parameter

$$\varepsilon = \ell_F/\Lambda. \qquad (2.246)$$

To derive the flamelet equations we normalize the independent variables x and t in the equations for the reactive scalars (1.81) by Λ and Λ/s_L^0, respectively,

2.13 Laminar flamelet equations for premixed combustion

and the velocities v by s_L^0:

$$x^* = x/\Lambda, \qquad t^* = t\, s_L^0/\Lambda, \qquad v^* = v/s_L^0. \tag{2.247}$$

The diffusivity D_i in (1.81) is normalized by $D = s_L^0 \ell_F = \varepsilon s_L^0 \Lambda$ such that

$$D_i^* = D_i/D. \tag{2.248}$$

The reactive scalars, the density, and the chemical source term are normalized, using suitable reference values $\psi_{i,\text{ref}}$ and ρ_{ref}, as

$$\psi_i^* = \psi_i/\psi_{i,\text{ref}}, \qquad \rho^* = \rho/\rho_{\text{ref}}, \qquad \omega_i^* = \omega_i \Lambda \big/ \left(s_L^0 \psi_{i,\text{ref}} \rho_{\text{ref}} \right). \tag{2.249}$$

Furthermore, G will be normalized by G' as

$$G^* = G/G'. \tag{2.250}$$

Then one obtains (1.81) in nondimensional form, which is written with the asterisk removed, as

$$\rho \frac{\partial \psi_i}{\partial t} + \rho v \cdot \nabla \psi_i = \varepsilon \nabla \cdot (\rho D_i \nabla \psi_i) + \omega_i. \tag{2.251}$$

It is important to note that the diffusive term is of order $O(\varepsilon)$ in these equations, which points at a thin flame structure at the large scales that were used in the normalization.

Following Keller and Peters (1994) we introduce a two-scale asymptotic expansion where x and t are the long scales and

$$\zeta = \varepsilon^{-1}(G(x,t) - G_0) \tag{2.252}$$

is a short range spatial variable. It resolves changes in the vicinity of the flame surface that occur over distances of the order of the flame thickness. In a rigorous asymptotic analysis the dependent variables should be expanded into a leading order and a first-order term as

$$\psi_i = \psi_i^0 + \varepsilon \psi_i^1 + \cdots. \tag{2.253}$$

Here, for simplicity, we will only consider the leading order solution, written without the suffix, and introduce a coordinate transformation where ψ_i is expressed as a function of the short and long spatial and temporal variables in the form

$$\psi_i = \psi_i(\zeta(x,t), \boldsymbol{\xi}, \tau). \tag{2.254}$$

Here the new coordinates $\boldsymbol{\xi}=\boldsymbol{x}$ replace the original long spatial coordinates $\boldsymbol{x}$ and the new time $\tau = t$ the original time t in the transformed coordinate system. The temporal and spatial derivatives in (2.251) may then be written as

$$\frac{\partial}{\partial t} = \frac{\partial \zeta}{\partial t}\frac{\partial}{\partial \zeta} + \frac{\partial}{\partial \tau} = \varepsilon^{-1}\frac{\partial G}{\partial t}\frac{\partial}{\partial \zeta} + \frac{\partial}{\partial \tau}, \tag{2.255}$$

$$\nabla = \nabla \zeta \frac{\partial}{\partial \zeta} + \nabla_\xi = \varepsilon^{-1}\nabla G\frac{\partial}{\partial \zeta} + \nabla_\xi. \tag{2.256}$$

When these transformations are introduced into (2.251) one obtains

$$\varepsilon\rho\frac{\partial \psi_i}{\partial \tau} + \varepsilon\rho\boldsymbol{v}\cdot\nabla_\xi \psi_i + \left[\rho\frac{\partial G}{\partial t} + \rho\boldsymbol{v}\cdot\nabla G\right]\frac{\partial \psi_i}{\partial \zeta}$$
$$= \nabla G \cdot \frac{\partial}{\partial \zeta}\left(\rho D_i \nabla G \frac{\partial \psi_i}{\partial \zeta}\right) + \varepsilon \nabla_\xi \cdot \left(\rho D_i \nabla G \frac{\partial \psi_i}{\partial \zeta}\right)$$
$$+ \varepsilon \nabla G \cdot \frac{\partial}{\partial \zeta}(\rho D_i \nabla_\xi \psi_i) + \varepsilon^2 \nabla_\xi \cdot (\rho D_i \nabla_\xi \psi_i) + \Omega_i. \tag{2.257}$$

Here, to retain the effect of chemical reactions it was necessary to rescale the chemical source term as

$$\varepsilon\,\omega_i = \Omega_i. \tag{2.258}$$

To justify a posteriori the choice of the large scale Λ, we analyze the first term on the r.h.s. of (2.257). If we take into account that the gradient ∇G in front of this term varies on the long scales only and therefore is independent of ζ, it can be pulled inside the brackets, where it can be combined with ∇G to yield $|\nabla G|^2$. If the nondimensional quantity $|\nabla G|^2$ is expressed in dimensional form again as

$$|\nabla G|^2\left(\frac{\Lambda}{G'}\right)^2 = \left(\frac{\sigma}{\bar{\sigma}}\right)^2 \tag{2.259}$$

where (2.245) was used, it is seen to be a quantity of order unity. This is required for a correct ordering of the different terms in the equation and therefore justifies the definition of Λ.

If only leading order terms are retained and ζ is replaced by G as independent variable using

$$d\zeta = \varepsilon^{-1}dG, \tag{2.260}$$

Equation (2.257) becomes, written in dimensional form again,

$$\left[\rho\frac{\partial G}{\partial t} + \rho\boldsymbol{v}\cdot\nabla G\right]\frac{\partial \psi_i}{\partial G} = \frac{\partial}{\partial G}\left(\rho D_i |\nabla G|^2 \frac{\partial \psi_i}{\partial G}\right) + \omega_i. \tag{2.261}$$

2.13 Laminar flamelet equations for premixed combustion

The term in square brackets represents the mass flow rate normal to the flame surface and is to leading order equal to (ρs_L^0) in the corrugated flamelets regime. Introducing this we obtain the steady flamelet equation

$$(\rho s_L^0)\sigma \frac{\partial \psi_i}{\partial G} = \frac{\partial}{\partial G}\left(\rho D_i \sigma^2 \frac{\partial \psi_i}{\partial G}\right) + \omega_i. \tag{2.262}$$

This equation uses the scalar G as independent variable and contains the gradient σ as an external parameter that varies only on the long scales. In that sense it resembles the flamelet equation for nonpremixed combustion, where Z is the independent variable and χ is the external parameter.

In contrast to Z, however, G varies from $G = -\infty$ to $G = +\infty$, if the laminar flamelet is sufficiently removed from boundaries such as solid walls. Then it is possible to rescale the scalar G by

$$x_n = \frac{G - G_0}{\sigma}, \tag{2.263}$$

where x_n is the normal distance to the instantaneous flame front (cf. Figure 2.16). Then one obtains (2.262) in the form

$$(\rho s_L^0)\frac{\partial \psi_i}{\partial x_n} = \frac{\partial}{\partial x_n}\left(\rho D_i \frac{\partial \psi_i}{\partial x_n}\right) + \omega_i. \tag{2.264}$$

These are the equations for a one-dimensional steady flame, equivalent to (2.2) and (2.3), from which the burning velocity eigenvalue s_L^0 can be calculated.

One could, of course, carry the asymptotic expansion to the first order to derive the response of the burning velocity to strain and curvature. These effects originate from the second term on the l.h.s. and the second term on the r.h.s. in (2.257), respectively. We will not do this here but refer to Matalon and Matkowsky (1982) and Keller and Peters (1994). Keller and Peters (1994) have also taken into account unsteady changes of pressure, which are, for instance, important in reciprocating engines. This was combined with the influences of heat loss and the temperature dependence of transport properties. Heat loss effects had previously been analyzed by Clavin and Nicoli (1985) and Clavin and Garcia (1983).

If the parameters describing strain or curvature become of order $O(\varepsilon^{-1})$ first-order effects will appear in the leading order flamelet equations. A very elegant derivation of the steady flamelet Equations (2.264) with stretch effects included at leading order has recently been presented by de Goey and ten Thije Boonkkamp (1999). Rather than solving (2.264) with a constant mass flux $(\rho u) = (\rho s_L^0)$ one may then use the counterflow geometry to impose a strain rate as an additional parameter on the flamelet structure. This also allows one to prescribe downstream boundary conditions that differ from the equilibrium values.

We now want to derive flamelet equations for the thin reaction zones regime. In that regime they should resolve the reaction zone of order ℓ_δ only. Since the laminar burning velocity is not of leading order anymore as shown in the discussion of (2.118), the normalization of the independent variables must differ from that in the corrugated flamelets regime. We still use the large length scale Λ to normalize the spatial coordinates but introduce the time $t_\Lambda = \Lambda^2/D$ to normalize the time t and the velocity vector v. Thus we have

$$x^* = x/\Lambda, \qquad t^* = t/t_\Lambda, \qquad v^* = vt_\Lambda/\Lambda. \tag{2.265}$$

The diffusivity D_i in (1.81) is normalized as in (2.248) and the reactive scalars and the density as in (2.249), but the nondimensional chemical source term is

$$\omega_i^* = \omega_i t_\Lambda/(\psi_{i,\text{ref}}\rho_{\text{ref}}). \tag{2.266}$$

With these normalizations the nondimensional form of (1.81) becomes the same as originally proposed with the asterisk removed:

$$\rho\frac{\partial \psi_i}{\partial t} + \rho v \cdot \nabla \psi_i = \nabla \cdot (\rho D_i \nabla \psi_i) + \omega_i. \tag{2.267}$$

As a small expansion parameter in the thin reaction zones regime we introduce the quantity

$$\varepsilon = \frac{\ell_\delta}{\Lambda} \tag{2.268}$$

and the corresponding short spatial variable defined by

$$\zeta = \varepsilon^{-1}(G(x,t) - G_0). \tag{2.269}$$

Here G is normalized by G' as in (2.250). In contrast from the previous scalings we introduce the short time variable

$$\tau = \varepsilon^{-2} t \tag{2.270}$$

to capture unsteady changes of the flamelet structure, which we expect to be more important in the thin reaction zones regime than in the corrugated flamelets regime, as noted at the end of Section 2.6. The reactive scalars are expanded as before and expressed as functions of the short and long spatial and temporal variables:

$$\psi_i = \psi_i(\zeta(x,t), \xi, \tau). \tag{2.271}$$

The temporal and spatial derivatives in (2.267) may then be written as

$$\frac{\partial}{\partial t} = \frac{\partial \zeta}{\partial t}\frac{\partial}{\partial \zeta} + \varepsilon^{-2}\frac{\partial}{\partial \tau} = \varepsilon^{-1}\frac{\partial G}{\partial t}\frac{\partial}{\partial \zeta} + \varepsilon^{-2}\frac{\partial}{\partial \tau}, \tag{2.272}$$

$$\nabla = \nabla\zeta\frac{\partial}{\partial \zeta} + \nabla_\xi = \varepsilon^{-1}\nabla G\frac{\partial}{\partial \zeta} + \nabla_\xi. \tag{2.273}$$

2.13 Laminar flamelet equations for premixed combustion

If these transformations are introduced into (2.267) one obtains

$$\rho \frac{\partial \psi_i}{\partial \tau} + \varepsilon^2 \rho \mathbf{v} \cdot \nabla_\zeta \psi_i + \varepsilon \left[\rho \frac{\partial G}{\partial t} + \rho \mathbf{v} \cdot \nabla G \right] \frac{\partial \psi_i}{\partial \zeta}$$

$$= \nabla G \cdot \frac{\partial}{\partial \zeta} \left(\rho D_i \nabla G \frac{\partial \psi_i}{\partial \zeta} \right) + \varepsilon \nabla_\xi \cdot \left(\rho D_i \nabla G \frac{\partial \psi_i}{\partial \zeta} \right)$$

$$+ \varepsilon \nabla G \cdot \frac{\partial}{\partial \zeta} (\rho D_i \nabla_\xi \psi_i) + \varepsilon^2 \nabla_\xi \cdot (\rho D_i \nabla_\xi \psi_i) + \Omega_i, \quad (2.274)$$

where the reaction rate has been rescaled as

$$\varepsilon^2 \omega_i = \Omega_i. \quad (2.275)$$

As before, we can pull the gradient ∇G in the first term on the r.h.s. inside the brackets. The leading order terms in (2.274) can then be collected to obtain

$$\rho \frac{\partial \psi_i}{\partial \tau} = \frac{\partial}{\partial \zeta} \left(\rho D_i |\nabla G|^2 \frac{\partial \psi_i}{\partial \zeta} \right) + \Omega_i, \quad (2.276)$$

which becomes, written in dimensional form again using $|\nabla G|^2 = \sigma^2$,

$$\rho \frac{\partial \psi_i}{\partial t} = \frac{\partial}{\partial G} \left(\rho D_i \sigma^2 \frac{\partial \psi_i}{\partial G} \right) + \omega_i. \quad (2.277)$$

This equation differs from (2.262), because it does not contain the propagation term, but an unsteady term instead. It only resolves the reaction zone that was identified with the inner layer at the end of Section 1.6. Boundary conditions matching the solution with the highly perturbed preheat zone and the oxidizer layer must complement the formulation of the problem. This will be considered in the next section.

In closing this section we want to combine (2.262) and (2.277) to obtain flamelet equations that are valid in both regimes, the corrugated flamelets and the thin reaction zones regime. Thereby, terms that may be of importance in either one of the two regimes will appear in a common equation. If the unsteady term originating from (2.276) is retained, the burning velocity is no longer a constant. We then replace the term in square brackets in (2.261) by $\rho s_L |\nabla G|$, where s_L is a fluctuating burning velocity that responds to outer perturbations, to obtain the common equation

$$\rho \frac{\partial \psi_i}{\partial t} + \rho s_L \sigma \frac{\partial \psi_i}{\partial G} = \frac{\partial}{\partial G} \left(\rho D_i \sigma^2 \frac{\partial \psi_i}{\partial G} \right) + \omega_i. \quad (2.278)$$

Göttgens et al. (2000) have considered unsteady flamelet equations for analyzing the response of the burning velocity to imposed fluctuations of the curvature. Although the curvature term was found to be of higher order in (2.257) its effect was included in the leading order equation to describe situations where local values of curvature are large. Performing numerical simulations with imposed harmonic fluctuations of the curvature Göttgens et al. found that not only does the burning velocity fluctuate, but also its mean changes. Those changes increase with the frequency of curvature fluctuations and depend on the Lewis number. Refering to the discussion on Lewis number effects on the burning velocity in the second and third paragraph after (2.200) it may sometimes seem appropriate to use a mean burning velocity obtained from unsteady flamelet calculations that replaces the burning velocity eigenvalue s_L^0 in (2.184) and in the subsequent correlations of the turbulent burning velocity.

2.14 Flamelet Equations in Premixed Turbulent Combustion

We now want to consider the properties of flamelets subjected to a turbulent flow field. In the corrugated flamelets regime the entire flame structure including the preheat zone is embedded within the smallest scale of turbulence, the Kolmogorov scale η, while in the thin reaction zones regime this is only true for the reaction zone of thickness ℓ_δ. Within a Kolmogorov eddy the flow and the scalar fields are quasi-laminar. If the flame surface $G(x, t)$ is placed at the location of the inner layer and averages conditioned on the distance from the surface[†] within the quasi-laminar regions in the vicinity of the surface are performed, no fluctuations of ψ_i and ω_i will appear and surface related averages of ψ_i and ω_i are equal to the instantaneous values. However, the gradient σ appearing in the flamelet Equations (2.262), (2.277), and (2.278) is an external fluctuating quantity. At each point in physical space the joint probability density of σ and G, conditioned at $G = G_0$, can be defined and a conditional probability $P(\sigma \mid G = G_0; x, t)$, as used before in (2.202), can be specified. We will use this again in the context of the presumed shape pdf approach.

The flamelet equations cannot be solved without information about the boundary values of ψ_i on both sides of the flamelet structure. In the corrugated flamelets regime, where the entire flamelet structure is embedded within a Kolmogorov eddy, boundary conditions specifying the unburnt and burnt state of the reactive scalars can be applied on either sides of the flamelet structure. Applying the boundary conditions at $G = \pm \infty$ presumes, however, that the flamelet is sufficiently removed from the boundaries of the flow field. If this is not the case, the particular value of G at which boundary conditions are to be applied and its statistical distribution enter into the flamelet model. Modeling

---

[†] We will call such averages surface related averages.

2.14 Flamelet equations in premixed turbulent combustion 153

problems of flame structures approaching the cylinder wall in a spark-ignition engine have been addressed by Ashurst (1995).

Similarly, if effects of strain or curvature are included in the leading order equations, fluctuations of these external parameters may lead to local extinction and possibly reignition events with rapid changes of the flamelet structure on the fast time scale. This also induces rapid fluctuations of the reactive scalars. If one assumes quasi-steady quenching conditions, as was done by Meneveau and Poinsot (1991), one can presume a joint pdf of G and σ and calculate the probability of finding burning or quenched flamelets.

In the thin reaction zones regime boundary conditions of the inner layer must be applied at the boundaries to the preheat zone and to the oxidation layer. We will first discuss those with respect to the preheat zone, since boundary conditions near the oxidation layer are less critical from a physical point of view, as discussed at the end of Section 1.6.

In the highly perturbed nonreacting preheat zone convection and diffusion are of the same order of magnitude. For an asymptotic analysis in that region we start from the equations for the reactive scalars (1.81) without the chemical source term. As a mental exercise we filter these equations over a region in physical space, which does not contain the reaction zone. The objective of this exercise is to envision scalar fields outside of the reaction zone that are smooth enough to allow for a two-scale asymptotic analysis as before.

We want to resolve the structure of the filtered turbulent preheat zone adjacent to the thin reaction zone. In the unburnt mixture this preheat zone extends over a region of the width of the mixing length ℓ_m. Since boundary conditions for the reactive scalar fields are imposed at inlet and outlet conditions of the entire turbulent flow region the large length scale is the length L of the flow region and the small length scale is ℓ_m. Therefore the small parameter ε is now the ratio

$$\varepsilon = \frac{\ell_m}{L}. \tag{2.279}$$

How do we imagine the filtering process? At first the G-field outside of $G(x, t) = G_0$ shall be reinitialized using $|\nabla G| = 1$ such that iso-scalar surfaces of G represent constant distance surfaces to the flame front. Filtering of the flame front shall proceed in the same way as illustrated in Figure 2.23 until it is smooth enough to contain only scales of order $O(\ell_m)$ and larger. All other G-iso-surfaces outside of the flame front shall be filtered in the same way. During the smoothing operation of the G-field the conditioning of the scalars ψ_i on particular values of G must be preserved, however, which implies that these fields must be reconstructed on the filtered G-field. Finally, the scalar fields outside $G(x, t) = G_0$ are filtered to make them smooth enough. Looking at Figure 2.13 one may imagine this filtering as a spatial averaging of the temperature profiles in the preheat zones to obtain smooth enough profiles that can be matched to the thin reaction zone.

The filtered equations are written as

$$\hat{\rho}\frac{\partial \hat{\psi}_i}{\partial t} + \hat{\rho}\hat{v} \cdot \nabla \hat{\psi}_i = \nabla \cdot (\hat{\rho}(\hat{D} + D_i)\nabla \hat{\psi}_i). \quad (2.280)$$

Similarly, the filtered G-equation is by analogy to (2.151)

$$\hat{\rho}\frac{\partial \hat{G}}{\partial t} + \hat{\rho}\hat{v} \cdot \nabla \hat{G} = (\hat{\rho}\hat{s}_T^0)|\nabla \hat{G}| - \hat{\rho}\hat{D}_T\hat{\kappa}|\nabla \hat{G}|. \quad (2.281)$$

In (2.280) and (2.281) $\hat{\rho}$ is the filtered mean density and $\hat{\psi}_i$ and $\hat{G}$ are filtered Favre averages defined by

$$\hat{\psi}_i = \frac{\widehat{\rho \psi_i}}{\hat{\rho}}, \quad \hat{G} = \frac{\widehat{\rho G}}{\hat{\rho}}, \quad (2.282)$$

$\hat{D}$ is a filtered turbulent diffusivity, which results from a gradient transport assumption applied to the nonreacting filtered quantities, $\hat{s}_T^0$ is the filtered burning velocity introduced in (2.169), and $\hat{\kappa}$ is the curvature of the filtered flame front.

In the two-scale asymptotic analysis we introduce a short range spatial variable $\hat{\zeta}$ defined as

$$\hat{\zeta} = \varepsilon^{-1}(\hat{G}(\mathbf{x}, t) - G_0), \quad (2.283)$$

which now resolves the mixing zone of size ℓ_m. The asymptotic analysis proceeds in the same way as in the derivation of the flamelet equations for the corrugated flamelets regime with the local instantaneous values of ρ, v, ψ_i, D_i, and G and s_L replaced by their respective filtered values. However, to capture unsteady effects that may be important in the filtered turbulent preheat zone, we rescale the time scale as

$$\tau = \varepsilon^{-1}t \quad (2.284)$$

such that the unsteady term becomes of leading order. One finally obtains, to leading order instead of (2.262),

$$\hat{\rho}\frac{\partial \hat{\psi}_i}{\partial t} + [(\hat{\rho}\hat{s}_T^0) - \hat{\rho}\hat{D}_T\hat{\kappa}]|\nabla \hat{G}|\frac{\partial \hat{\psi}_i}{\partial \hat{G}} = \frac{\partial}{\partial \hat{G}}\left[\hat{\rho}(\hat{D} + D_i)|\nabla \hat{G}|^2\frac{\partial \hat{\psi}_i}{\partial \hat{G}}\right]. \quad (2.285)$$

Except for the missing reaction term, this equation is very similar to (2.278).

In the vicinity of the instantaneous location of the reaction zone the filter width must be reduced down to the Kolmogorov scale in order not to average

2.14 Flamelet equations in premixed turbulent combustion

over reacting regions. Therefore the values of $\hat{\psi}_i$ become equal to ψ_i at the borderline between the thin reaction zone and the filtered turbulent preheat zone. The solution of (2.285) can then be matched to the flamelet Equation (2.278), thereby providing boundary conditions to the latter. However, rather than solving both equations separately in their region of validity and doing the matching process, we propose a common formulation for the entire domain. We had noted before in discussing (2.159) that the mass flux through the flame surface is inertial range invariant owing to continuity. In comparing (2.278) with (2.285) this statement may be generalized to curved flames to

$$[(\hat{\rho}\hat{s}_T^0) - \hat{\rho}\hat{D}_T\hat{\kappa}]|\nabla\hat{G}| = [(\rho s_L^0) - (\rho D)\kappa]\sigma, \qquad (2.286)$$

where the curvature effects contained in s_L have been expressed by replacing it by $s_L^0 - D\kappa$ following (2.121). Furthermore, by analogy to (2.168), one has for the filtered quantities

$$\hat{D}|\nabla\hat{G}|^2 \sim D\sigma^2. \qquad (2.287)$$

In contrast to (2.168) here the instantaneous value σ appears rather than its mean $\bar{\sigma}$. When surface related averages of the flamelet equations are taken, the mean value $\bar{\sigma}$ would appear.

One may write the term $(\hat{D} + D_i)|\nabla\hat{G}|^2$ in (2.285) as

$$(\hat{D} + D_i)|\nabla\hat{G}|^2 = \frac{\hat{D} + D_i}{\hat{D}} \frac{\hat{\chi}}{2}, \qquad (2.288)$$

where $\hat{\chi}$ represents the filtered scalar dissipation rate of the scalar G:

$$\hat{\chi} = 2\hat{D}|\nabla\hat{G}|^2 \sim 2D\sigma^2 = \chi, \qquad (2.289)$$

which is inertial range invariant. Setting $\hat{\chi}$ equal to χ one may write (2.288) as

$$(\hat{D} + D_i)|\nabla\hat{G}|^2 = \frac{\chi}{2\widehat{Le}_i}, \qquad (2.290)$$

where $\widehat{Le}_i$ is defined as

$$\widehat{Le}_i = \begin{cases} D/D_i & \text{in the thin reaction zone,} \\ \hat{D}/(\hat{D} + D_i) & \text{outside the reaction zone.} \end{cases} \qquad (2.291)$$

It is easily seen that $\widehat{Le}_i$ tends to unity outside of the thin reaction zone as $\hat{D}$ becomes larger than D_i.

Since the reaction rate ω_i in (2.278) is not affected by turbulence, it can be evaluated using the instantaneous values of ψ_i. Since notation is unimportant we combine (2.278) and (2.285) into a single equation

$$\rho\frac{\partial \psi_i}{\partial t} + \rho s_L \sigma \frac{\partial \psi_i}{\partial G} = \frac{\partial}{\partial G}\left(\frac{\rho\chi}{\widehat{Le}_i}\frac{\partial \psi_i}{\partial G}\right) + \omega_i. \qquad (2.292)$$

Here ρ, ψ_i, and G are interpreted as filtered mean values in the turbulent preheat zone; they become equal to the instantaneous values in the thin reaction zone.

Equation (2.292) has been constructed as an example of flamelet equations that are valid in both the corrugated flamelets regime and in the thin reaction zones regime. It was argued that for the latter the same structure of the equations is valid in the laminar flow region in the thin reaction zone as well as in the highly perturbed preheat zone. Similar arguments could be used for the oxidation layer that follows the thin reaction zone. There, however, a chemical source term would appear in the filtered equations of the reaction scalars. Even though the perturbations of the oxidation layer are expected to be small, as was argued at the end of Section 1.6, it is not evident that the filtered reaction rate can be approximated by evaluating it using the filtered mean values of the reacting scalars. This question could be addressed by analyzing DNS data in the thin reaction zones regime.

2.15 The Presumed Shape Pdf Approach

Once the reactive scalars ψ_i are obtained as a function of G and σ by solving the flamelet equations, one may calculate their means at any position x and time t in the flow field using the presumed pdf approach. Here G represents displacements of the instantaneous flame front with respect to the mean flame front in the direction normal to the latter, and σ takes fluctuations of its angle into account. Therefore the joint pdf of G and σ, written as in (2.202) must be determined. For the marginal pdf of G a Gaussian shape may be assumed to first approximation. Using Favre averages and relating the Favre mean $\tilde{G}$ to the normal coordinate x as $\tilde{G}(x) = G_0 + x - x_f$, we can write the Favre pdf as

$$\tilde{P}(G; x, t) = \frac{1}{\sqrt{2\pi \widetilde{(G''^2)}_0}} \exp\left(-\frac{(G - \tilde{G}(x))^2}{2\widetilde{(G''^2)}_0}\right). \qquad (2.293)$$

Here $\widetilde{(G''^2)}_0$ is the conditional variance evaluated at $\tilde{G}(x, t) = G_0$. For the conditional pdf of σ, by analogy to the scalar dissipation rate, to be discussed in Chapter 3, a lognormal distribution could be presumed. It could be determined, if its mean and the variance were known. However, since there is no information available that would allow us to estimate that variance, we presume a delta function pdf at the local mean value of $\langle \sigma \mid G = G_0 \rangle$.

Flamelet equations are typically solved in terms of the normal coordinate $x_n = (G - G_0)/\sigma$ and t. We fix the origin $G = G_0$ in the flamelet profiles at the inner layer and require that this coincides with the mean flame front position $x = x_f$. The Favre mean concentration $\tilde{\psi}_i$ of a reactive scalar may then be calculated from

$$\tilde{\psi}_i = \int_{-\infty}^{+\infty} \int_0^{\infty} \psi_i \left(\frac{G - G_0}{\sigma}, t \right) \delta(\sigma - \langle \sigma \mid G = G_0 \rangle) \tilde{P}(G; \boldsymbol{x}, t) \, d\sigma \, dG. \quad (2.294)$$

An integration over σ then leads to

$$\tilde{\psi}_i = \int_{-\infty}^{+\infty} \psi_i \left(\frac{G - G_0}{\langle \sigma \mid G = G_0 \rangle}, t \right) \tilde{P}(G; x, t) \, dG, \quad (2.295)$$

In evaluating (2.295), the dependence of the conditional gradient $\langle \sigma \mid G = G_0 \rangle$ on the position x within the flame brush must be known. As discussed at the end of Section 2.11 no model is available at this time to account for this dependence. It is therefore proposed that the conditional gradient be replaced by the mean gradient $\bar{\sigma}$.

Another concern is related to the use of a Gaussian distribution for the shape of the pdf. Given that only equations for the first two moments of G are available, it is an obvious choice, which allows us to express the pdf by a two-parameter presumed shape. In the limit $v'/s_L^0 \to \infty$ fluctuations around the mean flame position are expected to be symmetric. Nonsymmetries of the shape that show up in experimental data for not so large values of v'/s_L (cf. Figures 2.18 and 2.19) could be accounted for by considering a modeled equation for the third moment of G, but this is beyond the scope of this presentation.

2.16 Numerical Calculations of One-Dimensional and Multidimensional Premixed Turbulent Flames

Because of the many conceptual difficulties in formulating a consistent model for premixed turbulent combustion, there have been few attempts to calculate multidimensional turbulent flames. Most engineering approaches still rely on the eddy dissipation model, which, by construction, reproduces features of flame propagation, even though the physics behind the model is dubious and excessive tuning of the modeling constants is required.

Among the more fundamentally based models the BML model has been validated in one-dimensional configurations. Calculating one-dimensional unsteady turbulent flames in a closed vessel, Cant and Bray (1988) have compared a model for the chemical source term of the type (2.51) with an Eddy-Break-Up Model. They found that, to obtain agreement between the two models, the Eddy-Break-Up constant must be varied considerably from one case to the other. In a series of papers, Bray, Champion, and Libby studied premixed flames in stagnating turbulence. In Bray et al. (1991) they presented a similarity transformation for obtaining a one-dimensional equation for the Favre mean progress variable. In Bray et al. (1992) they solved this equation, having identified three different regions that were matched to each other. They applied the gradient flux approximation (2.48) but predicted a production of turbulent kinetic energy caused by coupling between the mean pressure gradient and density inhomogeneities. In Bray et al. (1998) comparisons between theoretical predictions and five experimental data sets for impinging and opposed flows are discussed. The predictions are based on an asymptotic formulation involving the relative intensity of turbulence at the exit plane of the jet as a small parameter. To leading order in the asymptotic treatment the scalar flux term disappears in the equation for the Favre mean progress variable, leaving a convective–reactive balance. Calculated profiles of the mean progress variable and the mean axial velocity compare favorably with experiments.

Wu and Bray (1997) applied the Coherent Flame Model to calculate the counterflow flame of Kostiuk (1991). The similarity transformation was used to reduce all equations, including the one for the flame surface density, to a one-dimensional form. The gradient transport approximation was used throughout. To yield good agreement with experiments, modeling of the kinematic restoration term in the flame surface density equation, the last term in (1.155), required using a strain rate dependence.

The Coherent Flame Model was also used in a series of papers to predict turbulent flame propagation in spark-ignition engines. Boudier et al. (1992) have added a laminar flame ignition model to predict the radius of the first laminar flame kernel. A criterion based on flame stretch is used for the transition to fully turbulent combustion modeled by a modified ε-equation. The model is validated by comparison with average flame contours in an experimental engine with central ignition. Baritaud et al. (1996) apply the Coherent Flame Model to the prediction of combustion and pollutant formation in a stratified charge spark-ignition engine. Partial equilibrium for the radicals is used in the postflame region and finite rate chemistry calculations are applied to predict concentrations of NO (using the Zeldovich mechanism) and CO. The ability of the model to reproduce trends of exhaust gas emissions from a 4-valve spark-ignition engine is demonstrated. A similar study is carried out by Duclos et al. (1996) to predict the effects of fuel distribution, residual gas distribution, and spark location. An Extended Coherent Flame Model (ECFM) has been presented recently by Duclos et al. (1999). It is validated by comparing measurements and computations on the gasoline direct injection engine by Mitsubishi.

Two-dimensional calculations of the counterflow experiments by Cheng and Shepherd (1991) were performed by Lindstedt and Váos (1999) using second moment closure for the Reynolds stress tensor and the scalar flux of the progress

2.16 One- and multidimensional premixed turbulent flames

variable. The production term in the ε-equation was modified to account for velocity divergence and influences of the mean pressure gradient. An algebraic model for the mean reaction rate is adopted. Comparison with experiments is encouraging, showing good agreement for the mean progress variable and the mean axial velocity. As a conclusion the superiority of second moment closure over gradient transport models is emphasized. The same second moment method was also applied by Lindstedt and Váos (1999) to the oblique flames by Gulati and Driscoll (1986) as well as to the counterflow flames by Kostiuk (1991) and Mastorakos (1993) (cf. also Mastorakos et al., 1994). Gradient diffusion and second moment closure are compared, demonstrating again the superiority of the latter.

The counterflow flame of Mastorakos (1993) has also been used by Jones and Prasetyo (1996) to validate a model based on the pdf transport equation. A four-step reduced mechanism was used to model the chemistry. While the profiles of the mean and r.m.s. velocities along the centerline, as well as most of the mean reactive scalars, show very good agreement with experiments, the concentrations of carbon monoxide is considerably underpredicted in the lean flame considered. Two mixing models have been compared in this study. It turns out that only the coalescence–dispersion model generated plausible results, while the flame extinguished, when the IEM model was used.

Hůlek and Lindstedt (1996a) consider the influence of closure approximations on the turbulent burning velocity in the pdf transport equation model. They propose adding to the classical model a flamelet acceleration term involving the mean pressure gradient, and they find that this term is important for predicting countergradient diffusion in the corrugated flamelets regime, while it is unimportant in the thin reaction zones regime. In Hůlek and Lindstedt (1996b) they calculate turbulent burning velocities using this method for various turbulence intensities and length scales. Their results, when plotted over the Damköhler number, show a trend similar to that of the turbulent burning velocities in Figure 2.24, but with significantly smaller values.

A full second-order Reynolds stress model, including equations for the temperature and its variance, was presented by Bradley et al. (1994). The heat release rate was evaluated by assuming that the flame brush consists of an array of stretched laminar flamelets, which are subjected to flame stretch. A joint pdf of the progress variable and the stretch rate was introduced but, by assuming a sudden extinction at a critical stretch rate, the expression for the mean heat release rate could be reduced to

$$\bar{q} = p_b \int_0^1 q(c) P(c)\,dc. \qquad (2.296)$$

Here $q(c)$ is the heat release rate of the laminar flamelet and p_b is the probability of burning. For the latter an empirical approximation in terms of the Karlovitz stretch factor and the Lewis number was derived. The heat release rate $q(c)$ was approximated as being proportional to $c^a(1-c)^b$ and the pdf $P(c)$ was presumed to be a beta function. Turbulent burning velocities were predicted that are in good agreement with the correlation (2.200) over a wide range of parameters. It is interesting to note that in the limit $Da \to \infty$, corresponding to $p_b = 1$, the constant b_1 was calculated as 2.45 instead of 2.0 as in Table 2.1. The turbulent flame brush thickness was determined from the distribution of the mean heat release rate over the flame brush, which corresponds to $P(x)$ in (2.127). In the limit $p_b \to 1$, the constant b_2 becomes 2.46 instead of 1.78 in Table 2.1.

Two-dimensional calculations using the mean G-equation with a presumed burning velocity $s_T^0 = s_L^0 + v'$ are presented by Smiljanovski et al. (1997). They use a novel in-cell reconstruction technique for an accurate calculation of fluxes across the cell boundaries, when the flame front crosses the cell. The method was applied to the problem of turbulent flame acceleration in a semiclosed duct containing a series of obstacles. When the premixed gas was ignited at the closed end of the duct, combustion-induced gas expansion at the mean turbulent flame front generated an axial flow in the duct, which, upon passage over the obstacles, generated turbulence in their wakes. This led to a dramatic acceleration of the flame propagation process, which, in terms of propagation length, was in very good agreement with experimental data.

A first application of the G-equation to spark-ignition engines is due to Keller-Sornig (1996). In his dissertation he solved the equations for the mean and the variance of G and an equation for the flame surface area ratio $\bar{\sigma}$. This latter equation had a production term and a kinematic restoration term, but no dissipation term. The dissertation clearly showed the potential of the level set approach for modeling turbulent flame propagation in engines. This set the basis for the dissertation of Dekena (1998) who used this G-equation model to simulate flame propagation in a direct injection gasoline engine (cf. also Dekena and Peters, 1999). The equations for $\widetilde{G}$ and $\widetilde{G''^2}$ were implemented into a source version of the FIRE code, which was available at Volkswagen, but instead of the $\bar{\sigma}$-equation, an algebraic relation for the turbulent burning velocity was used. As an example of flame propagation in a homogeneous charge engine, Figure 2.26 shows the comparison of different combustion models, as a side view into the combustion chamber. Panel a shows a simulation using the Eddy Dissipation Model where an acceleration in the vicinity of the walls is observed. This unphysical behavior is often found in using the Eddy Dissipation Model or similar Eddy Break-Up models. The pdf model shown in panel b is the one

2.16 One- and multidimensional premixed turbulent flames

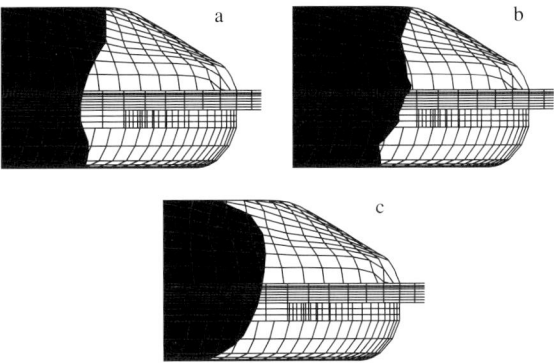

Figure 2.26. Simulation of turbulent flame propagation in a homogeneous charge spark-ignition engine. Three turbulent combustion models were compared by Dekena (1998). Panel a shows calculations using the Eddy Dissipation Model, panel b shows results from using a pdf transport equation model and panel c shows results from using the G-equation model. (Reprinted with permission by M. Dekena.)

implemented in the FIRE code. It shows wriggles of the flame front, which are presumably caused by the limited number of 100 particles per cell. The G-equation flamelet model shown in panel c generates a smooth mean front that proceeds the fastest in the center.

There are many formulations based on the mean progress variable that try to improve upon the Eddy Dissipation Model. A model by Zimont and Lipatnikov (1995), for instance, mixes ideas from the level set approach with a formulation for the mean progress variable and proposes the equation

$$\bar{\rho}\frac{\partial \tilde{c}}{\partial t} + \bar{\rho}\tilde{v} \cdot \nabla \tilde{c} = \nabla \cdot (\rho D_t \nabla \tilde{c}) + (\bar{\rho} s_T)|\nabla \tilde{c}| \qquad (2.297)$$

with boundary conditions $\tilde{c} = 0$ in the unburnt mixture and $\tilde{c} = 1$ in the burnt gas. The model uses an empirical expression for the turbulent burning velocity, as mentioned in Section 2.11.

Clearly Equation (2.297) has no solution for a steady planar flame, for which it would reduce to

$$\bar{\rho} s_T \frac{\partial \tilde{c}}{\partial x} \neq \frac{\partial}{\partial x}\left(\bar{\rho} D_t \frac{\partial \tilde{c}}{\partial x}\right) + (\bar{\rho} s_T)\left|\frac{\partial \tilde{c}}{\partial x}\right|. \qquad (2.298)$$

In the unsteady case the turbulent diffusion term appearing in (2.297) will decrease the gradient of the mean progress variable in the direction normal to the flame with time. This will lead to an indefinite broadening of the region where $0 < \tilde{c} < 1$, which is unphysical, because combustion will then change the flow field far ahead of the turbulent flame. The model also misses another important property of the level set approach, namely that the turbulent burning velocity is only defined at the flame front and has no physical meaning outside. Zimont et al. (1998) use the model for calculations of flame propagation in a gas turbine combustion chamber, claiming it to be numerically robust and efficient.

A similar model based on a LES version of (2.297) has been proposed by Weller et al. (1998). They do not prescribe the turbulent burning velocity but solve an equation for the flame surface area ratio instead, which they call the subgrid flame wrinkling. That equation resembles (2.182) without the dissipation term but its modeling is entirely different. Curiously enough, a modeled transport equation for the filtered laminar burning velocity also is solved, which apparently is needed to model effects of time-varying strain. The model is applied to the highly unstable combustion pattern behind a backward facing step and is compared with experimental data for this configuration.

2.17 A Numerical Example Using the Presumed Shape Pdf Approach

The set of differential equations for $\tilde{G}$, $\widetilde{G''^2}$, and $\bar{\sigma}_t$, namely (2.151), (2.152), and (2.184), together with the flamelet equations and the presumed pdf approach are sufficient to determine the Favre averaged mean values of all reactive scalars and the mean density in the flow field. The Favre averaged velocities and the mean pressure may, in addition, be calculated from the continuity equation and the Navier–Stokes equations, (1.34) and (1.35), respectively, the turbulent kinetic energy $\tilde{k}$ from (1.39), and the dissipation $\tilde{\varepsilon}$ from (1.40). Other turbulence models for the flow field and the turbulence quantities, such as Reynolds-stress models or LES could alternatively be used.

Before discussing the boundary conditions and the numerical results of the example calculation, we want to point to a specific numerical feature associated with the level set approach. The flow field outside of the mean flame surface $\tilde{G}(x, t) = G_0$ will convect iso-scalar surfaces with speeds different from that of the flame surface, which changes the scalar gradient $|\nabla \tilde{G}|$ and therefore the propagation speed of these iso-scalar surfaces according to (2.151). This problem can be circumvented by requiring that at each time step the condition

$$|\nabla \tilde{G}| = 1 \tag{2.299}$$

is satisfied outside the surface $\tilde{G}(x, t) = G_0$. A numerical procedure for doing this is called reinitialization as already mentioned in Example 1 in Section 2.5. A method proposed by Sussman et al. (1994) solves after each time step in the entire flow field the equation

$$\frac{\partial g}{\partial t} = \text{sign}(\tilde{G}(x, t) - G_0)(1 - |\nabla g|), \tag{2.300}$$

2.17 A numerical example using the presumed shape pdf approach

which has characteristics that originate at the flame surface and propagate the unity gradient information from there into the surrounding field. Starting from the initial condition $g(\mathbf{x}, t = t_0) = \tilde{G}(\mathbf{x}, t)$, the g-field is calculated until for large times the stationary solution $g_\infty(\mathbf{x})$ is reached. Then $\tilde{G}(\mathbf{x}, t)$ is set equal to $g_\infty(\mathbf{x})$ for $\tilde{G} \neq G_0$ at that time step such that it satisfies (2.299), while the surface $\tilde{G}(\mathbf{x}, t) = G_0$ remains unchanged. This numerical process is time consuming. A more efficient procedure that assumes the propagation velocity normal to the adjacent iso-scalar surfaces to be equal to that of the front has recently been proposed by Adalsteinsson and Sethian (1999). A similar method was used for laminar flame calculations by Terhoeven (1998). A overview of numerical techniques related to the level set approach may be found in Klein (1999).

Here we want to present some results for a turbulent Bunsen flame on a slot burner similar to the laminar flame shown in Figure 2.17. The burner had a width of $b = 12$ mm and was surrounded by sinter metal plates of 30 mm on both sides for pilot flames. The fuel was methane with an equivalence ratio $\phi = 0.78$ for the main flame and $\phi = 1$ for the pilot flames. A combined planar Rayleigh-OH-LIF technique, described in Plessing et al. (1998b), was used to visualize temperature and OH concentrations. Typical pictures of instantaneous temperature and OH fields are shown in the upper row of Figure 2.27. Taking averages over 1,000 imagines and defining the flame front as an iso-temperature surface at 1,000 K, we can obtain the probability density $P(G = G_0)$, shown on the picture on the lower left of Figure 2.27. Average OH concentrations taken over 1,000 imagines are shown in the picture on the lower right. Because of averaging over unburnt and burnt gas conditions, the peak values in the mean OH concentration is, of course, lower than the peak values in the instantaneous figure above. The picture on the lower right also shows vectors indicating the magnitude and direction of the mean velocity, obtained from particle image velocimetry (PIV) measurements.

Calculations of this flame were performed in the dissertation by Herrmann (2000). Zero gradient boundary conditions were used for all scalars at all boundaries except at the burner exit plane $x = 0$. For $y < \pm b/2$ the axial velocity was assumed constant and equal to $\bar{u} = 5.8$ m/s. The turbulent kinetic energy k was 1.5 m^2/s^2 and the dissipation was $\varepsilon = 540$ m^2/s^3, as obtained from the PIV measurements. The value of ε was calculated from the decay of k over a small distance in the axial direction, assuming homogeneous turbulence at the burner outlet. Boundary conditions for the $\tilde{G}$-equation at $x = 0$ are $\partial^2 \tilde{G}/\partial x^2 = 0$ for the unburnt mixture at $y < \pm b/2$ and for the co-flow $y > \pm b/2 + h$. Here $h = 1.5$ mm is the width of the burner rim. Along a vertical line of 6 mm the condition $\tilde{G} = G_0$ was imposed to anchor the flame. Since the flame is supposed to start with the laminar flame thickness ℓ_F, which then develops into the

164 2. *Premixed turbulent combustion*

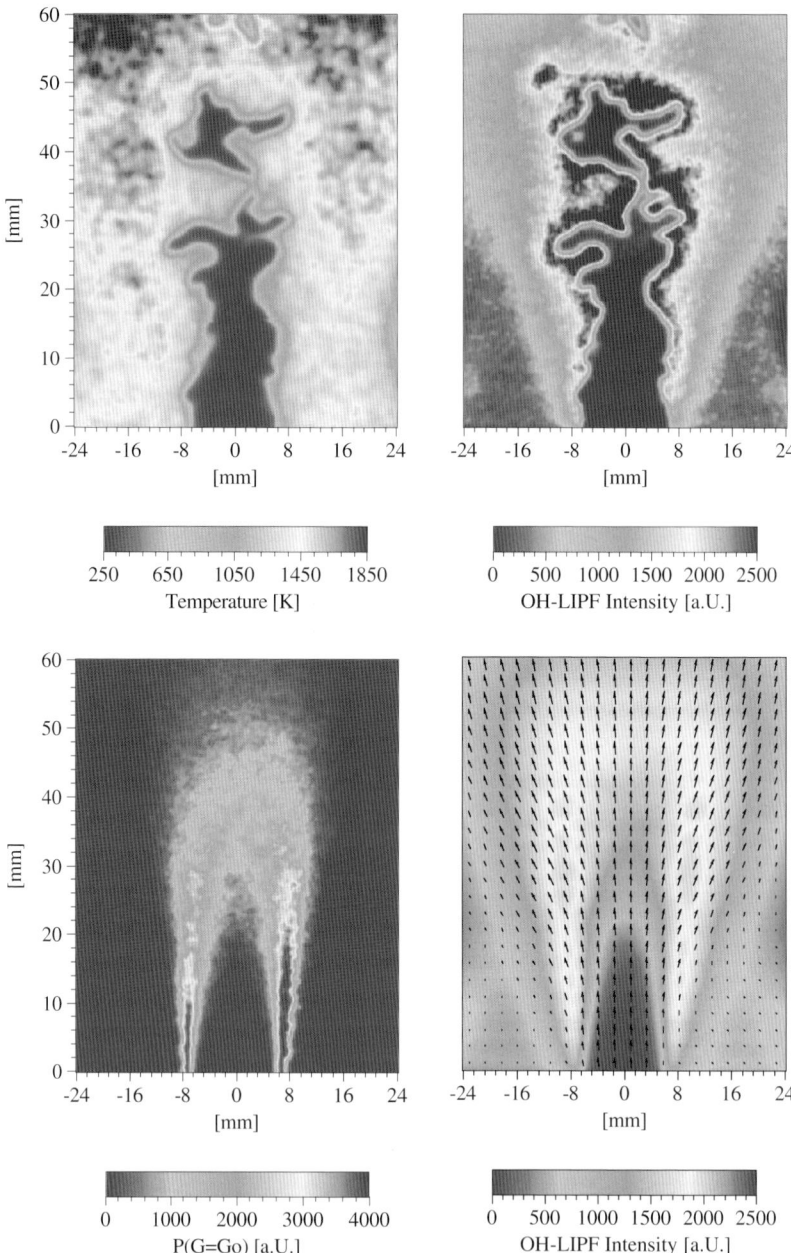

Figure 2.27 (see Color Plate II, Top). Two-dimensional laser measurements of scalar and velocity fields in a Bunsen flame on a slot burner. These results were kindly provided by T. Plessing and A. Joedicke. The pictures are explained in the text. (Reprinted with permission by T. Plessing and A. Joedicke.)

2.17 A numerical example using the presumed shape pdf approach

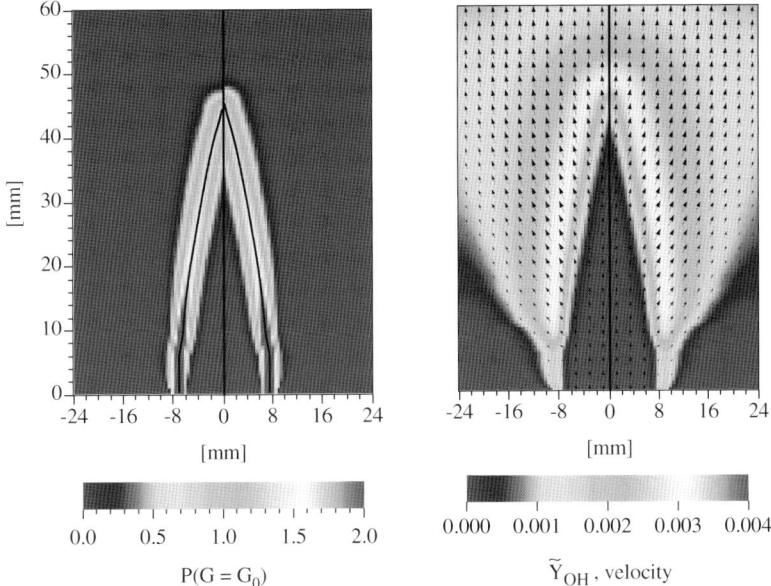

Figure 2.28 (see Color Plate I, Right). Mean scalar and velocity fields obtained by Herrmann (2000) for the Bunsen flame on a slot burner. These pictures are to be compared with those on the lower row in Figure 2.27. Details are explained in the text. (Reprinted with permission by M. Herrmann.)

turbulent flame brush thickness, boundary conditions for the $\widetilde{G''^2}$-equation at $x = 0$ were $\ell_{F,t} = \ell_F$ in the unburnt mixture and in the co-flow and zero gradient on the burner rim. For the $\bar{\sigma}_t$-equation the value $\bar{\sigma}_t = 0.01$ was imposed at $x = 0$ over the entire length of the inlet. Reinitialization was applied to the $\tilde{G}$-field but it was not found to be necessary for the $\widetilde{G''^2}$ and $\bar{\sigma}_t$ fields, because they turned out to be sufficiently smooth in the vicinity of the mean flame surface. For simplicity the gradient transport approximation has been used throughout in solving the $\widetilde{G''^2}$- and $\bar{\sigma}_t$-equations.

Figure 2.28 shows on the left the flame surface $\tilde{G}(x, y) = G_0$ as a solid line, superimposed on the probability density $P(G = G_0)$ of finding the instantaneous flame surface. The latter has been calculated using the presumed shape pdf approach. The width of the green region corresponds to the flame brush thickness. The picture on the right shows the mean mass fraction of OH in the flame. These two pictures should be compared to the pictures in the lower row of Figure 2.27.

There are two solutions plotted in each of the pictures in Figure 2.28: The l.h.s. corresponds to a solution obtained by solving the $\bar{\sigma}_t$-Equation (2.184) in the entire flow field, while the solution on the r.h.s. only uses the algebraic balance of the source terms (2.186). We see that the flame shape calculated with the algebraic balance is nearly the same as the one obtained by solving (2.184). Therefore, at least for the turbulent flame considered here, there seems little need to solve the differential equation for $\bar{\sigma}_t$.

For the flame shape and the flame brush thickness the agreement is clearly not satisfactory. By analyzing many of the single images that have led to the picture on the lower left of Figure 2.27 it becomes evident that the instantaneous flame fronts on both sides fluctuate significantly in the lateral direction. This is probably due to shear layer instabilities, which are enhanced by instabilities induced by gas expansion. These instabilities cannot be captured by the RANS calculations that led to the picture on the left of Figure 2.28, although these calculations were done using an unsteady code, where it was observed that the solution did not reach a steady state. The remaining unsteady oscillations of the flame front, although quite small, were averaged to produce Figure 2.28. This seems to indicate that the example presented here is only marginally stable at the RANS level. The fluctuating nature of the flame fronts would probably become more apparent, if LES was applied.

The mean OH mass fraction field shown in the picture on the right in Figure 2.28 was obtained by using the presumed pdf approach, namely (2.295). Premixed flamelet profiles were calculated in a counterflow geometry, setting the burnt gas temperatures to values between 1,400 K and 1,966 K as boundary conditions. This temperature range corresponds to the temperatures measured in the co-flow for increasing distances from the burner. The strain rate a of the counterflow in the libraries was varied between $a = 100/s$ and $a = 500/s$, which corresponds to the range of the values of $\tilde{\varepsilon}/\tilde{k}$ encountered in the flame brush. For the calculation of the reactive scalar fields an interpolation between different flamelets in the library was performed to match the local values of $\tilde{\varepsilon}/\tilde{k}$ to the strain rate in the library and to match the co-flow temperature in the Bunsen flame to the burnt gas temperature of the flamelets.

To further analyze the conditions at the flame front, the flame surface area ratio $\bar{\sigma}$, the local Damköhler number, and the local Karlovitz number are shown in Figure 2.29 along the normalized arclength s/L, where s is measured along the mean flame surface $\tilde{G}(x, y) = G_0$ from the burner rim to the flame tip L. We see that $\bar{\sigma}$, obtained here from the algebraic balance (2.186), varies slightly, owing to variations of the turbulence along the flame surface. The Damköhler number and the Karlovitz number at the flame surface $\tilde{G}(x, y) = G_0$ shown in Figure 2.29 are relatively small, indicating that the flame lies toward the left and close to the line $Ka = 1$ in the regime diagram (Figure 2.8).

In Figure 2.30 the three terms in the algebraic balance (2.186) are plotted along the normalized arclength. It is interesting to note that the production term and the scalar dissipation term are the dominant terms in this balance.

2.17 A numerical example using the presumed shape pdf approach   167

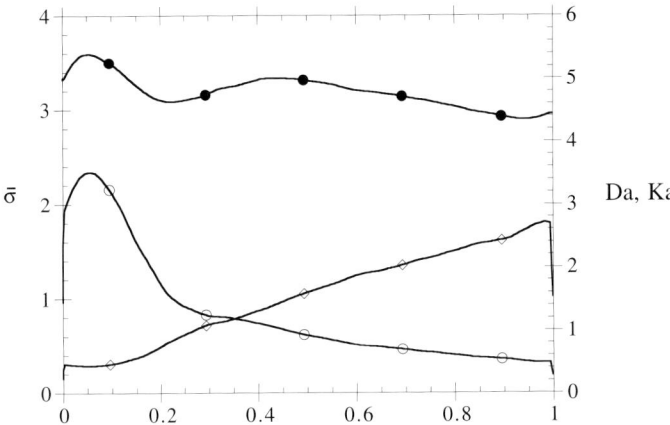

Figure 2.29. Calculated values of the flame surface area ratio $\bar{\sigma}$ –•–•–, the Damköhler number –◇–◇–, and the Karlovitz number –○–○– along the normalized arclength of the mean flame surface in a turbulent Bunsen flame on the slot burner. (Reprinted with permission by M. Hermann.)

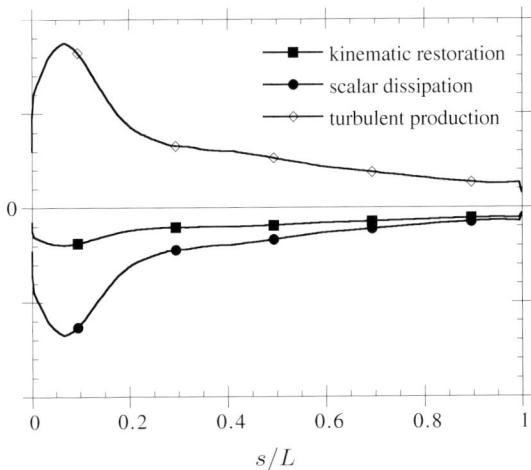

Figure 2.30. The relative magnitude of production, kinematic restoration, and scalar dissipation in the algebraic balance (2.186) used to calculate the flame surface area ratio in the Bunsen flame on a slot burner. (Reprinted with permission by M. Hermann.)

Close to the burner rim kinematic restoration amounts to about 30% of scalar dissipation. Further downstream at $s/L = 0.5$, where the Karlovitz number becomes less than unity and therefore the conditions at the flame front change from the thin reaction zones regime to the corrugated flamelets regime, the relative importance of kinematic restoration increases, but scalar dissipation continues to be the larger one of the two sink terms. This shows that, as far as modeling of the turbulent burning velocity is concerned, scalar dissipation is important even in the corrugated flamelets regime.

2.18 Concluding Remarks

In this chapter premixed turbulent combustion was analyzed essentially on the basis of the level set approach. This formulation appears natural for a propagating surface – but seems also unavoidable. The most important quantity characterizing premixed combustion is the laminar burning velocity. Since the governing equations for the reactive scalars do not contain the laminar burning velocity explicitly, it was useful to identify it as in (2.115) as the sum of displacement speeds, namely those due to normal diffusion and to reaction. Using the laminar burning velocity for local flame propagation implicitly assumes scale separation between chemistry occuring in thin layers and mixing in the inertial range of turbulence. The introduction of a thin propagating front naturally leads to the level set approach.

The advantage of this formulation is that it circumvents the unresolved problems that arise with any attempts to model premixed turbulent combustion on the basis of equations for moments of reactive scalars. It was shown in Chapter 1 that those problems are not only related to the chemical source term but also to the turbulent transport term. On the contrary, the level set equation does not contain a turbulent transport term and no chemical source term. Therefore classical closure models based on the eddy cascade hypothesis can be applied. Two-point closure for premixed combustion thereby follows the route of turbulence modeling for nonreactive scalars. Since the laminar burning velocity s_L is the only molecular quantity that appears in the level set equation, a new inertial range invariant quantity, called kinematic restoration, appears in the variance equation as a sink term in the corrugated flamelets regime. As shown in Section 2.12, gas expansion competes with kinematic restoration. It is argued, however, that in most situations kinematic restoration can suppress instabilities induced by gas expansion on scales smaller than the integral length scale. This allows us to introduce the turbulent burning velocity as a well-defined mean quantity, which can be related to other turbulent mean quantities.

There are two major disadvantages of the level set approach, however. First, it is unappealing: In combustion one is used to thinking in terms of balance equations for mass, momentum, energy, and chemical species. The scalar G appears as an artificial quantity, and it is somewhat artificial, indeed. It is calculated by a field equation in three-dimensional space, but it is only defined on a two-dimensional surface, namely the flame surface defined by $G(x, t) = G_0$.

2.18 Concluding remarks

The same property is carried over to the mean flame surface which is defined by $\tilde{G}(x, t) = G_0$. Outside of that surface the turbulent burning velocity and therefore the solution is not well defined. To prevent a feedback of the unphysical G-field outside $\tilde{G}(x, t) = G_0$, a numerical reinitialization technique must be applied. The second disadvantage of the level set approach therefore concerns the additional numerical effort associated with this reinitialization procedure. The procedure is time consuming but unavoidable for a robust and reliable numerical code.

To determine influences of finite rate chemistry on the reactive scalars, flamelet equations can be derived. In the most fundamental case they are identical to the equations for a steady laminar premixed flame. Unsteady effects and influences of strain and curvature as well as nonadiabatic downstream boundary conditions may also be included. Using the presumed shape pdf approach mean values of the reactive scalars may then be determined in premixed turbulent flames.

3

Nonpremixed turbulent combustion

3.1 Introduction

In many combustion applications fuel and oxidizer enter separately into the combustion chamber where they mix and burn during continuous interdiffusion. This process is called nonpremixed combustion.

A typical example is combustion in furnaces, which are operated under nonpremixed conditions mainly for safety reasons. Fuel is supplied, for instance, by jets of gaseous fuel, which entrain enough air from the surroundings so that all the fuel can be burned within a certain distance from the nozzle. That distance is called the flame length. Other fuels used in furnaces are coal dust injected with air as a carrier gas, or liquid fuel that is injected as a spray. Since mixing and combustion in jets and sprays occur simultaneously, the formation of large volumes of unburnt flammable mixture can be avoided. In a practical application this requires a control system to make sure that each of the flames in a furnace is burning as long as fuel is supplied.

Other applications of nonpremixed combustion include diesel engines and gas turbines. In diesel engines the air is compressed by the piston before a liquid fuel spray is injected into the combustion chamber. The hot compressed air is entrained into the spray, leading to liquid fuel breakup, evaporation, and finally to auto-ignition. During the combustion phase, at first the premixed fraction of the gas is rapidly consumed, but then combustion takes place under nonpremixed conditions. During this phase most of the formation of NO_x and soot is taking place, but it also provides the necessary conditions for soot oxidation.

In aircraft gas turbine engines nonpremixed combustion occurs in the swirl-stabilized combustion zone downstream of the spray injector. Combustion products formed in that zone are convected downstream and then are diluted by secondary air to reduce the temperature at the exit of the combustor to values

3.1 Introduction

that are bearable for the turbine blade material. In modern stationary gas turbines the fuel is often prevaporized and partially premixed before entering the gas turbine combustion chamber. Therefore models used for partially premixed combustion (Chapter 4) are more relevant for describing the flame propagation and combustion processes occuring in these engines.

A last example of nonpremixed combustion is fire. If the fuel is a solid or a liquid, it will first be gasified by radiative heat flux from the fire before mixing with the surrounding air. The mixing process is often dominated by buoyancy rather than by forced convection as in jets and sprays used in furnaces and engines. A well-known and generally pleasant example of a buoyancy dominated nonpremixed flame, to be discussed below, is the candle flame. If, however, one has to consider a fire hazard where the fuel is a pressurized gas that issues at a high velocity into the surroundings, the mixing process is momentum controlled.

Nonpremixed combustion is sometimes called diffusive combustion or combustion in diffusion flames since diffusion is the rate-controlling process. The time needed for convection and diffusion, both being responsible for turbulent mixing, is typically much larger than the time needed for combustion reactions to occur. The assumption of infinitely fast chemistry (or local chemical equilibrium) therefore appears to be appropriate. It is an assumption that introduces an important simplification, since it eliminates all parameters associated with finite-rate chemical kinetics from the analysis.

Over many years, since the pioneering work of Burke and Schumann (1928) the fast chemistry assumption has been used in nonpremixed combustion to predict global properties such as the flame length of a jet diffusion flame. The essential feature for the calculation of the latter is the introduction of a chemistry-independent "conserved scalar" variable called the mixture fraction. All scalars such as temperature, concentrations, and density are then uniquely related to the mixture fraction. For a review of the conserved scalar approach see Bilger (1980).

There are, however, situations in turbulent flows where local diffusion time scales are not so large compared to the rate-determining combustion reactions. The fast chemistry assumption is then not valid and nonequilibrium effects must be taken into account. If the diffusion time scale becomes locally of the same order of magnitude as the chemical time scale of the combustion reactions, even local quenching may occur. A further reduction of diffusion time scales then leads to lift-off and eventually blow-off of the entire turbulent flame. (Lifted diffusion flames will be addressed in detail in Chapter 4.) But already in globally stable flames diffusion may interact selectively with the different chemical processes occuring in the system. For instance, the reduction of the maximum temperature by finite-rate chemistry influences the strongly temperature dependent NO formation rate considerably whereas soot oxidation is very much faster in the presence of OH radicals, which requires a sufficiently high temperature. The trade-off between NO_x and soot emissions in diesel engines, therefore, is a consequence of this adverse response to nonequilibrium temperature changes. If soot formation is not an issue, technical combustion is best accomplished at moderately low temperature conditions so that the main combustion reactions can be brought to completion, while NO_x formation can be kept at a minimum.

3.2 The Mixture Fraction Variable

A very important quantity for the description of nonpremixed combustion is the mixture fraction Z (sometimes also denoted by f or ξ) which plays a role similar to that of the scalar G in premixed combustion in determining the flame surface. Therefore, before going into the description of current modeling approaches, we will present various definitions of the mixture fraction.

The definition of the mixture fraction is best derived for a homogeneous system in the absence of diffusion. Then, by writing the global reaction equation for complete combustion of a hydrocarbon fuel $C_m H_n$, for instance, as

$$\nu'_F C_m H_n + \nu'_{O_2} O_2 \rightarrow \nu''_{CO_2} CO_2 + \nu'''_{H_2O} H_2O \qquad (3.1)$$

one defines the stoichiometric coefficients of fuel and oxygen as ν'_F and $\nu'_{O_2} = (m + n/4)\nu'_F$, respectively. The reaction equation relates the changes of mass fractions of oxygen dY_{O_2} and fuel dY_F to each other by

$$\frac{dY_{O_2}}{\nu'_{O_2} W_{O_2}} = \frac{dY_F}{\nu'_F W_F}, \qquad (3.2)$$

where the W_is are the molecular weights. For a homogeneous system this equation may be integrated to obtain

$$\nu Y_F - Y_{O_2} = \nu Y_{F,u} - Y_{O_2,u}, \qquad (3.3)$$

where $\nu = \nu'_{O_2} W_{O_2}/\nu'_F W_F$ is the stoichiometric oxygen-to-fuel mass ratio and the subscript u denotes the initial conditions in the unburnt mixture. The mass fractions Y_F and Y_{O_2} correspond to any state of combustion between the unburnt and the burnt state. If the diffusivities of fuel and oxidizer are equal, (3.3) is also valid for spatially inhomogeneous systems such as a diffusion flame.

A mixture is called stoichiometric if the fuel-to-oxygen ratio is such that both fuel and oxygen are entirely consumed after combustion to CO_2 and H_2O

3.2 The mixture fraction variable

is completed. The condition for a stoichiometric mixture requires that the ratio of the concentrations $[X_i] = \rho Y_i / W_i$ of oxygen and fuel in the unburnt mixture is equal to the ratio of the stoichiometric coefficients:

$$\left. \frac{[X_{O_2}]_u}{[X_F]_u} \right|_{st} = \frac{\nu'_{O_2}}{\nu'_F}, \tag{3.4}$$

or in terms of mass fractions

$$\left. \frac{Y_{O_2,u}}{Y_{F,u}} \right|_{st} = \frac{\nu'_{O_2} W_{O_2}}{\nu'_F W_F} = \nu. \tag{3.5}$$

In a two-feed system,† where subscript 1 denotes the fuel stream with mass flux $\dot{m}_1$ and subscript 2 denotes the oxidizer stream with mass flux $\dot{m}_2$ into the system, the mixture fraction Z is defined at any location in the system as the local ratio of the mass flux originating from the fuel feed to the sum of both mass fluxes:

$$Z = \frac{\dot{m}_1}{\dot{m}_1 + \dot{m}_2}. \tag{3.6}$$

Both fuel and oxidizer streams may contain inert substances such as nitrogen. If the system is homogeneous or if equal diffusivities of fuel, oxygen, and inert substances are assumed in an inhomogeneous system, the local mass fraction $Y_{F,u}$ of fuel in the unburnt mixture is related to the mixture fraction Z as

$$Y_{F,u} = Y_{F,1} Z, \tag{3.7}$$

where $Y_{F,1}$ denotes the mass fraction of fuel in the fuel stream. Similarly, since $1 - Z$ represents the mass fraction of the oxidizer stream locally in the unburnt mixture, one obtains the local mass fraction of oxygen as

$$Y_{O_2,u} = Y_{O_2,2}(1 - Z), \tag{3.8}$$

where $Y_{O_2,2}$ represents the mass fraction of oxygen in the oxidizer stream ($Y_{O_2,2} = 0.232$ for air). Introducing (3.7) and (3.8) into (3.3) and integrating between the unburnt and any other state of combustion, one can relate the mass fractions of fuel and oxygen to the mixture fraction as

$$Z = \frac{\nu Y_F - Y_{O_2} + Y_{O_2,2}}{\nu Y_{F,1} + Y_{O_2,2}}. \tag{3.9}$$

---

† If more than two feeds enter into a combustion chamber, the concept of a single mixture fraction can no longer be used. It can formally be extended to multiple mixture fraction variables, but then it becomes less attractive for modeling because the resulting flamelet equations are defined in a multi-dimensional mixture fraction space.

For a stoichiometric mixture the r.h.s of (3.3) vanishes by definition, since both fuel and oxygen are zero, such that

$$\nu Y_F - Y_{O_2} = 0. \tag{3.10}$$

Therefore the stoichiometric mixture fraction is

$$Z_{st} = \left[1 + \frac{\nu Y_{F,1}}{Y_{O_2,2}}\right]^{-1}. \tag{3.11}$$

For pure fuels ($Y_{F,1} = 1$) mixed with air, the stoichiometric mixture fraction is, for instance, 0.0284 for H_2, 0.055 for CH_4, 0.0635 for C_2H_4, 0.0601 for C_3H_8, and 0.072 for C_2H_2. It is worth keeping in mind that these numbers are all much smaller than unity, which implies that in terms of mass ratios, a large amount of oxidizer is needed to completely consume the fuel.

The mixture fraction can be related to the commonly used equivalence ratio ϕ, which is defined as the fuel-to-air ratio in the unburnt mixture normalized by that of a stoichiometric mixture:

$$\phi = \frac{Y_{F,u}/Y_{O_2,u}}{\left(Y_{F,u}/Y_{O_2,u}\right)_{st}} = \frac{\nu Y_{F,u}}{Y_{O_2,u}}. \tag{3.12}$$

Introducing (3.7) and (3.8) into (3.12) leads with (3.11) to

$$\phi = \frac{Z}{1-Z} \frac{(1-Z_{st})}{Z_{st}}. \tag{3.13}$$

This suggests that the mixture fraction can be interpreted as a normalized fuel-to-air equivalence ratio.

There is a more general way to define the mixture fraction, namely as a quantity related to chemical elements, rather than to the equivalence ratio. While the mass of chemical species may change due to chemical reactions, the mass of elements is conserved. Denoting a_{ij} as the number of atoms of element j in a molecule of species i, and W_j as the molecular weight of that atom, one may write the mass of all atoms j in the system as

$$m_j = \sum_{i=1}^{n} \frac{a_{ij} W_j}{W_i} m_i. \tag{3.14}$$

The mass fraction of element j is then

$$Z_j = \frac{m_j}{m} = \sum_{i=1}^{n} \frac{a_{ij} W_j}{W_i} Y_i, \tag{3.15}$$

where $j = 1, 2, \ldots, n_e$, and n_e is the total number of elements in the system. Adding Equations (1.62) for the mass fractions Y_i in the same way as in (3.15) one obtains

$$\rho \frac{\partial Z_j}{\partial t} + \rho v \cdot \nabla Z_j = -\nabla \cdot \left(\sum_{i=1}^{n} \frac{a_{ij} W_j}{W_i} j_i \right), \quad (3.16)$$

where no chemical source term appears, since

$$\sum_{i=1}^{n} a_{ij} W_i \nu_{ik} = 0 \quad (3.17)$$

for each element j in any reaction k. This shows that the element mass fraction is conserved during combustion. If a binary diffusion flux as in (1.63) is used and if all diffusivities are equal, $D_i = D$, the balance equation for the element mass fraction becomes

$$\rho \frac{\partial Z_j}{\partial t} + \rho v \cdot \nabla Z_j = \nabla \cdot (\rho D \nabla Z_j). \quad (3.18)$$

Let Z_C, Z_H, and Z_O denote the element mass fractions of C, H, and O, and W_C, W_H, and W_O their molecular weights, respectively. Setting, for simplicity of notation, the stoichiometric coefficient ν'_F of the global reaction (3.1) equal to unity, one gets the element mass fractions

$$\frac{Z_C}{m W_C} = \frac{Z_H}{n W_H} = \frac{Y_{F,u}}{W_F}, \quad Z_O = Y_{O_2,u}. \quad (3.19)$$

From (3.5) it follows that the coupling function

$$\beta = \frac{Z_C}{m W_C} + \frac{Z_H}{n W_H} - 2 \frac{Z_O}{\nu'_{O_2} W_{O_2}} \quad (3.20)$$

vanishes at stoichiometric conditions. This corresponds to the original definition of Burke and Schumann (1928) of a conserved scalar. It can be normalized to vary between 0 and 1 as

$$Z = \frac{\beta - \beta_2}{\beta_1 - \beta_2} \quad (3.21)$$

to obtain Bilger's (1988) definition of the mixture fraction:

$$Z = \frac{Z_C/(m W_C) + Z_H/(n W_H) + 2(Y_{O_2,2} - Z_O)/(\nu'_{O_2} W_{O_2})}{Z_{C,1}/(m W_C) + Z_{H,1}/(n W_H) + 2 Y_{O_2,2}/(\nu'_{O_2} W_{O_2})}. \quad (3.22)$$

This formula is often used to determine the mixture fraction from experimental or numerical data of mass fractions that are available. However, in experiments mass fractions of minor species are usually not available and in most numerical calculations the diffusivities are not all equal to each other. Therefore the definition (3.22) implicitly contains unspecified additional influences. Its advantage is, however, that it reduces to (3.11) if the elements C, H, and O are in stoichiometric proportions.

If all diffusivities D_i are equal to D, it follows from (3.18), (3.20), and (3.21) that Z satisfies the balance equation

$$\rho \frac{\partial Z}{\partial t} + \rho \boldsymbol{v} \cdot \nabla Z = \nabla \cdot (\rho D \nabla Z). \tag{3.23}$$

This equation, together with its boundary conditions $Z = 1$ in the fuel stream and $Z = 0$ in the oxidizer stream, can also be postulated, independently of any combinations of element mass fractions, to provide a further and alternative definition of the mixture fraction. The diffusion coefficient D in (3.23) then is arbitrary, but since the maximum temperature determines the location of the reaction zone, diffusion of enthalpy is the most important transport process in mixture fraction space. It has been discussed in Chapter 1 that under certain simplifying conditions the enthalpy equation (1.75) takes the same form as (3.23), where D is the thermal diffusivity. Therefore we choose the thermal diffusivity as the diffusion coefficient in the mixture fraction equation.

3.3 The Burke–Schumann and the Equilibrium Solutions

The previous section has indicated that the mixture fraction is a normalized conserved scalar representing the chemical elements locally available. To show the restrictions that the presence of chemical elements impose on the combustion process we will first derive the Burke–Schumann solution and then the equilibrium solution as a function of the mixture fraction. In the limit of an infinitely fast one-step irreversible reaction of the kind (3.1) there exists an infinitely thin nonequilibrium layer at $Z = Z_{st}$. Outside of this layer the mass fractions of fuel and oxygen are either zero or due to (3.9) they are piecewise linear functions of Z:

$$Y_F = Y_{F,1} \frac{Z - Z_{st}}{1 - Z_{st}}, \quad Y_{O_2} = 0, \quad \text{for} \quad Z \geq Z_{st},$$
$$Y_{O_2} = Y_{O_2,2} \left(1 - \frac{Z}{Z_{st}}\right), \quad Y_F = 0, \quad \text{for} \quad Z \leq Z_{st}. \tag{3.24}$$

The mass fractions of product species may be written similarly. If the assumptions leading to the simplified temperature Equation (1.77) are valid and all Lewis numbers are equal to unity, coupling between the temperature and the

3.3 The Burke–Schumann and the equilibrium solutions

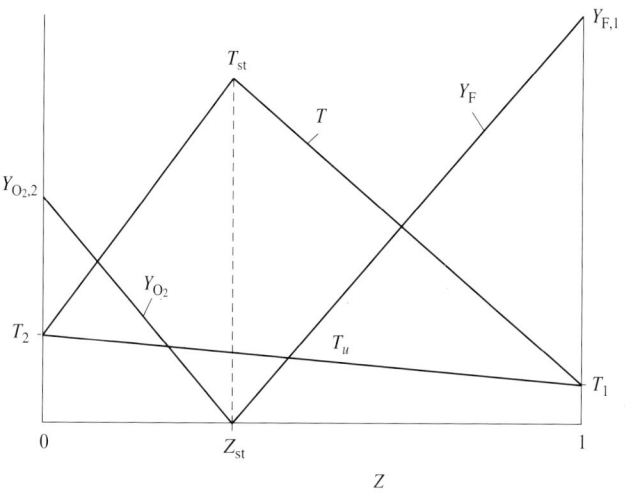

Figure 3.1. The Burke–Schumann solution as a function of mixture fraction.

mass fractions using (1.91) shows that the temperature is also a piecewise linear function of Z:

$$T(Z) = \begin{cases} T_u(Z) + \dfrac{QY_{F,1}}{c_p \nu'_F W_F} Z, & Z \leq Z_{st}, \quad (3.25) \\ T_u(Z) + \dfrac{QY_{O,2}}{c_p \nu'_{O_2} W_{O_2}} (1-Z), & Z \geq Z_{st}. \quad (3.26) \end{cases}$$

The Burke–Schumann solution showing the profiles of $Y_F(Z)$, $Y_{O_2}(Z)$, and $T(Z)$ for complete combustion is plotted as a function of mixture fraction in Figure 3.1. For the unburnt mixture, the temperature Equation (1.77) has the same form as the mixture fraction Equation (3.23), leading to the linear relation

$$T_u(Z) = T_2 + Z(T_1 - T_2), \quad (3.27)$$

which is also shown in Figure 3.1.

It is evident that before combustion, if equal diffusivities of fuel, oxygen and inerts are assumed, the mass fractions of fuel and oxygen and the temperature can uniquely be related to the mixture fraction. These are given by (3.7), (3.8), and (3.27), respectively. After combustion a different kind of relation among the mass fractions of fuel and oxygen, temperature, and the mixture fraction exists. For the case of a one-step irreversible reaction with infinitely fast chemistry these are given by (3.24), (3.25), and (3.26) and are shown in Figure 3.1. If infinitely fast but reversible reactions are assumed, all species are in chemical

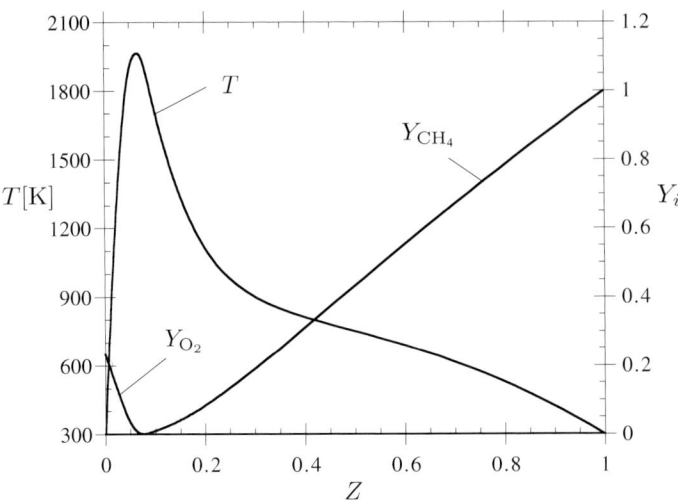

Figure 3.2. The equilibrium solution for methane–air combustion as a function of mixture fraction.

equilibrium at each value of the mixture fractions. If the enthalpy equation (1.75) takes the same form as the mixture fraction equation (3.23), the enthalpy becomes a linear function of the mixture fraction. The equilibrium temperature calculated under these conditions is the adiabatic flame temperature for each mixture fraction. This equilibrium solution is shown in Figure 3.2 for methane combustion. It differs from the Burke–Schumann solution, in particular for $Z \geq Z_{st}$. It also is a function of mixture fraction only.

Since the mixture fraction is uniquely related to the fuel–air equivalence ratio ϕ, one may also choose the equivalence ratio to present the structure of diffusion flames. Sivathanu and Faeth (1990) have plotted experimental data of mass fractions of major species (N_2, O_2, fuel, CO_2, H_2O, CO, and H_2) and temperature as functions of the local equivalence ratio for various steady laminar hydrocarbon–air diffusion flames, denoting that relation as "generalized state relationship." They found for a given fuel nearly the same dependencies on ϕ for all major species for fuel-lean conditions, whereas the profiles departed from each other for near-stoichiometric and fuel-rich conditions. This shows that the profiles in real flames are not in chemical equilibrium and that an additional nonequilibrium parameter is needed to represent these profiles.

3.4 Nonequilibrium Flames

We assume the mixture fraction field to be given in the flow field as a function of time by the solution of (3.23). Then the surface of the stoichiometric mixture

3.4 Nonequilibrium flames

can be determined from

$$Z(x, t) = Z_{st}. \tag{3.28}$$

Since the temperature is the highest in the vicinity of that surface, combustion chemistry is fast. Therefore there will be a thin reaction zone in the vicinity of the surface of stoichiometric mixture, where the fuel and the oxygen are depleted and radicals and products are formed. In that region also the highest temperatures will occur.

In Section 1.11, the location of the flame surface was defined as that of the inner layer. For simplicity, we approximate this by the iso-surface of the stoichiometric mixture. Then Equation (3.28) defines the flame surface in non-premixed combustion in a similar way in terms of the mixture fraction field as $G(x, t) = G_0$ defines the flame surface in premixed combustion in terms of the G-field. An important difference, however, is that the mixture fraction Z is well defined in the entire flow field, while G is not.

The surface of the stoichiometric mixture is shown schematically in Figure 3.3 for a laminar candle flame. The flow entraining the air into the flame is driven by buoyancy rather than by forced convection as in a jet flame. The paraffin of the candle first melts because of the radiative heat flux received from the flame; it mounts by capillary forces into the wick where it then evaporates to become paraffin vapor, a gaseous fuel. As the fuel vapor is transported toward the surface of the stoichiometric mixture it heats up and begins to pyrolize and to form soot under fuel rich conditions. Soot particles strongly radiate at a typical yellow color. This region is also shown in Figure 3.3. The soot particles are convected to the surface of the stoichiometric mixture where they are

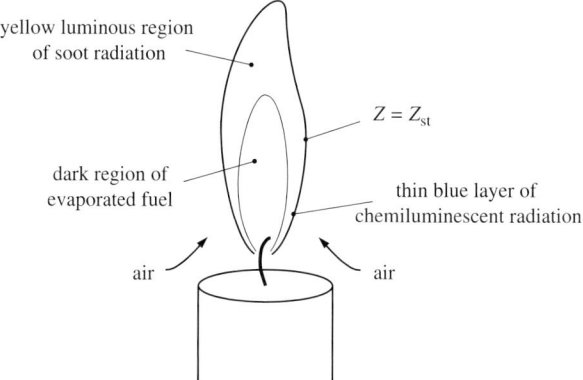

Figure 3.3. The candle flame as the classical example of a laminar diffusion flame.

abruptly oxidized within the thin reaction zone, mainly by OH radicals, which have a strong maximum concentration in that zone. Therefore there is typically a sharp boundary to the yellow luminous region in the laminar candle flame. If, however, flame stretch or radiative heat loss causes the temperature of the reaction zone to fall below a temperature of typically 1,300 K, which corresponds to the crossover temperature of chain-branching and chain-breaking reactions, identified as the inner layer temperature in the footnote in Section 1.6, soot particles can break through the surface of the stoichiometric mixture, generating a sooting candle flame. Kent and Wagner (1984), for instance, found that soot was emitted from gaseous laminar diffusion flames if the particle temperature dropped below about 1,300 K. At the leading edge of the surface of the stoichiometric mixture shown in Figure 3.3 there appears a thin blue layer, which is caused by chemiluminiscence, mainly of CH and C_2^* radicals. This provides visual evidence of the existence of a thin reaction zone in a candle flame.

As was already discussed in Section 1.11, an important quantity in nonpremixed combustion is the instantaneous scalar dissipation rate defined by

$$\chi = 2D|\nabla Z|^2. \tag{3.29}$$

It has the dimension 1/s and may be interpreted as the inverse of a characteristic diffusion time. Once the solution of (3.23) is known, χ can be calculated at each location in the flow field. The instantaneous local value of the scalar dissipation rate at stoichiometric mixture will be denoted by

$$\chi_{st} = 2D_{st}|\nabla Z|_{st}^2. \tag{3.30}$$

Two Example Solutions of the Mixture Fraction Equation

Example 1: The Counterflow Diffusion Flame. The counterflow geometry is very often used in experimental and numerical studies of diffusion flames because it leads to an essentially one-dimensional diffusion flame structure. Figure 3.4 shows a flame that has been established between an oxidizer stream from above and a gaseous fuel stream from below.

We consider a steady two-dimensional (planar or axially symmetric) counterflow configuration. There exists an exact solution in terms of a similarity coordinate, which results in a set of one-dimensional equations (cf. for instance Dixon-Lewis et al., 1984 or Peters and Kee, 1987). Here we will not introduce a similarity coordinate, but use the y coordinate directly. We introduce the ansatz

3.4 Nonequilibrium flames

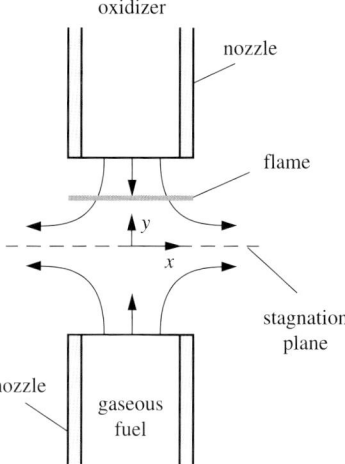

Figure 3.4. A schematic illustration of the experimental configuration for counterflow diffusion flames for gaseous fuels.

$u = Ux$ to obtain the following governing equations:

Continuity
$$\frac{d(\rho v)}{dy} + (j+1)\rho U = 0, \tag{3.31}$$

Momentum
$$\rho v \frac{dU}{dy} = -\rho U^2 + P + \frac{d}{dy}\left(\mu \frac{dU}{dy}\right), \tag{3.32}$$

Mixture fraction
$$\rho v \frac{dZ}{dy} = \frac{d}{dy}\left(\rho D \frac{dZ}{dy}\right), \tag{3.33}$$

Reactive scalars
$$\rho v \frac{d\psi_i}{dy} = \frac{d}{dy}\left(\rho D_i \frac{d\psi_i}{dy}\right) + \omega_i. \tag{3.34}$$

Here $j = 0$ applies for the planar configuration and $j = 1$ for the axially symmetric configuration. The velocity in the y direction is denoted by v and the gradient of the velocity u in the x direction by U. The parameter P represents the axial pressure gradient and is related to the strain rate a by

$$P = \rho_\infty a^2. \tag{3.35}$$

The inverse of the strain rate a is the characteristic time scale of the problem. It is prescribed if two potential flows coming from $y = +\infty$ and $y = -\infty$ are considered. If, however, the oxidizer and the fuel streams issue from a counterflow burner closer to the flame, nonslip boundary conditions $U = 0$ must be imposed on both sides. This is called the plug flow configuration. The continuity and momentum equations, forming a third-order system of ordinary differential equations, must then satisfy four boundary conditions. This is only possible if P and consequently the strain rate a is calculated as an eigenvalue of the problem.

As an introduction to the flamelet concept to be presented in Sections 3.11–3.13, it is useful to show that, for a one-dimensional system, flamelet equations can easily be derived by replacing the spatial coordinate y by the mixture fraction as a new independent variable. Introducing the transformation

$$\frac{d}{dy} = \frac{dZ}{dy}\frac{d}{dZ} \tag{3.36}$$

into (3.34) one obtains

$$\rho v \frac{dZ}{dy}\frac{d\psi_i}{dZ} = \frac{d}{dy}\left(\rho D_i \frac{dZ}{dy}\right)\frac{d\psi_i}{dZ} + \rho D_i \left(\frac{dZ}{dy}\right)^2 \frac{d^2\psi_i}{dZ^2} + \omega_i. \tag{3.37}$$

For simplicity we will consider here the special case $D_i = D$, leaving the more general case $D_i \neq D$, known as differential diffusion, to Section 3.12. Then it is immediately seen that in view of (3.33) the term on the l.h.s. in (3.37) and the first term on the r.h.s. cancel. One then obtains the steady state flamelet equation

$$\rho D \left(\frac{dZ}{dy}\right)^2 \frac{d^2\psi_i}{dZ^2} + \omega_i = 0, \tag{3.38}$$

which is equivalent to (1.139) for $Le_i = 1$, since the term in front of the second derivative can be identified as $\rho \chi / 2$. For a one-dimensional system the scalar dissipation rate χ is a function of the spatial coordinate y, which again can be replaced by the mixture fraction coordinate Z, by inverting the solution $Z(y)$, as will be shown in the following.

If one assumes that the flow velocities of both streams are sufficiently large and sufficiently removed from the stagnation plane, the flame is embedded between two potential flows, one coming from the oxidizer and one from the fuel side. Prescribing the potential flow velocity gradient in the oxidizer stream by $a = -\partial v_\infty / \partial y$, the boundary conditions in the oxidizer stream are

$$y \to \infty: \quad v_\infty = -ay, \quad U_\infty = a. \tag{3.39}$$

3.4 Nonequilibrium flames

Equal stagnation point pressure for both streams requires the boundary conditions in the fuel stream to be

$$y \to -\infty: v_{-\infty} = -(\rho_\infty/\rho_{-\infty})^{1/2} ay, \quad (3.40)$$
$$U_\infty = a\rho_\infty/\rho_{-\infty}.$$

By definition, the boundary conditions of the mixture fraction equation are $Z = 0$ for $y \to +\infty$ and $Z = 1$ for $y \to -\infty$.

An integral of the mixture fraction equation may be obtained by separation of variables. If Z' denotes $\rho D \, dZ/dy$, (3.33) may be transformed to $d(\ln Z')/dy = v/D$. Integrating this twice leads to

$$Z = c_1 I(y) + c_2, \quad (3.41)$$

where

$$I(y) = \int_0^y \frac{1}{\rho D} \exp\left\{\int_0^y \frac{v}{D} dy\right\} dy \quad (3.42)$$

and c_1 and c_2 are constants of integration. Applying the boundary conditions $Z = 0$ at $y \to \infty$ and $Z = 1$ at $y \to -\infty$ one obtains

$$Z = \frac{I(\infty) - I(y)}{I(\infty) - I(-\infty)}. \quad (3.43)$$

An approximate solution is obtained by assuming a constant density in the momentum equation. Then $U = a$ satisfies (3.32) and the integral $I(y)$ becomes

$$I(y) = \int_0^y \frac{1}{\rho^2 D} \exp\left\{-(j+1)a \int_0^{y'} \frac{1}{\rho^2 D} y' dy'\right\} dy', \quad (3.44)$$

where $y' = \int_0^y \rho \, dy$. An often used approximation in analytic studies of laminar flames is to set $\rho^2 D = \rho_\infty^2 D_\infty = $ const. This approximation is sometimes called the Chapman gas approximation. It is justified by the observation that D varies with the temperature typically as $T^{1.7}$, while ρ is inversely proportional to the temperature. With this approximation the nondimensional coordinate

$$\eta = \left(\frac{(j+1)a}{D_\infty}\right)^{1/2} \int_0^y \frac{\rho}{\rho_\infty} dy \quad (3.45)$$

may be introduced and one obtains from (3.43) with (3.44) the solution

$$Z = \frac{1}{2} \operatorname{erfc}(\eta/\sqrt{2}), \quad (3.46)$$

where erfc(x) = $1 - \text{erf}(x)$ is the complementary error function. This may be inverted and the scalar dissipation rate (3.29) may then be calculated as a function of Z:

$$\chi(Z) = \frac{a(j+1)}{\pi} \exp(-\eta^2(Z))$$
$$= \frac{a(j+1)}{\pi} \exp(-2[\text{erfc}^{-1}(2Z)]^2), \qquad (3.47)$$

where $\text{erfc}^{-1}(x)$ is the inverse (not the reciprocal) of erfc(x). An approximation that modifies (3.47) by accounting for nonconstant density in the solution of the momentum equation is due to Kim and Williams (1997).

Example 2: The One-Dimensional Unsteady Laminar Mixing Layer. As a second example we investigate the unsteady laminar mixing layer. This example describes a situation where infinite amounts of fuel and oxidizer, initially separated at $x = 0$, are suddenly allowed to interdiffuse thereby generating a mixing region that grows with time. For the special case of equal diffusivities $D_i = D$ the problem is governed by the equations

$$\rho \frac{\partial Z}{\partial t} = \frac{\partial}{\partial x}\left(\rho D \frac{\partial Z}{\partial x}\right), \qquad (3.48)$$

$$\rho \frac{\partial \psi_i}{\partial t} = \frac{\partial}{\partial x}\left(\rho D_i \frac{\partial \psi_i}{\partial x}\right) + \omega_i. \qquad (3.49)$$

Replacing in this unsteady case as before y by Z as the new independent spatial coordinate, while keeping the time, now written as τ, the same in the transformed system, one obtains the transformation rules

$$\frac{\partial}{\partial y} = \frac{\partial Z}{\partial y}\frac{\partial}{\partial Z}, \qquad \frac{\partial}{\partial t} = \frac{\partial Z}{\partial t}\frac{\partial}{\partial Z} + \frac{\partial}{\partial \tau}. \qquad (3.50)$$

When these are introduced into (3.49), one obtains by a similar procedure as in the first example the unsteady flamelet equations

$$\rho \frac{\partial \psi_i}{\partial \tau} = \rho D \left(\frac{\partial Z}{\partial y}\right)^2 \frac{\partial^2 \psi_i}{\partial Z^2} + \omega_i. \qquad (3.51)$$

This is the same as (1.139) for $Le_i = 1$.

The scalar dissipation rate $\chi = 2D(\partial Z/\partial y)^2$ may again be expressed as a function of Z. Writing the initial and boundary conditions for (3.48) as

$$t = 0: Z = 1 \quad \text{for } x < 0, \qquad Z = 0 \quad \text{for } x > 0;$$
$$t > 0: Z = 1 \quad \text{for } x \to -\infty, \qquad Z = 0 \quad \text{for } x \to +\infty \qquad (3.52)$$

and introducing the assumption

$$\rho^2 D = \rho_\infty^2 D_\infty \qquad (3.53)$$

one obtains in terms of the similarity coordinate

$$\eta = (2D_\infty t)^{-1/2} \int_0^x \frac{\rho}{\rho_\infty} dx \qquad (3.54)$$

formally the same solution as (3.46),

$$Z = \frac{1}{2}\operatorname{erfc}(\eta/\sqrt{2}). \qquad (3.55)$$

The scalar dissipation rate, however, is now

$$\chi(Z, t) = \frac{1}{2\pi t} \exp(-\eta^2(Z))$$
$$= \frac{1}{2\pi t} \exp(-2[\operatorname{erfc}^{-1}(2Z)]^2). \qquad (3.56)$$

This shows that in a one-dimensional mixing layer the scalar dissipation rate decreases with time. As discussed by Peters (1984) lateral perturbations of that layer would lead to an increase of χ. Bish and Dahm (1995) have performed a similar analysis with a strain term added to (3.48) and (3.49). They call the resulting model the Strained Dissipation and Reaction Layer (SDRL) model and construct images of OH and O mass fraction fields from measured fields of the scalar dissipation into. They point at the effect of strain in forming thin dissipation layers in turbulent flows.

Equation (3.56) shows the same functional dependence of χ on Z for the unsteady mixing layer as was found for the counterflow configuration in (3.47). Both example configurations represent typical mixing fields in turbulent combustion: To a moving observer

- the counterflow configuration appears in the highly strained region between two large vortices and
- an unsteady mixing layer appears in the rolled-up region within a vortex.

This is illustrated in Figure 3.5 for an unstable shear layer behind a splitter plate. Since these are quite different mechanisms of mixing, it is appealing to use the functional dependence $\chi(Z)$ in (3.47) and (3.56) as a model for more general turbulent flows. This model will be used in Section 3.12 to relate the conditional to the mean scalar dissipation rate.

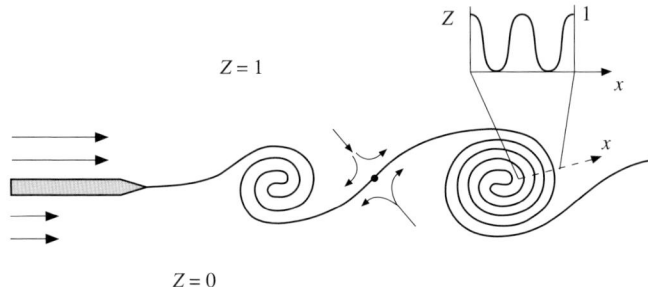

Figure 3.5. Mixing in a shear layer behind a splitter plate.

3.5 Numerical and Asymptotic Solutions of Counterflow Diffusion Flames

If the system (3.31)–(3.34) is solved numerically with appropriate boundary conditions using detailed or reduced chemical mechanisms, the entire structure of a counterflow diffusion flame can be determined. Many such calculations with detailed and reduced chemical mechanisms are presented in Peters and Rogg (1993). Previous work (cf. Dixon-Lewis et al., 1984) had been devoted, for instance, to calculating a methane–air diffusion flame in the stagnation point boundary layer investigated by Tsuji and Yamaoka (1971). That flame has also been used by Peters and Kee (1987) to validate the four-step reduced mechanism (1.92). Figure 3.6 shows temperature profiles, plotted against the mixture fraction defined by (3.22), for two different strain rates a from that paper. We see that the temperature decreases with increasing strain rate, which is equivalent to an increasing scalar dissipation rate according to (3.47). The physics behind this behavior is the following: With increasing strain rates the mixture fraction gradient becomes steeper in physical space and the temperature profile becomes narrower. This leads to enhanced heat conduction out of the reaction zone into the fuel-rich and fuel-lean regions on either side of the diffusion flame. In the flamelet equations (3.38), this effect is represented by an increase of the scalar dissipation rate χ, which acts like an increased diffusion coefficient and thereby enhances diffusive transport in mixture fraction space. If the strain rate and therefore the scalar dissipation rate is increased further, one approaches a situation where heat production by chemical reaction can no longer balance the heat conduction out of the reaction zone. At that point the flame will extinguish.

The scalar dissipation rate $\chi(Z)$ may be parameterized by its value at stoichiometric mixture χ_{st}. At extinction (or quenching) that value is

$$\chi_q = \chi_{st,ext} \qquad (3.57)$$

For counterflow methane flames Chelliah et al. (1993) found values around

3.5 Numerical and asymptotic solutions

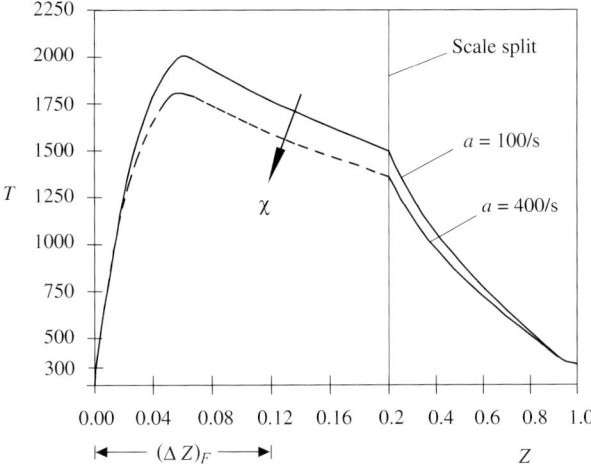

Figure 3.6. Temperature profiles of methane–air diffusion flames for $a = 100/s$ and $a = 400/s$ as a function of mixture fraction. From Peters and Kee (1987). (Reprinted by permission of Elsevier Science from "The computation of stretched laminar methane–air diffusion flames using a reduced four-step mechanism," by N. Peters and R. J. Kee, Combustion & Flame, Vol 68, pp 17–29, Copyright 1987 by The Combustion Institute.)

$\chi_q = 18/s$ and extinction strain rates around $a_q = 500/s$, depending on whether potential flow or plug flow boundary conditions were used. Steady state solutions of diffusion flames may also be represented by an S-shaped curve as shown in Figure 1.1, where extinction occurs at $Da = Da_Q$.

The canonical paper analyzing the structure of counterflow diffusion flames is by Liñán (1974). The analysis is based on a one-step irreversible reaction which is first-order with respect to fuel and to oxygen and has a large activation energy. Four different regimes are treated: the diffusion flame regime, the premixed flame regime, the partial burning regime, and the ignition regime. Among these, only the diffusion flame regime and the ignition regime lead to equations for the reactive–diffusive layer that have stable solutions over the entire range of parameters. The analysis of the diffusion flame regime describes first-order nonequilibrium corrections to the Burke–Schumann solution and, owing to the large activation energy, predicts the possibility of quenching of diffusion flames. For $Z_{st} < 0.5$ this is preceded by leakage of the fuel through the reaction zone. Peters (1980, 1983) has used this analysis to set the basis for the flamelet concept for nonpremixed combustion. Chung and Law (1983) and Cuenot and Poinsot (1996) have presented solutions for the same one-step reaction allowing for Lewis numbers different from unity. A recent study of Cheatham and Matalon (2000) provides a general formulation in physical space, not restricted to a particular geometry, for distinct and arbitrary Lewis numbers.

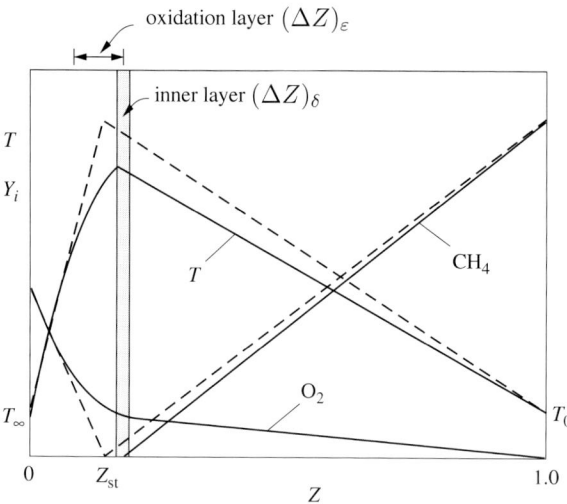

Figure 3.7. Illustration of the asymptotic structure of the methane–air diffusion flame based on the reduced four-step mechanism. The dashed line is the Burke–Schumann solution. From Seshadri and Peters (1988). (Reprinted by permission of Elsevier Science from "Asymptotic structure and extinction of methane–air diffusion flames," by K. Seshadri and N. Peters, Combustion & Flame, Vol 73, pp 23–44, Copyright 1988 by The Combustion Institute.)

An asymptotic analysis by Seshadri and Peters (1988) based on the four-step model (1.92) for methane flames shows a close correspondence between the different layers identified in the premixed flame in Section 1.6, and those in the diffusion flame. The structure obtained from the asymptotic analysis is schematically shown in Figure 3.7. The outer structure of the diffusion flame is the Burke–Schumann solution governed by the overall one-step reaction $CH_4 + 2O_2 \rightarrow CO_2 + 2H_2O$, with the flame sheet positioned at $Z = Z_{st}$. The nonequilibrium structure consists of a thin H_2–CO oxidation layer of thickness $O(\varepsilon)$ on the lean side and a thin inner layer of thickness $O(\delta)$ slightly toward the rich side of $Z = Z_{st}$. Beyond this layer the rich side is chemically inert because all radicals are consumed by the fuel. Comparison of the diffusion flame structure with that of a premixed flame shows that the rich part of the diffusion flame corresponds to the upstream preheat zone of the premixed flame while its lean part corresponds to the downstream oxidation layer. In contrast to the one-step asymptotic analysis the four-step kinetics show leakage of oxygen rather than fuel through the reaction zone, which agrees with the experimental data of Tsuji and Yamaoka (1971). The maximum temperature calculated by the four-step asymptotic analysis is shown in Figure 3.8 as a function of the inverse of the scalar dissipation rate. This plot corresponds to the upper branch of the S-shaped curve shown in Figure 1.1. The calculations agree well with

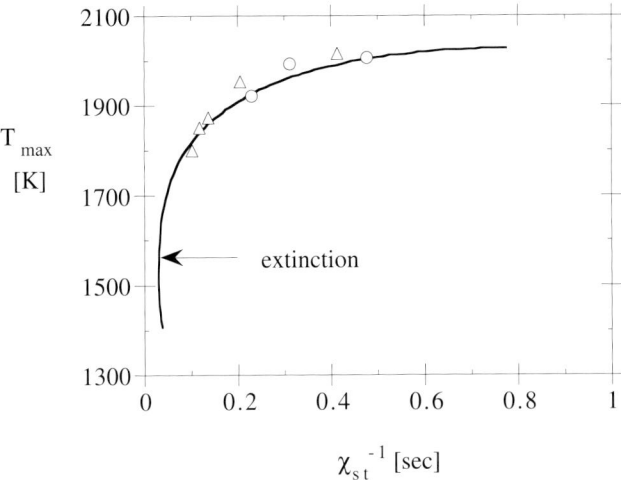

Figure 3.8. Maximum flame temperature plotted as a function of the inverse of the scalar dissipation rate at stoichiometric conditions. △, numerical calculations by Peters and Kee (1987) using the four-step mechanism; ○, experimental data. From Seshadri and Peters (1988). (Reprinted by permission of Elsevier Science from "Asymptotic structure and extinction of methane–air diffusion flames," by K. Seshadri and N. Peters, Combustion & Flame, Vol 73, pp 23–44, Copyright 1988 by The Combustion Institute.)

numerical and experimental data and also show the vertical slope of $T_{\max}$ versus χ_{st}^{-1} at extinction, where $\chi_{st} = \chi_q$.

Peters (1991) has shown that χ_q can be related to the flame time $t_F = D/s_L^2$ of a stoichiometric premixed flame as

$$\chi_q = \frac{Z_{st}^2(1 - Z_{st})^2}{t_F}. \tag{3.58}$$

This relation demonstrates the fundamental equivalence between extinction of a diffusion flame and propagation of a premixed flame. In a premixed flame, there is a balance between the heat generated by chemical reaction and heat conduction into the preheat zone, which leads to a so-called deflagration wave. In a diffusion flame the same balance exists at extinction between heat generation and heat conduction toward both sides of the flame. Based on this correspondence one could take the view, which we do not advocate here, however, that a premixed flame is an extinction wave, governed by heat loss toward the unburnt mixture.

While a premixed laminar flame allows only for a single time scale, namely t_F, in a diffusion flame there is a competition between at least two time scales: the one imposed by the flow, expressed in terms of χ_{st} (or in terms of the velocity gradient a as for instance in a counterflow diffusion flame), and the chemical time scale t_c of the rate-determining combustion reaction. At extinction both time scales are equal. The product $Da = (\chi_{st} t_c)^{-1}$ is a Damköhler number that appears as a parameter in a diffusion flame problem. In a premixed flame, in contrast, the flame time is equal to the chemical time and the corresponding Damköhler number is always equal to unity.

3.6 Regimes in Nonpremixed Turbulent Combustion

In nonpremixed combustion, in contrast to premixed combustion, there is no characteristic velocity scale such as the laminar burning velocity, and therefore there is no flame thickness defining a characteristic length scale. As noted above there is, however, the strain rate a, locally imposed by the flow field, which is the inverse of a characteristic time. This can be used together with the diffusion coefficient D_{st} to define a diffusion thickness

$$\ell_D = \left(\frac{D_{st}}{a}\right)^{1/2}. \tag{3.59}$$

There also exists the gradient of the mixture fraction field $|\nabla Z|_{st}$, which relates the thickness ℓ_D, defined in physical space, to a diffusion thickness $(\Delta Z)_F$ in mixture fraction space:

$$(\Delta Z)_F = |\nabla Z|_{st} \ell_D. \tag{3.60}$$

Using the definition (3.30) of the scalar dissipation rate χ_{st} one then obtains

$$(\Delta Z)_F = \left(\frac{\chi_{st}}{2a}\right)^{1/2}. \tag{3.61}$$

In Peters (1991) $\chi(Z)$ from (3.47) is expanded for small values of Z. If this is used in (3.61) one finds that (ΔZ_F) is approximately two times Z_{st}. Using this as a definition,

$$(\Delta Z)_F = 2Z_{st}, \tag{3.62}$$

one sees that the diffusion thickness in mixture fraction space covers the reaction zone and the surrounding diffusion layers as shown in Figure 3.6. It therefore corresponds to the flame thickness ℓ_F in premixed flames, which also includes the convective–diffusive preheat zone.

In addition to $(\Delta Z)_F$ one may define a reaction zone thickness $(\Delta Z)_R$ in mixture fraction space. In contrast to the case of premixed flames, as argued at the end of Section 1.6, here we will define the reaction zone as containing both the fuel consumption and the oxidation layer. Since the fuel consumption layer is much thinner than the oxidation layer, it is the width of the latter that determines the reaction zone thickness $(\Delta Z)_R$. The oxidation layer thickness $(\Delta Z)_\varepsilon$ is shown in Figure 3.7. We relate the thickness $(\Delta Z)_R = (\Delta Z)_\varepsilon$ to the flame thickness as

$$(\Delta Z)_R = \varepsilon (\Delta Z)_F. \tag{3.63}$$

3.6 Regimes in nonpremixed turbulent combustion

From Equation (57) in Seshadri and Peters (1988) it can be seen that in the four-step methane–air diffusion flame the oxidation layer thickness $(\Delta Z)_\varepsilon$ is proportional to $\chi_{st}^{1/4}$. (It is proportional to $\chi_{st}^{1/3}$ for a diffusion flame with one-step kinetics; cf. Liñán, 1974.) Scaling ε with its value ε_q at extinction

$$\frac{\varepsilon}{\varepsilon_q} = \left(\frac{\chi_{st}}{\chi_q}\right)^{1/4} \tag{3.64}$$

leads to

$$\frac{(\Delta Z)_R}{(\Delta Z)_F} = \varepsilon_q \left(\frac{\chi_{st}}{\chi_q}\right)^{1/4}. \tag{3.65}$$

From Figure 5 of Seshadri and Peters (1988), where $(\Delta Z)_\varepsilon$ (denoted there as ε) is plotted as a function of χ_{st}^{-1}, one obtains $(\Delta Z)_\varepsilon = 0.0175$, leading with $(\Delta Z)_F = 0.11$ to $\varepsilon_q = 0.16$.

In a turbulent diffusion flame the characteristic thicknesses $(\Delta Z)_F$ and $(\Delta Z)_R$ in mixture fraction space must be compared with mixture fraction fluctuations defined by the r.m.s of the variance

$$Z' = (\widetilde{Z''^2})^{1/2}. \tag{3.66}$$

Since comparisons are most meaningful at mean stoichiometric mixture, the relevant fluctuation is denoted by

$$Z'_{st} = (\widetilde{Z''^2})^{1/2}_{st}, \tag{3.67}$$

where the index st stands for the value of the variance at the mean mixture fraction $\tilde{Z}(x,t) = Z_{st}$.

The regime diagram for nonpremixed turbulent combustion shown in Figure 3.9 is a log–log plot of the ratio $Z'_{st}/(\Delta Z)_F$ versus the time scale ratio $\chi_q/\tilde{\chi}_{st}$. Here $\tilde{\chi}_{st}$ is the conditional Favre mean scalar dissipation rate defined by (3.157) in Section 3.12 below. For large mixture fraction fluctuations, where $Z'_{st} > (\Delta Z)_F$, fluctuations in mixture fraction space extend to sufficiently lean and rich mixtures such that the diffusion layers surrounding the reaction zone are separated. For small mixture fraction fluctuations, where $Z'_{st} < (\Delta Z)_F$, which may be due to intense mixing or to partial premixing of the fuel stream, the reaction zones are connected. Therefore the criterion $Z'_{st}/(\Delta Z)_F = 1$, represented by the horizontal line in Figure 3.9, distinguishes between two regimes. If mixture fraction fluctuations are larger than $(\Delta Z)_F$ one has separated flamelets; otherwise connected flame zones occur.

In Figure 3.9 the line $Z'_{st}/(\Delta Z)_R = 1$ is also plotted. It corresponds to mixture fraction fluctuations that are equal to the thickness of the reaction zone. From (3.65) it follows that the slope of that line is $-1/4$ if, for simplicity,

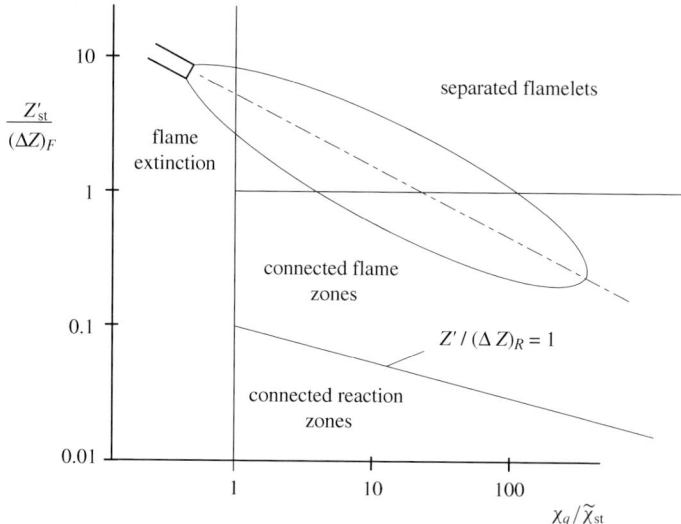

Figure 3.9. Regimes in nonpremixed turbulent combustion.

χ_{st} is set equal to $\tilde{\chi}_{st}$. Below this line mixture fraction fluctuations are smaller than the reaction zone thickness and even the reaction zones are connected. This means that the mixture fraction field is nearly homogeneous.

A second criterion defining combustion regimes is that related to extinction,

$$\tilde{\chi}_{st} = \chi_q, \tag{3.68}$$

which appears as a vertical line in Figure 3.9. It separates the regime of flame extinction from the regimes of separated flamelets, connected flame zones, and connected reaction zones.

The shape of a diffusion flame in Figure 3.9 illustrates how local conditions along the contour of mean stoichiometric mixture in a jet diffusion flame would fit into the different regimes of nonpremixed turbulent combustion. Since Z'_{st} decreases as x^{-1} and $\tilde{\chi}_{st}$ approximately as x^{-2} along the surface of mean stoichiometric mixture in a turbulent jet flame (cf. Peters and Williams, 1983), in Figure 3.9 the dash-dotted line corresponds to a line with a slope $-1/2$. The flame extinction regime corresponds to the region close to the nozzle where local extinction will happen frequently. This will eventually lead to lift-off of the diffusion flame, which will be discussed in detail in Section 4.1. At $\chi_q/\tilde{\chi}_{st} = 1$ one enters into the regime of separated flamelets. Further downstream, when mixture fraction fluctuations have decayed to values of the order of $(\Delta Z)_F$, one enters into the connected flame zones regime.

3.6 Regimes in nonpremixed turbulent combustion

One may, as Bilger (1988) does, consider a criterion for the validity of the laminar flamelet concept (and thereby of scale separation in general) for turbulent diffusion flames. As for the thin reaction zone regime in premixed turbulent combustion, the flamelet concept requires that the reaction zone of thickness ℓ_R (cf. the discussion at the end of Section 1.6) be embedded within the smallest scale of turbulence, the Kolmogorov scale η. Similar to the relation (3.60) for the diffusion thickness, the thickness ℓ_R is related to the reaction zone thickness in mixture fraction space $(\Delta Z)_R$ by the mixture fraction gradient. Therefore the criterion for the validity of the flamelet concept for diffusion flames is

$$\ell_R = \frac{(\Delta Z)_R}{|\nabla Z|_{\text{st}}} < \eta. \tag{3.69}$$

Introducing (3.65), (3.61), (3.30), and (3.58) one obtains

$$\ell_R \approx \varepsilon_q \frac{(\Delta Z)_F^{1/2}}{Z_{\text{st}}^{1/2}(1-Z_{\text{st}})^{1/2}} \left(\frac{2D_{\text{st}}^2 \cdot t_F}{a}\right)^{1/4} < \eta. \tag{3.70}$$

If one neglects Z_{st} compared to unity in this equation and sets $(\Delta Z)_F = 2Z_{\text{st}}$ and $D_{\text{st}} t_F = \ell_F^2$ one obtains

$$\varepsilon_q \ell_F \left(\frac{8}{a\, t_F}\right)^{1/4} < \eta. \tag{3.71}$$

It is appropriate for this estimate, which is evaluated at the dissipation scales, to set the imposed strain rate a equal to the inverse of the Kolmogorov time scale t_η. Then the criterion for the validity of the flamelet concept becomes

$$8\varepsilon_q^4 Ka < 1, \tag{3.72}$$

where the definition (2.26),

$$Ka = t_F/t_\eta = \ell_F^2/\eta^2, \tag{3.73}$$

has been used. It is easily seen that with the estimate $\varepsilon_q = 0.16$ mentioned above for the methane–air flames analysed by Seshadri and Peters (1988) the criterion (3.72) allows for Karlovitz numbers up to 190 for the flamelet concept to be valid. This is of the same order of magnitude as the estimate $Ka = 100$ for the thin reaction zones regime in premixed turbulent combustion. In an assessment of flamelet modeling for turbulent jet diffusion flames Lentini (1994) has plotted trajectories corresponding to local values of $v'/\hat{s}_L$ and $\ell/\hat{\ell}_F$ on a diagram similar to Figure 2.8. Here $\hat{s}_L = (\nu/t_F)^{1/2}$ and $\hat{\ell}_F = (\nu t_F)^{1/2}$, with t_F defined by (3.58), were introduced as the appropriate laminar velocity and the corresponding laminar length scale for nonpremixed combustion. The trajectories for three flames with nozzle exit Reynolds numbers of 8,500, 10,000, and 40,000 fall well below a Karlovitz number of 10 in that diagram.

3.7 Modeling Nonpremixed Turbulent Combustion

Models in nonpremixed turbulent combustion are often based on the presumed shape pdf approach to be presented in Section 3.8. This requires knowledge of the Favre mean mixture fraction $\tilde{Z}$ and its variance $\widetilde{Z''^2}$ at position x and time t. Averaging of (3.23) leads to the equation for the Favre mean mixture fraction $\tilde{Z}$:

$$\bar{\rho}\frac{\partial \tilde{Z}}{\partial t} + \bar{\rho}\tilde{v}\cdot\nabla\tilde{Z} = -\nabla\cdot(\overline{\rho v''Z''}). \tag{3.74}$$

The molecular diffusivity D in (3.23) is much smaller than the turbulent diffusivity D_t and has therefore been neglected in (3.74). Since Z is a nonreacting scalar, the gradient transport assumption (1.107) can be used:

$$\widetilde{v''Z''} = -D_t\nabla\tilde{Z}. \tag{3.75}$$

In addition to the mean mixture fraction we need an equation for the Favre variance $\widetilde{Z''^2}$, which is obtained similarly to (1.110) as

$$\bar{\rho}\frac{\partial \widetilde{Z''^2}}{\partial t} + \bar{\rho}\tilde{v}\cdot\nabla\widetilde{Z''^2} = -\nabla\cdot(\overline{\rho v''Z''^2}) + 2\bar{\rho}D_t(\nabla\tilde{Z})^2 - \bar{\rho}\tilde{\chi}, \tag{3.76}$$

where (3.75) has been used. For the turbulent flux in this equation the gradient transport assumption

$$\widetilde{v''Z''^2} = -D_t\nabla\widetilde{Z''^2} \tag{3.77}$$

can also be used. In (3.76) the mean scalar dissipation rate appears, which is defined as

$$\tilde{\chi} = 2D\widetilde{|\nabla Z''|^2} \tag{3.78}$$

and will be modeled similarly to (1.114) as

$$\tilde{\chi} = c_\chi \frac{\tilde{\varepsilon}}{\tilde{k}}\widetilde{Z''^2}, \tag{3.79}$$

where the time scale ratio c_χ is assumed to be a constant. Jones (1994) suggests a value of $c_\chi = 1.0$ while Janicka and Peters (1982) found that a value of $c_\chi = 2.0$ would predict the decay of scalar variance in an inert jet of methane very well. Overholt and Pope (1996) and Juneja and Pope (1996) performing DNS studies of one and two passive scalar mixing find an increase of c_χ with Reynolds number and steady state values around 2.0 and 3.0, respectively. In the numerical simulations of diesel engine combustion, to be presented below, a value of $c_\chi = 2.0$ has been used.

It has been pointed out, however, that the scalar dissipation rate should depend in a more subtle way on the characteristics of the turbulent flow field than through an algebraic relation like (3.79). Several authors (cf. Pope, 1983b and Jones and Musonge, 1988, for instance) proposed a modeled transport equation for the scalar dissipation rate $\tilde{\chi}$ itself. The model of Jones and Musonge (1988) reads (cf. Jones, 1994)

$$\bar{\rho}\frac{\partial \tilde{\chi}}{\partial t} + \bar{\rho}\tilde{v} \cdot \nabla \tilde{\chi} = \nabla \cdot \left(C_0 \bar{\rho} \frac{\tilde{k}}{\tilde{\varepsilon}} \widetilde{v''v''} \cdot \nabla \tilde{\chi} \right) - C_1 \bar{\rho} \frac{\tilde{\chi}^2}{\widetilde{Z''^2}} - C_2 \bar{\rho} \frac{\tilde{\varepsilon}}{\tilde{k}} \tilde{\chi}$$
$$- C_3 \bar{\rho} \frac{\tilde{\varepsilon}}{\tilde{k}} \widetilde{v''Z''} \cdot \nabla \tilde{Z} - C_4 \bar{\rho} \frac{\tilde{\chi}}{\tilde{k}} \widetilde{v''v''} : \nabla \tilde{v}. \quad (3.80)$$

Note that this equation has features in common with the modeled equation (2.184) for $\bar{\sigma}_t$, if in the latter only terms pertaining to the thin reaction zones regime are considered. This is easily seen if $\tilde{\chi}$ is replaced in favor of $D\tilde{\sigma}_t^2$ and Z by G in the entire equation (3.80). Furthermore, $\tilde{\varepsilon}/\tilde{k}$ must be replaced in the term containing C_3 by $D\tilde{\sigma}_t^2/\widetilde{G''^2}$ and the gradient flux approximation must be used. The term containing C_1 then corresponds to the dissipation term in (2.184) and the terms proportional to C_3 and C_4 to the production terms due to turbulence and to mean velocity gradients, respectively. The difference between (3.80) and (2.184) in the modeling of the turbulent transport terms remains, since Z is a diffusive scalar while G is not.

In many cases, as in turbulent jet diffusion flames in air, zero gradient boundary conditions, except at the inlet, can be imposed. If the simplifying assumptions mentioned after (1.75) in Section 1.5 can be introduced the enthalpy h can be related to the mixture fraction by the linear coupling relation

$$h = h_2 + Z(h_1 - h_2), \quad (3.81)$$

which also holds for the mean values

$$\tilde{h} = h_2 + \tilde{Z}(h_1 - h_2), \quad (3.82)$$

and no additional equation for the enthalpy is required. In (3.81) and (3.82) h_2 is the enthalpy of the air and h_1 that of the fuel.

A more general formulation is needed, if different boundary conditions have to be applied for $\tilde{Z}$ and $\tilde{h}$ or heat loss from radiation or unsteady pressure changes must be accounted for. Then an equation for the Favre mean enthalpy $\tilde{h}$ as an additional variable must be solved. An equation for $\tilde{h}$ can be obtained from (1.75) by averaging:

$$\bar{\rho}\frac{\partial \tilde{h}}{\partial t} + \bar{\rho}\tilde{v}\cdot\nabla\tilde{h} = \frac{\partial \bar{p}}{\partial t} + \nabla\cdot\left(\bar{\rho}\,D_t\nabla\tilde{h}\right) + \overline{q_R}. \tag{3.83}$$

As in (1.75) the terms containing the mean spatial pressure gradient have been neglected in this equation by applying the limit of zero Mach number, where fast acoustic waves are rapidly homogenizing the pressure field. The term describing temporal mean pressure changes $\partial \bar{p}/\partial t$ has been retained, because it is important for the modeling of combustion in internal combustion engines operating under nonpremixed conditions, such as the diesel engine. The mean volumetric heat exchange term $\overline{q_R}$ must also be retained in many applications where radiative heat exchange has an influence on the local enthalpy balance. As an example, Marracino and Lentini (1997) have used the Stretched Laminar Flamelet Model to calculate the radiative heat flux from buoyant turbulent methane–air diffusion flames. Temperature changes due to radiation within the flamelet structure also have a strong influence on the prediction of NO_x formation (cf. Pitsch et al., 1998). Changes of the mean enthalpy also occur as the result of convective heat transfer at the boundaries or the evaporation of a liquid fuel. As in (3.74) and (3.76) the transport term containing the molecular diffusivity has been neglected in (3.83) as being small compared to the turbulent transport term. Effects due to nonunity Lewis numbers have also been neglected.

No equation for enthalpy fluctuations is presented here, because in nonpremixed turbulent combustion, it is often assumed that fluctuations of the enthalpy are mainly due to mixture fraction fluctuations and are described by those.

3.8 The Presumed Shape pdf Approach

Equations (3.74)–(3.77) can be used to calculate the mean mixture fraction and the mixture fraction variance at each point of the turbulent flow field, provided that the density field is known. In addition, of course, equations for the turbulent flow field, the Reynolds stress equations (or the equation for the turbulent kinetic energy $\tilde{k}$), and the equation for the dissipation $\tilde{\varepsilon}$ must be solved.

If the assumption of fast chemistry is introduced and the coupling between the mixture fraction and the enthalpy (3.81) can be used, the Burke–Schumann solution or the equilibrium solution relates all reactive scalars to the local mixture fraction (cf. Section 3.3). Using these relations the easiest way to obtain mean values of the reactive scalars is to use the presumed shape pdf approach. This is called the Conserved Scalar Equilibrium Model. In this approach a

3.8 The presumed shape pdf approach

suitable two-parameter probability density function is "presumed" in advance, thereby fixing the functional form of the pdf by relating the two parameters to the known values of $\tilde{Z}$ and $\widetilde{Z''^2}$ at each point of the flow field.

Since in a two-feed system the mixture fraction Z varies between $Z = 0$ and $Z = 1$, the beta function pdf is widely used for the Favre pdf in nonpremixed turbulent combustion. The beta function pdf has the form

$$\tilde{P}(Z; \mathbf{x}, t) = \frac{Z^{\alpha-1}(1-Z)^{\beta-1}}{\Gamma(\alpha)\Gamma(\beta)} \Gamma(\alpha + \beta). \quad (3.84)$$

Here Γ is the gamma function. The two parameters α and β are related to the Favre mean $\tilde{Z}(\mathbf{x}, t)$ and variance $\widetilde{Z''^2}(\mathbf{x}, t)$ by

$$\alpha = \tilde{Z}\gamma, \qquad \beta = (1 - \tilde{Z})\gamma, \quad (3.85)$$

where

$$\gamma = \frac{\tilde{Z}(1-\tilde{Z})}{\widetilde{Z''^2}} - 1 \geq 0. \quad (3.86)$$

The beta function is plotted for different combinations of the parameters $\tilde{Z}$ and γ in Figure 3.10. It can be shown that in the limit of very small $\widetilde{Z''^2}$ (large γ) it approaches a Gaussian distribution. For $\alpha < 1$ it develops a singularity at $Z = 0$ and for $\beta < 1$ a singularity at $Z = 1$. Despite its surprising flexibility, it is unable to describe distributions with a singularity at $Z = 0$ or $Z = 1$ and an additional intermediate maximum in the range $0 < Z < 1$. For such shapes, which have been found in jets and shear layers, a composite model has been developed by

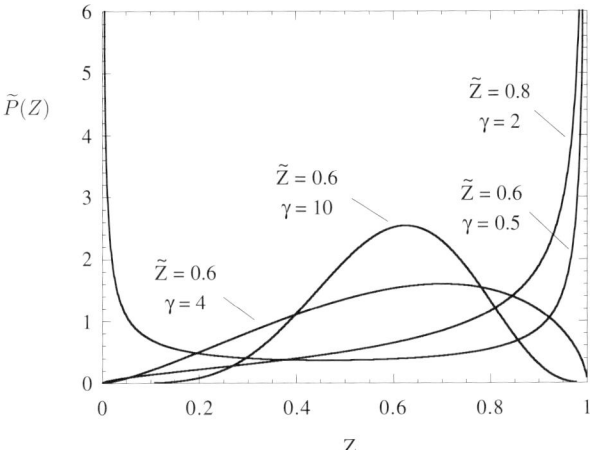

Figure 3.10. Shapes of the beta-function pdf for different parameters $\tilde{Z}$ and γ.

Effelsberg and Peters (1983). It identifies three different contributions to the pdf: a fully turbulent part, an outer flow part, and a part related to the viscous superlayer between the outer flow and the fully turbulent flow region. The model shows that the intermediate maximum is due to the contribution from the fully turbulent part of the scalar field.

By the presumed shape pdf approach, means of any quantity that depends only on the mixture fraction can be calculated. For instance, the mean value of ψ_i can be obtained from

$$\tilde{\psi}_i(\boldsymbol{x}, t) = \int_0^1 \psi_i(Z)\tilde{P}(Z; \boldsymbol{x}, t)\, dZ. \qquad (3.87)$$

A further quantity of interest is the mean density $\bar{\rho}$. Since Favre averages are considered, one must take the Favre average of ρ^{-1}, which leads to

$$\widetilde{\rho^{-1}} = \frac{1}{\bar{\rho}} = \int_0^1 \rho^{-1}(Z)\tilde{P}(Z)\, dZ. \qquad (3.88)$$

With (3.84)–(3.88) and the Burke–Schumann solution or the equilibrium solution the Conserved Scalar Equilibrium Model for nonpremixed combustion is formulated. It is based on a closed set of equations that do not require any further chemical input other than the assumption of infinitely fast chemistry. It may therefore be used as an initial guess in a calculation where the Burke–Schumann solution or the equilibrium solution later on is replaced by the solution of the flamelet equations to account for nonequilibrium effects.

3.9 Turbulent Jet Diffusion Flames

In many applications fuel enters into the combustion chamber as a turbulent jet, with or without swirl. To provide an understanding of the basic properties of jet diffusion flames, we will consider here the easiest case, the axisymmetric jet flame without buoyancy, for which we can obtain approximate analytical solutions. This will enable us to determine, for instance, the flame length. The flame length is defined as the distance from the nozzle to the point on the centerline of the flame where the mean mixture fraction is equal to Z_{st}. The flow configuration and the flame contour of a vertical jet diffusion flame are shown schematically in Figure 3.11.

We consider a fuel jet issuing from a nozzle with diameter d and exit velocity u_0 into quiescent air. The indices 0 and ∞ denote conditions at the nozzle and in the ambient air, respectively. Using Favre averaging and the boundary layer assumption we obtain a system of two-dimensional axisymmetric equations, in

3.9 Turbulent jet diffusion flames

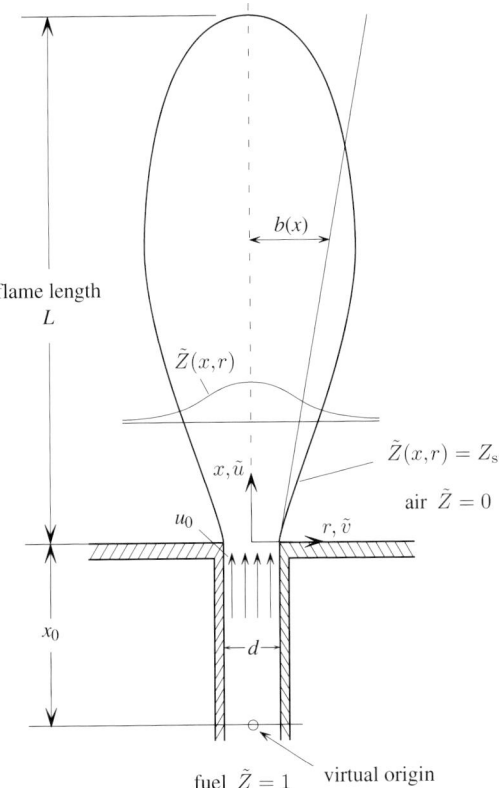

Figure 3.11. Schematic representation of a vertical jet flame into quiescent air.

terms of the axial coordinate x and the radial coordinate r:

continuity

$$\frac{\partial}{\partial x}(\bar{\rho}\tilde{u}r) + \frac{\partial}{\partial r}(\bar{\rho}\tilde{v}r) = 0, \tag{3.89}$$

momentum in x direction

$$\bar{\rho}\tilde{u}r\frac{\partial \tilde{u}}{\partial x} + \bar{\rho}\tilde{v}r\frac{\partial \tilde{u}}{\partial r} = \frac{\partial}{\partial r}\left(\bar{\rho}\nu_t r \frac{\partial \tilde{u}}{\partial r}\right), \tag{3.90}$$

mean mixture fraction

$$\bar{\rho}\tilde{u}r\frac{\partial \tilde{Z}}{\partial x} + \bar{\rho}\tilde{v}r\frac{\partial \tilde{Z}}{\partial r} = \frac{\partial}{\partial r}\left(\frac{\bar{\rho}\nu_t r}{Sc_t} \frac{\partial \tilde{Z}}{\partial r}\right). \tag{3.91}$$

We have again neglected molecular as compared to turbulent transport terms. Turbulent transport was modeled by the gradient flux approximation. For the scalar flux we have replaced D_t by introducing the turbulent Schmidt number $Sc_t = v_t/D_t$.

For simplicity, we will not consider equations for $\tilde{k}$ and $\tilde{\varepsilon}$ or the mixture fraction variance but seek an approximate solution by introducing a model for the turbulent viscosity v_t. Details may be found in Peters and Donnerhack (1981). The dimensionality of the problem may be reduced by introducing the similarity transformation

$$\eta = \frac{\bar{r}}{\zeta}, \qquad \bar{r}^2 = 2\int_0^r \frac{\bar{\rho}}{\rho_\infty} r\, dr, \qquad \zeta = x + x_0, \qquad (3.92)$$

which contains a density transformation defining the density weighted radial coordinate $\bar{r}$. The new axial coordinate ζ starts from the virtual origin of the jet located at $x = -x_0$. Introducing a stream function ψ by

$$\bar{\rho}\tilde{u}r = \partial\psi/\partial r, \qquad \bar{\rho}\tilde{v}r = -\partial\psi/\partial x \qquad (3.93)$$

we can satisfy the continuity equation (3.89). The axial and radial velocity components may now be expressed in terms of the nondimensional stream function defined by

$$F(\eta) = \psi/\rho_\infty v_{tr}\zeta \qquad (3.94)$$

as

$$\tilde{u} = v_{tr} F_\eta/\eta\zeta, \qquad \bar{\rho}\tilde{v}r = -\rho_\infty v_{tr}(F - F_\eta \eta). \qquad (3.95)$$

Here v_{tr} is the eddy viscosity of a constant density jet, here used as a reference value. It has been fitted (cf. Peters and Donnerhack, 1981) to experimental data as

$$v_{tr} = \frac{u_0 d}{70}. \qquad (3.96)$$

For the mixture fraction the ansatz

$$\tilde{Z} = \tilde{Z}_{CL}(\zeta)\omega(\eta) \qquad (3.97)$$

is introduced, where $\tilde{Z}_{CL}$ stands for the Favre mean mixture fraction on the centerline.

For a jet into still air a similarity solution exists. Introducing (3.92)–(3.97) into (3.89)–(3.91) one obtains the following equations, valid in the similarity region of the jet:

$$-\frac{\partial}{\partial \eta}\left(\frac{FF_\eta}{\eta}\right) = \frac{\partial}{\partial \eta}\left[C\eta\frac{\partial}{\partial \eta}\left(\frac{F_\eta}{\eta}\right)\right],$$
$$-\frac{\partial}{\partial \eta}(F\omega) = \frac{\partial}{\partial \eta}\left(\frac{C}{Sc_t}\eta\frac{\partial \omega}{\partial \eta}\right). \qquad (3.98)$$

3.9 Turbulent jet diffusion flames

Here the Chapman–Rubesin parameter has been introduced:

$$C = \frac{\bar{\rho}^2 v_t r^2}{\rho_\infty^2 v_{tr} \bar{r}^2}. \tag{3.99}$$

To derive an analytical solution we must assume that C is a constant in the entire jet. With a constant value of C one obtains from (3.98) the solutions

$$F(\eta) = C\gamma^2\eta^2/(1+(\gamma\eta)^2/4),$$
$$\omega(\eta) = (1+(\gamma\eta)^2/4)^{-2Sc_t}. \tag{3.100}$$

The axial velocity profile then is obtained from (3.95) as

$$\tilde{u} = \frac{2C\gamma^2 v_{tr}}{\zeta}\left(1+\frac{(\gamma\eta)^2}{4}\right)^{-2}, \tag{3.101}$$

where the jet spreading parameter

$$\gamma^2 = \frac{3\cdot 70^2}{64}\frac{\rho_0}{\rho_\infty C^2} \tag{3.102}$$

is obtained from the requirement of integral momentum conservation along the axial direction. Similarly, conservation of the mixture fraction integral across the jet yields the mixture fraction on the centerline

$$\tilde{Z}_{CL} = \frac{70(1+2Sc_t)}{32}\frac{\rho_0}{\rho_\infty C}\frac{d}{\zeta} \tag{3.103}$$

such that the mixture fraction profile is given by

$$\tilde{Z} = \frac{2.19(1+2Sc_t)d}{x+x_0}\frac{\rho_0}{\rho_\infty C}\left(1+\frac{(\gamma\eta)^2}{4}\right)^{-2Sc_t}. \tag{3.104}$$

From this equation the flame length L can be calculated by setting $\tilde{Z} = Z_{st}$ at $x = L, r = 0$:

$$\frac{L+x_0}{d} = \frac{2.19(1+2Sc_t)}{Z_{st}}\frac{\rho_0}{\rho_\infty C}. \tag{3.105}$$

Experimental data by Hawthorne et al. (1949) suggest that the flame length L should scale as

$$\frac{L+x_0}{d} = \frac{5.3}{Z_{st}}\left(\frac{\rho_0}{\rho_{st}}\right)^{1/2}. \tag{3.106}$$

This fixes the turbulent Schmidt number as $Sc_t = 0.71$ and the Chapman–Rubesin parameter as

$$C = \frac{(\rho_0 \rho_{st})^{1/2}}{\rho_\infty}. \tag{3.107}$$

When this is introduced into (3.101) and (3.102) one obtains the centerline velocity as

$$\frac{\tilde{u}_{CL}}{u_0} = \frac{6.56\,d}{x+x_0}\left(\frac{\rho_0}{\rho_{st}}\right)^{1/2}. \qquad (3.108)$$

The distance of the virtual origin from $x=0$ may be estimated by setting $\tilde{u}_{CL} = u_0$ at $x=0$ in (3.108) so that

$$x_0 = 6.56\,d\left(\frac{\rho_0}{\rho_{st}}\right)^{1/2}. \qquad (3.109)$$

As an example for the flame length, we set the molecular weight at the stoichiometric mixture equal to that of nitrogen, thereby estimating the density ratio ρ_0/ρ_{st} from

$$\frac{\rho_0}{\rho_{st}} = \frac{W_0}{W_{N_2}}\frac{T_{st}}{T_0}. \qquad (3.110)$$

The flame length may then be calculated from (3.106) with $Z_{st} = 0.055$ as $L \sim 200\,d$.

In large turbulent diffusion flames buoyancy influences the turbulent flow field and thereby the flame length. To derive a scaling law for that case, Peters and Göttgens (1991) have integrated the boundary layer equations for momentum and mixture fraction for a vertical jet flame over the radial direction in order to obtain first-order differential equations in terms of the axial coordinate for cross-sectional averages of the axial velocity and the mixture fraction. Since turbulent transport disappears entirely as the result of averaging, an empirical model for the entrainment coefficient β is needed; this relates the half-width b of the jet to the axial coordinate as

$$b(x) = \beta x. \qquad (3.111)$$

By comparison with the similarity solution for a nonbuoyant jet β was determined as

$$\beta = 0.23\left(\frac{\rho_{st}}{\rho_0}\right)^{1/2}. \qquad (3.112)$$

For the buoyant jet flame, the following closed-form solution for the flame length can be derived:

$$\left(\frac{3}{4}\beta Fr^* - \frac{1}{8}\right)\left(\frac{\beta L}{d_{eff}}\right)^2 + \left(\frac{\beta L}{d_{eff}}\right)^5 = \frac{3\beta\alpha_1^2}{16 Z_{st}^2}Fr^*. \qquad (3.113)$$

Here Fr^* is the modified Froude number

$$Fr^* = Fr\left(\frac{\rho_0}{\rho_\infty}\right)^{1/2}\frac{\rho_\infty}{\alpha_2(\rho_\infty - \rho_{st})}, \qquad Fr = \frac{u_0^2}{gd}, \qquad (3.114)$$

3.10 Experimental data from turbulent jet diffusion flames

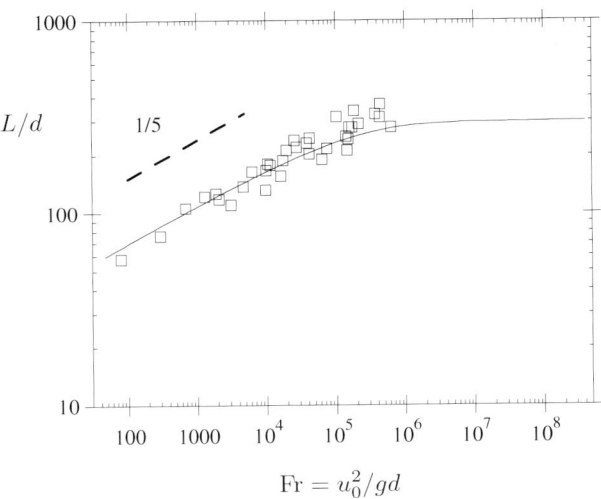

Figure 3.12. Dimensionless flame length, L/d, versus Froude number for propane–air flames. Comparison is made with experimental data of Sønju and Hustad (1984).

and

$$d_{\text{eff}} = d \left(\frac{\rho_0}{\rho_\infty} \right)^{1/2} \qquad (3.115)$$

is the effective nozzle diameter. The constants in (3.113) and (3.114) are determined as $\alpha_1 = 1 + 2Sc_t$ and $\alpha_2 = 1.0$. Details of the derivation may be found in Peters and Göttgens (1991).

The flame length of propane flames calculated from (3.113) is compared with measurements from Sønju and Hustad (1984) in Figure 3.12. For Froude numbers smaller than 10^5 the data show a Froude number scaling as $Fr^{1/5}$, which corresponds to a balance of the second term on the l.h.s. with the term on the r.h.s. in (3.113). For Froude numbers larger than 10^6 the flame length becomes Froude number independent and equal to the value calculated from (3.105).

3.10 Experimental Data from Turbulent Jet Diffusion Flames

Although the flame length may be calculated on the basis of the fast chemistry assumption using the solution for the mean mixture fraction field alone, more details on scalars are needed if one wants to determine chemical effects and pollutant formation in turbulent jet flames. For that purpose we want to discuss data taken locally within turbulent jet flames by nonintrusive laser-diagnostic techniques. There is a large body of experimental data on single point measurements using laser Rayleigh and Raman scattering techniques combined with laser-induced fluorescence (LIF). Since a comprehensive review on the subject by

Masri et al. (1996) is available, it suffices to present as an example the results by Barlow et al. (1990) obtained by the combined Raman–Rayleigh–LIF technique. This paper has set a landmark not only because of the diagnostics that were used but also because of the interpretation of the chemical structure in terms of laminar flamelet profiles. The fuel stream of the two flames that were investigated consisted of a mixture of 78 mole% H_2 and 22 mole% argon, the nozzle inner diameter d was 5.2 mm, and the co-flow air velocity was 9.2 m/s. The resulting flame length was approximately $L = 60\,d$. Two cases of exit velocities were analyzed, but only case B with $u_0 = 150$ m/s will be considered here.

The stable species H_2, O_2, N_2, and H_2O were measured using Raman scattering at a single point with light from a flash-lamp pumped dye laser. In addition, quantitative OH radical concentrations from LIF measurements were obtained by using the instantaneous one-point Raman data to calculate quenching corrections for each laser shot. The correction factor was close to unity for stoichiometric and moderately lean conditions but increased rapidly for very lean and moderately rich mixtures. The temperature was calculated for each laser shot by adding number densities of the major species and using the ideal gas law. The mixture fraction was calculated from the stable species concentrations using (3.22). An ensemble of one-point, one-time Raman-scattering measurements of major species and temperature is plotted against mixture fraction in Figure 3.13.

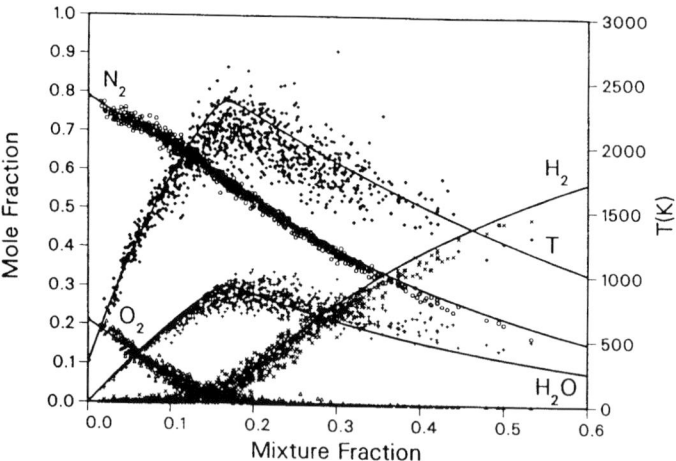

Figure 3.13. Ensemble of Raman scattering measurements of major species concentrations and temperatures at $x/d = 30$, $r/D = 2$. The solid curves show equilibrium conditions. From Barlow et al. (1990). (Reprinted by permission of Elsevier Science from "Effect of the Damköhler number on super-equilibrium OH concentration in turbulent nonpremixed jet flames," by R.S. Barlow, R.W. Dibble, J.-Y. Chen and R.P. Lucht, Combustion & Flame, Vol 82, pp 235–251, Copyright 1990 by The Combustion Institute.)

3.10 Experimental data from turbulent jet diffusion flames

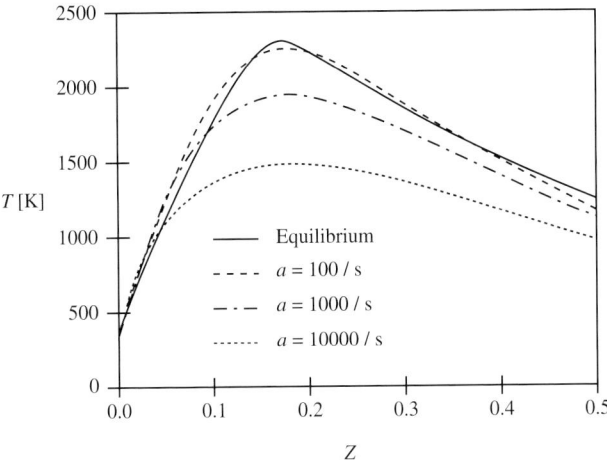

Figure 3.14. Temperature profiles from flamelet calculations at different strain rates. From Barlow et al. (1990). (Reprinted by permission of Elsevier Science from "Effect of the Damköhler number on super-equilibrium OH concentration in turbulent non-premixed jet flames," by R.S. Barlow, R.W. Dibble, J.-Y. Chen and R.P. Lucht, Combustion & Flame, Vol 82, pp 235–251, Copyright 1990 by The Combustion Institute.)

They were taken at $x/d = 30$, $r/d = 2$ in the case B flame. Also shown are calculations based on the assumption of chemical equilibrium.

The overall agreement between the experimental data and the equilibrium solution is quite good. This is often observed for hydrogen flames where chemistry typically is very fast. In contrast, hydrocarbon flames at high exit velocities and small nozzle diameters are likely to exhibit local extinction and nonequilibrium effects. A discussion on localized extinction observed in turbulent jet flames may be found in Masri et al. (1996).

Figure 3.14 shows temperature profiles versus mixture fraction calculated for counterflow diffusion flames at different strain rates. These steady state flamelet profiles display the characteristic decrease of the maximum temperature with increasing strain rates (which corresponds to decreasing Damköhler numbers) as shown schematically by the upper branch of the S-shaped curve in Figure 1.1. The strain rates vary here between $a = 100/s$, which is close to chemical equilibrium, and $a = 10,000/s$.

Data of OH concentrations are shown in Figure 3.15. They are to be compared to flamelet calculations in Figure 3.16 for the different strain rates mentioned before. It is evident from Figure 3.15 that the local OH concentrations exceed those of the equilibrium profile by a factor up to 3. The flamelet calculations in Figure 3.16 show an increase of the maximum values by a factor of 3 already at the low strain rates $a = 100/s$ and $a = 1,000/s$. The maximum value of

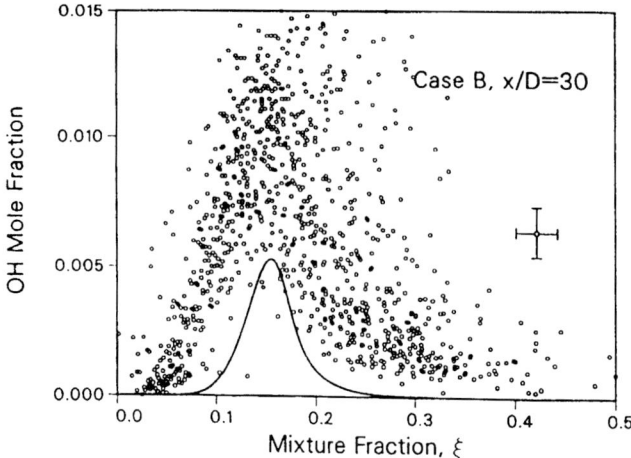

Figure 3.15. Ensemble of LIF measurements of OH concentrations at $x/d = 30$, $r/D = 2$. The solid curve shows the equilibrium solution. From Barlow et al. (1990). (Reprinted by permission of Elsevier Science from "Effect of the Damköhler number on super-equilibrium OH concentration in turbulent nonpremixed jet flames," by R.S. Barlow, R.W. Dibble, J.-Y. Chen and R.P. Lucht, Combustion & Flame, Vol 82, pp 235–251, Copyright 1990 by The Combustion Institute.)

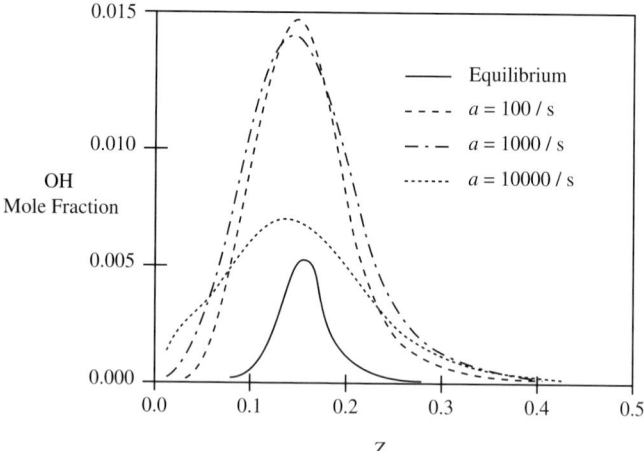

Figure 3.16. OH mole fractions from flamelet calculations at different strain rates. From Barlow et al. (1990). (Reprinted by permission of Elsevier Science from "Effect of the Damköhler number on super-equilibrium OH concentration in turbulent nonpremixed jet flames," by R.S. Barlow, R.W. Dibble, J.-Y. Chen and R.P. Lucht, Combustion & Flame, Vol 82, pp 235–251, Copyright 1990 by The Combustion Institute.)

$a = 10,000/\text{s}$ shown in Figure 3.16 is close to extinction and does not represent conditions in the turbulent hydrogen flame considered here.

In summary, it may be concluded that one-point, one-time experimental data in turbulent flames, when plotted as a function of mixture fraction, show strong similarities with laminar flamelet profiles in counterflow diffusion flames. Nonequilibrium effects are evident in both cases and lead to an increase of radical concentrations and a decrease of temperatures. This has an important influence on NO_x formation in turbulent diffusion flames, as will be discussed in Section 3.14.

3.11 Laminar Flamelet Equations for Nonpremixed Combustion

For nonpremixed combustion we will use a procedure similar to that used to derive the laminar flamelet equation for premixed combustion, namely a two-scale asymptotic analysis. This is different from the previous presentation in Peters (1984), where a local coordinate transformation and boundary layer arguments were used. A characteristic length scale in physical space representing flame surface corrugations is defined by

$$\Lambda = \frac{Z'_{st}}{|\nabla Z|_{st}}. \tag{3.116}$$

Here Z'_{st} is the r.m.s. of mixture fraction fluctuations defined by (3.67) and $|\nabla Z_{st}|$ is obtained from

$$|\nabla Z|_{st} = \left(\frac{\tilde{\chi}_{st}}{2 D_{st}}\right)^{1/2}, \tag{3.117}$$

where $\tilde{\chi}_{st}$ is defined by (3.157) below. If we define by analogy to (1.53) a Taylor scale λ_Z of the mixture fraction field by replacing the mixture fraction gradient $|\nabla Z|_{st}$ by Z'_{st}/λ_Z, we find that the length scale Λ corresponds to a Taylor scale of the mixture fraction field.

For the two-scale asymptotic expansion a small parameter ε defined here as

$$\varepsilon = \frac{\ell_R}{\Lambda} \tag{3.118}$$

is introduced where ℓ_R is defined by (3.69). We normalize the independent variables x and t and the velocity v in the equation for the reactive scalars by Λ and the time $t_\Lambda = \Lambda^2/D_{st}$, respectively:

$$x^* = x/\Lambda, \quad t^* = t/t_\Lambda, \quad v^* = v t_\Lambda/\Lambda. \tag{3.119}$$

The diffusivity is normalized as

$$D_i^* = D_i/D_{st}. \tag{3.120}$$

Using suitable reference values $\psi_{i,\mathrm{ref}}$ and ρ_{ref} the reactive scalars, the density, and the chemical source term are in nondimensional form

$$\psi_i^* = \psi_i/\psi_{i,\mathrm{ref}}, \quad \rho^* = \rho/\rho_{\mathrm{ref}}, \quad \omega_i^* = \omega_i t_\Lambda/(\psi_{i,\mathrm{ref}}\rho_{\mathrm{ref}}). \tag{3.121}$$

Furthermore, the mixture fraction is normalized with Z'_{st} as

$$Z^* = Z/Z'_{\mathrm{st}}. \tag{3.122}$$

Then the nondimensional version of (1.81) takes, with the asterisk removed for notational convenience, the same form as before:

$$\rho\frac{\partial\psi_i}{\partial t} + \rho\mathbf{v}\cdot\nabla\psi_i = \nabla\cdot(\rho D_i\nabla\psi_i) + \omega_i. \tag{3.123}$$

The short scale in the two-scale asymptotic analysis is defined by

$$\zeta = \varepsilon^{-1}(Z(\mathbf{x},t) - Z_{\mathrm{st}}). \tag{3.124}$$

This short-range coordinate covers the vicinity normal to the instantaneous flame surface, while the flow and the mixing field further away is described by the long-range spatial coordinates $\mathbf{x}$ and the long time t. In the new coordinate system $\boldsymbol{\xi}$ replaces $\mathbf{x}$ and the short time scale

$$\tau = \varepsilon^{-2} t \tag{3.125}$$

is introduced to describe rapid temporal changes of the flamelet structure in the vicinity of the surface of stoichiometric mixture.

The reactive scalars are expanded into a leading order and a first-order term as

$$\psi_i = \psi_i^0 + \varepsilon\psi_i^1 + \cdots. \tag{3.126}$$

Here, we will only consider the leading order solution, written without the suffix, and express it as a function of the short and the long spatial and temporal variables in the form

$$\psi_i = \psi_i(\zeta(\mathbf{x},t), \boldsymbol{\xi}, \tau). \tag{3.127}$$

The temporal and spatial derivatives in the original equations may then be written as

$$\frac{\partial}{\partial t} = \frac{\partial \zeta}{\partial t}\frac{\partial}{\partial \zeta} + \varepsilon^{-2}\frac{\partial}{\partial \tau} = \varepsilon^{-1}\frac{\partial Z}{\partial t}\frac{\partial}{\partial \zeta} + \varepsilon^{-2}\frac{\partial}{\partial \tau},$$

$$\nabla = \nabla\zeta\frac{\partial}{\partial \zeta} + \nabla_\xi = \varepsilon^{-1}\nabla Z\frac{\partial}{\partial \zeta} + \nabla_\xi. \tag{3.128}$$

3.11 Laminar flamelet equations for nonpremixed combustion

When these transformations are introduced into (3.123) one obtains in the vicinity of stoichiometric mixture

$$\rho \frac{\partial \psi_i}{\partial \tau} + \varepsilon^2 \rho \boldsymbol{v} \cdot \nabla_\xi \psi_i + \varepsilon \left[\rho \frac{\partial Z}{\partial t} + \rho \boldsymbol{v} \cdot \nabla Z \right] \frac{\partial \psi_i}{\partial \zeta}$$

$$= \nabla Z \cdot \frac{\partial}{\partial \zeta} \left(\rho D_i \nabla Z \frac{\partial \psi_i}{\partial \zeta} \right) + \varepsilon \nabla_\xi \cdot \left(\rho D_i \nabla Z \frac{\partial \psi_i}{\partial \zeta} \right)$$

$$+ \varepsilon \nabla Z \cdot \frac{\partial}{\partial \zeta} (\rho D_i \nabla_\xi \psi_i) + \varepsilon^2 \nabla_\xi \cdot (\rho D_i \nabla_\xi \psi_i) + \varepsilon^2 \omega_i. \quad (3.129)$$

To retain the effect of chemical reactions the chemical source term $\varepsilon^2 \omega_i$ must be of same order as the other leading order terms. We set

$$\varepsilon^2 \omega_i = \Omega_i, \quad (3.130)$$

where Ω_i is of order unity, indicating that the chemical reaction rate ω_i is large. Furthermore, since ∇Z varies only on the long scales, it may be pulled inside the parentheses in the first term on the r.h.s. of (3.129). Introducing this and (3.130) into (3.129) and collecting the leading order terms only, one obtains

$$\rho \frac{\partial \psi_i}{\partial \tau} = \frac{\partial}{\partial \zeta} \left(\rho D_i |\nabla Z|^2 \frac{\partial \psi_i}{\partial \zeta} \right) + \Omega_i. \quad (3.131)$$

If the nondimensional quantity $D_i |\nabla Z|^2$ is rewritten in dimensional form again,

$$\frac{D_i}{D_{st}} |\nabla Z|^2 \left(\frac{\Lambda}{Z'_{st}} \right)^2 = \frac{1}{Le_i} \frac{\chi}{\chi_{st}}, \quad (3.132)$$

the r.h.s. turns out to be of order unity. This should be the case in the asymptotic expansion, thereby justifying the definition of Λ. Since χ varies on the long scales only, the term $D_i |\nabla Z|^2$ can also be taken outside the parentheses in (3.131). Using

$$d\zeta = \varepsilon^{-1} dZ \quad (3.133)$$

we can rewrite (3.131) in dimensional form again to obtain the classical flamelet equations

$$\rho \frac{\partial \psi_i}{\partial t} = \frac{\rho}{Le_i} \frac{\chi}{2} \frac{\partial^2 \psi_i}{\partial Z^2} + \omega_i. \quad (3.134)$$

Because of the asymptotic expansion the instantaneous scalar dissipation rate should be that at stoichiometric position, namely χ_{st} defined by (3.30). In one-dimensional flow situations, such as those discussed in Section 3.4, one does not need the asymptotic expansion to obtain flamelet equations like (3.134). This suggests that there is a more general dependence of χ on mixture fraction, as given by (3.47), for instance. We will return to this discussion in the next section. As noted before, χ acts as an external parameter that is imposed on the flamelet structure by the mixture fraction field.

In deriving (3.134) the fast time $\tau = \varepsilon^{-2} t$ has been introduced to describe strong temporal changes of the flamelet structure. If the flamelet is convected by a mean flow, the time coordinate in (3.134) may be interpreted as a Lagrangian time. In a situation where one or more velocity components are large, one could also have introduced the rescaled velocity

$$\mathbf{V} = \varepsilon^2 \mathbf{v}, \tag{3.135}$$

which then would make the second term on the l.h.s. of (3.129) a leading order term to be retained in the flamelet equations. Written in dimensional form this would result in

$$\rho \frac{\partial \psi_i}{\partial t} + \rho \mathbf{v} \cdot \nabla \psi_i = \frac{\rho}{Le_i} \frac{\chi}{2} \frac{\partial^2 \psi_i}{\partial Z^2} + \omega_i \tag{3.136}$$

instead of (3.134).

Among the first-order terms in (3.129) the second term on the r.h.s. may be expressed as

$$\varepsilon \nabla \cdot (\rho D_i \nabla Z) \frac{\partial \psi_i}{\partial \zeta} + \varepsilon \rho D_i \nabla Z \cdot \nabla_\xi \left(\frac{\partial \psi_i}{\partial \zeta} \right), \tag{3.137}$$

where the ∇_ξ-operator in the first term was replaced by the ∇-operator. This term and the third term on the l.h.s. of (3.129) may be combined to give

$$\varepsilon \left[\rho \frac{\partial Z}{\partial t} + \rho \mathbf{v} \cdot \nabla Z - \nabla \cdot (\rho D_i \nabla Z) \right] \frac{\partial \psi_i}{\partial \zeta}. \tag{3.138}$$

Using (3.23) where D has been defined as the thermal diffusivity, the term in square brackets in this equation vanishes for the temperature equation. In the species equations, however, this term does not vanish, but, following Pitsch and Peters (1998a), it may be expressed in a more convenient form. We first use (3.23) to replace the first two terms in the square brackets of (3.138) by the diffusive term. Then, since a second derivative of z with respect to x, $d^2 z/dx^2$, may be transformed with $y = dz/dx$ to $dy/dx = dy/dz \cdot dz/dx = (dy^2/dz)/2$, we obtain, with y defined as $\rho D \nabla Z$ and the assumption of constant Lewis numbers for the term in square brackets in (3.138),

$$\left[\left(1 - \frac{1}{Le_i} \right) \nabla \cdot (\rho D \nabla Z) \right] = \frac{1}{4 \rho D} \left(1 - \frac{1}{Le_i} \right) \frac{\partial (\rho^2 D \chi)}{\partial Z}. \tag{3.139}$$

As in the examples in Section 3.4 we assume $\rho^2 D$ to be constant and move this term outside the parentheses in the derivative on the r.h.s. of (3.139). Retaining only this first-order term in (3.129), we can then write the flamelet equations for the chemical species in dimensional form as

$$\rho \frac{\partial \psi_i}{\partial t} + \frac{1}{4} \left(1 - \frac{1}{Le_i} \right) \rho \frac{\partial \chi}{\partial Z} \frac{\partial \psi_i}{\partial Z} = \frac{\rho}{Le_i} \frac{\chi}{2} \frac{\partial^2 \psi_i}{\partial Z^2} + \omega_i. \tag{3.140}$$

3.11 Laminar flamelet equations for nonpremixed combustion

As compared to (3.134) the additional term accounts for differential diffusion effects and appears as a convective term in mixture fraction space, where the derivative of the scalar dissipation rate with respect to Z acts like a velocity. This term is particularly important for reactive scalars with a very large diffusivity such as hydrogen radicals enhancing soot formation. (cf. Pitsch et al., 2000).

An alternative interpretation of the term on the l.h.s. of (3.139) is in terms of curvature. If the definition (2.111) is used for the iso-mixture fraction surface $Z(\boldsymbol{x}, t) = Z_{st}$, the second derivative becomes

$$\nabla \cdot (\rho D \nabla Z) = -\rho D |\nabla Z| \kappa + \boldsymbol{n} \cdot \nabla (\rho D \boldsymbol{n} \cdot \nabla Z). \quad (3.141)$$

If the curvature κ becomes locally large of order $O(\varepsilon^{-1})$, normal diffusion can be neglected and the first term on the r.h.s. of (3.141) can be inserted into the flamelet equations using κ as an additional parameter. Recent experiments by Yoshida and Takagi (1998) on the effect of flame curvature generated by a microjet show, for instance, local extinction and reignition in a laminar counterflow diffusion flame.

It should be noted, however, that the simple binary flux approximation used in (3.123) with binary Lewis numbers different from unity conflicts with the requirement that the sum of the diffusion fluxes must be zero. To satisfy this requirement, correction terms must be included and these lead to a much more complicated expression than (3.140). Species equations containing those expressions were derived by Pitsch and Peters (1998a). Those equations should be used to correctly account for differential diffusion effects in the flamelet equations.

There are a number of papers that analyze differential diffusion in nonpremixed systems. Yeung and Pope (1993), Kronenburg and Bilger (1997), and Nilsen and Kosály (1997) used DNS to gain a better understanding of the underlying physics. Kerstein et al. (1995) had predicted that the Reynolds number dependence of the variance of the difference between two element mass fractions should be proportional to $Re^{-1/2}$. This was not confirmed by Nilsen and Kosály, at least not in the limited range of turbulent Reynolds numbers ($Re = 33$–326) achieved in their simulations. Laser Raman scattering measurements by Smith et al. (1995) in turbulent jet flames of H_2/CO mixtures with air had shown that differential diffusion effects were present at all jet exit Reynolds numbers measured ($Re_j = 1,000$–$30,000$). The effect was also found in turbulent methane–air diffusion flames investigated by Barlow and Frank (1998) and Bergmann et al. (1998). For the latter experiment Pitsch (1999) has provided an interesting explanation based on unsteady flamelet modeling. He included the differential diffusion terms in the flamelet equations only in the laminar flow region close to the nozzle and introduced the $Le_i = 1$ assumption further downstream. As the result of differential diffusion in the early part of the jet, the element mass fractions at a given mixture fraction become different from those corresponding to the equal diffusivity approximation. Because the value of the scalar dissipation rate decreases rapidly as one moves downstream in the jet, diffusion in mixture fraction space is frozen as the flamelet is convected downstream. Therefore the elemental composition generated in the early part of the jet prevails even in the far downstream region.

3.12 Flamelet Equations in Nonpremixed Turbulent Combustion

A turbulent flow and mixing field will impose on flamelets its unsteady random nature. However, as long as the reaction zone ℓ_R is thinner than a Kolmogorov eddy, combustion occurs in the quasi-laminar flow within that eddy and (3.134) can be used within the thickness $(\Delta Z)_R$ in mixture fraction space. Therefore, within that thickness fluctuations of ψ_i and the reaction rate ω_i need not be considered, whereas the scalar dissipation rate χ fluctuates and its statistical distribution must be specified. If we parameterize χ by χ_{st} and use the presumed shape pdf approach for both Z and χ_{st}, together with statistical independence (cf. Peters, 1984), the joint Favre pdf of Z and χ_{st} becomes

$$P(Z, \chi_{st}; \mathbf{x}, t) = P(\chi_{st}; \mathbf{x}, t) P(Z; \mathbf{x}, t). \tag{3.142}$$

If the flamelet equations are used, the reactive scalar profiles are functions of Z and χ_{st}. By analogy to the means in premixed combustion, calculated from (2.294), their Favre means in nonpremixed combustion are to be calculated from

$$\tilde{\psi}_i(\mathbf{x}, t) = \int_0^1 \int_0^\infty \psi_i(Z, t, \chi_{st}) P(\chi_{st}; \mathbf{x}, t) \tilde{P}(Z; \mathbf{x}, t) \, d\chi_{st} \, dZ. \tag{3.143}$$

An alternative approach, similar to the CMC method discussed in Section 1.12, is to replace the pdf of the dissipation rate by a delta function at the conditional Favre mean value $\tilde{\chi}_Z$. This amounts to replacing χ in (3.134) by $\tilde{\chi}_Z$:

$$\rho \frac{\partial \psi_i}{\partial t} = \frac{\rho}{Le_i} \frac{\tilde{\chi}_Z}{2} \frac{\partial^2 \psi_i}{\partial Z^2} + \omega_i. \tag{3.144}$$

Solving these equations with $\tilde{\chi}_Z$ prescribed gives reactive scalars that are functions only of Z and t. In contrast to (3.143) the Favre mean value of ψ_i is obtained by integrating over the pdf of the mixture fraction only:

$$\tilde{\psi}_i = \int_0^1 \psi_i(Z, t, \tilde{\chi}_Z) \tilde{P}(Z; \mathbf{x}, t) \, dZ. \tag{3.145}$$

If similar arguments as for the dissipation rate are used for the velocity v in (3.136), the close relation between Conditional First Moment Closure and the flamelet concept becomes evident.

There remains the question of whether the terms that have been neglected to leading order in the flamelet equations can also be neglected in the turbulent

3.12 Flamelet equations in nonpremixed turbulent combustion

mixing field outside of the thin reaction zone $(\Delta Z)_R$. We want to address this question by introducing suitably filtered equations outside of the surface of the stoichiometric mixture. The arguments are the same as in Section 2.14 of the previous chapter for the thin reaction zones regime and will not be repeated here. If the scalar fields are filtered over a region in physical space, which does not contain the reaction zone, the Favre filtered equations for the reactive scalars and the mixture fraction are

$$\hat{\rho}\frac{\partial \hat{\psi}_i}{\partial t} + \hat{\rho}\hat{v}\cdot\nabla\hat{\psi}_i = \nabla\cdot(\hat{\rho}(\hat{D}+D_i)\nabla\hat{\psi}_i), \qquad (3.146)$$

$$\hat{\rho}\frac{\partial \hat{Z}}{\partial t} + \hat{\rho}\hat{v}\cdot\nabla\hat{Z} = \nabla\cdot(\hat{\rho}(\hat{D}+D_i)\nabla\hat{Z}), \qquad (3.147)$$

where $\hat{\rho}$ is the filtered mean density and $\hat{\psi}_i$ and $\hat{Z}$ are the Favre filtered reactive scalar and mixture fraction, respectively. As before, the hat replaces the bar for the filtered mean density; otherwise it replaces the tilde for Favre filtered quantities. The gradient transport assumption has been used for these nonreacting scalar fields and therefore the turbulent diffusivity $\hat{D}$, originating from the convective terms, is the same for all reactive scalars. The sum of the diffusivities $(\hat{D} + D_i)$ approaches $\hat{D}$ for increasing filter width Δ since $\hat{D}$ becomes much larger than D_i.

Equation (3.146) is to be solved in the nonreacting part of the flow field outside of the reaction zone. Boundary conditions for the lean part are to be imposed at the oxidizer inflow where $\hat{Z} = 0$ and at the location of the reaction zone where $\hat{Z} = Z_{st}$. For the rich part, the boundary conditions of (3.146) are at $\hat{Z} = Z_{st}$ and at the fuel inflow where $\hat{Z} = 1$.

Now we perform, starting from (3.146) and (3.147), a two-scale asymptotic analysis similar to the one used to derive the flamelet equations. Again two characteristic length scales can be identified. Since the boundary conditions at $\hat{Z} = 0$ and $\hat{Z} = 1$ are imposed at the boundaries of the entire flow, the large length scale is the length L of the turbulent flow region (the width of the jet, for instance). The small length scale is a mixing length scale ℓ_m that resolves a relatively thin turbulent mixing layer in the vicinity of $\hat{Z} = Z_{st}$. As in premixed turbulent combustion the thickness ℓ_m of that layer is obtained by equating the chemical time t_c with the turnover time $t_n = (\ell_n^2/\tilde{\varepsilon})^{1/3}$ to obtain a mixing length ℓ_m by setting $\ell_n = \ell_m$:

$$\ell_m = (\varepsilon t_c^3)^{1/2}. \qquad (3.148)$$

This definition is similar to (2.35) except that t_c is not a fixed quantity as is the time t_q. In a diffusion flame the instantaneous chemical time depends on the instantaneous scalar dissipation rate. For the steady state counterflow diffusion flame it was shown at the end of Section 3.5 that there is a balance between diffusion and reaction such that the chemical time scale can be related to the inverse of the scalar dissipation rate χ_q at extinction. We can generalize (3.58) to define the instantaneous chemical time t_c as

$$t_c = \frac{Z_{st}^2 (1-Z_{st})^2}{\chi_{st}} \tag{3.149}$$

in order to define ℓ_m.

The reaction zone is embedded within two turbulent mixing layers of thickness ℓ_m, one to the lean and one to the rich side of Z_{st}; both are thin compared to the large length scale L. In the vicinity of the reaction zone a filter width Δ smaller than ℓ_m must be chosen to resolve the structure of the mixing layer, while larger filters can be used further away from that zone to obtain sufficiently smooth scalar fields. Smoothness is necessary to prevent strong gradients that would invalidate the asymptotic analysis.

For this analysis the small parameter is defined as

$$\varepsilon = \frac{\ell_m}{L}. \tag{3.150}$$

Using this parameter and performing a two-scale asymptotic expansion in the same way as for premixed turbulent combustion, one obtains instead of (3.134)

$$\hat{\rho} \frac{\partial \hat{\psi}_i}{\partial t} = \hat{\rho} \frac{(\hat{D} + D_i)}{\hat{D}} \frac{\hat{\chi}}{2} \frac{\partial^2 \hat{\psi}_i}{\partial \hat{Z}^2}. \tag{3.151}$$

Here the filtered scalar dissipation rate

$$\hat{\chi} = 2\hat{D}|\nabla \hat{Z}|^2 \tag{3.152}$$

has been introduced.

Since the asymptotic derivation has been performed in the vicinity of $Z = Z_{st}$, $\hat{\chi}$ should be the conditional value $\hat{\chi}_{st}$. For a one-dimensional flow configuration, however, one can perform an analysis similar to that done in the two examples discussed in Section 3.4. Therefore $\hat{\chi}$ should depend on $\hat{Z}$ outside of $Z(x,t) = Z_{st}$ in a similar way as in (3.47) and (3.56). We will use this information below for a model of the conditional mean scalar dissipation rate $\tilde{\chi}_{st}$.

Except for the chemical source term Equation (3.151) has a form similar to (3.134). These equations differ, in addition, in the apparent Lewis number defined by

$$\widehat{Le}_i = \begin{cases} Le_i, & \text{in laminar flow regions;} \\ \hat{D}/(\hat{D}+D_i), & \text{in turbulent flow regions.} \end{cases} \tag{3.153}$$

Refering to results by Pitsch (1999) it seems appropriate to use $Le_i = 1$ whenever the flow is fully turbulent. Flamelet calculations by Oevermann (1997) and by

3.12 Flamelet equations in nonpremixed turbulent combustion

Pitsch et al. (1998) show that a Lewis number of unity gives the best results for the temperature and all the major species when compared with the hydrogen–air flame of Pfuderer et al. (1996).

In principle, one would have to solve (3.134) and (3.151) separately within their domain of validity in mixture fraction space. The two solutions of (3.151) on both sides of the reaction zone can then be matched to the solution of (3.134) in the reaction zone. This provides boundary conditions to (3.134) in a similar way as in the asymptotic analysis by Liñán (1974) (cf. also the discussion in Peters, 1980). For arbitrary chemistry such a procedure, however, is impractical. It is important to realize that $\hat{\chi}$ in (3.151) remains of the same order of magnitude independently of the filter width that has been used since as a dissipation rate it is inertial range invariant. Since notation is unimportant, we therefore can combine (3.134) and (3.151) into a single equation by replacing $\hat{Z}$ by Z, $\hat{\rho}$ by ρ, and $\hat{\psi}_i$ by ψ_i in (3.151), giving

$$\rho \frac{\partial \psi_i}{\partial t} = \frac{\rho}{\widehat{Le_i}} \frac{\chi}{2} \frac{\partial^2 \psi_i}{\partial Z^2} + \omega_i \qquad (3.154)$$

for the entire domain $0 < Z < 1$ with the understanding that ρ, ψ_i, and Z are instantaneous quantities within the reaction zone but filtered mean values outside of it.

Ferreira (1996) has compared OH concentration measurements from Barlow and Carter (1994) with two different solutions of the flamelet equations. The first is calculated from the laminar counterflow diffusion flame configuration at different strain rates. This calculation included effects of differential diffusion. The strain rate was set equal to $\tilde{\varepsilon}/\tilde{k}$ locally in the turbulent flame. The other case is a solution of the flamelet equations in mixture fraction space, where the Lewis numbers were unity and χ was replaced by the Favre mean value $\tilde{\chi}$, determined from (3.79) with $c_\chi = 2.0$. The OH mole fractions at stoichiometric mixture fraction from Ferreira (1994) are plotted against the normalized axial distance in the jet in Figure 3.17. Only the use of the flamelet equations with χ as a parameter and $Le_i = 1$ shows a good agreement.

There remains the question of how the pdf $P(\chi_{st}; x, t)$ needed in (3.143) should be modeled. Peters (1984) has argued that the pdf of χ_{st} should be a lognormal pdf. There is some limited experimental evidence (cf. Effelsberg and Peters, 1988 and Nandula et al., 1994) that the variance of the lognormal distribution of χ_{st} should be close to unity. Using DNS Yeung et al. (1990) find from area weighted averages of χ a variance of 0.68 and a value of 1.3 for unweighted averages.

If the presumed shape pdf approach is used, the double integration (3.143) to be performed at every grid point is time consuming. For time-dependent

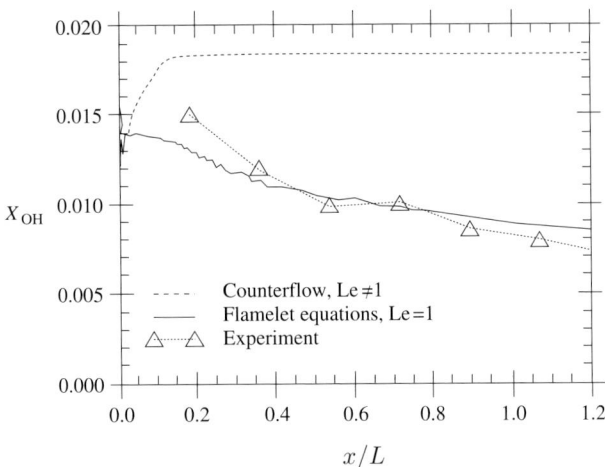

Figure 3.17. OH mole fraction along the mean contour of stoichiometric mixture fraction. (Reprinted with permission from J. C. Fereira.)

flamelet calculations this becomes virtually impossible, because then the time history of χ_{st} must be taken into account. If a conditional average over flamelet structures is performed, only the conditional mean scalar dissipation rate $\tilde{\chi}_Z$ appears in the flamelet equations, while ψ_i and ω_i are to be interpreted as conditional means. If, in addition, the dependence of χ on Z as in the one-dimensional flow problems in Section 3.4 is retained, the flamelet equations become equal to (3.144) and mean values, to be obtained from (3.145), involve only a single integration.

Then there remains the problem of how to model the conditional scalar dissipation rate $\hat{\chi}_Z$. One way to solve that problem is to use the procedure by Janicka and Peters (1982) where $\tilde{\chi}_Z$ is determined from the pdf transport equation for the mixture fraction as outlined in Section 1.11. Another possibility is to construct $\tilde{\chi}_Z$ from a model. Cook et al. (1997) assume that the same dependence on Z as in (3.47) or (3.56) is valid. They relate the conditional scalar dissipation rate $\tilde{\chi}_Z$ to that at a fixed value, say Z_{st}, by

$$\tilde{\chi}_Z = \tilde{\chi}_{st} \frac{f(Z)}{f(Z_{st})}, \qquad (3.155)$$

where $f(Z)$ is the exponential term in (3.47) and $\tilde{\chi}_{st}$ is the conditional mean scalar dissipation rate at $Z = Z_{st}$. Then, with the presumed pdf $\tilde{P}(Z)$ being known, the unconditional average can be written as

$$\tilde{\chi} = \int_0^1 \tilde{\chi}_Z \tilde{P}(Z)\,dZ = \tilde{\chi}_{st} \int_0^1 \frac{f(Z)}{f(Z_{st})} \tilde{P}(Z)\,dZ. \qquad (3.156)$$

Therefore, using the model (3.79) for $\tilde{\chi}$, the conditional Favre mean scalar dissipation rate $\tilde{\chi}_{st}$ can be expressed as

$$\tilde{\chi}_{st} = \frac{\tilde{\chi} f(Z_{st})}{\int_0^1 f(Z)\tilde{P}(Z)\,dZ}, \qquad (3.157)$$

which is to be used in (3.155) to calculate $\tilde{\chi}_Z$. Cook et al. (1997) also emphasize that $\tilde{\chi}_Z$ should vanish for $Z = 0$ and $Z = 1$ as in the counterflow model (3.47).

With $\tilde{\chi}_Z$ known one may solve the flamelet equations in mixture fraction space numerically. This has the advantage that boundary conditions are to be imposed at the fixed values at $Z = 0$ and $Z = 1$ rather than asymptotically at $y = +\infty$ and $y = -\infty$ as in the counterflow diffusion flame problem. In some cases, where Z does not extend to either $Z = 0$ or to $Z = 1$, one must impose the boundary conditions at the minimum or maximum value of Z. The model for the conditional scalar dissipation rate then must be modified accordingly. Details may be found in Pitsch (1998).

Calculations performed in mixture fraction space can also be used to determine the scalar dissipation rate χ_q at extinction. As in the counterflow geometry, numerical continuation techniques (cf. Giovangigli and Smooke, 1987) can be used to determine the extinction point Q in Figure 3.8. At that point $\chi_q = \chi_{st,ext}$ and the temperature should have decreased considerably from its adiabatic value.

Similarly, the flamelet equations can also be used to describe ignition in a nonpremixed system. If fuel and oxidizer are initially at the unburnt temperature $T_u(Z)$, as was shown in Figure 3.1, but the scalar dissipation rate is still large enough, so that heat loss out of the reaction zone exceeds the heat release by chemical reactions, a thermal runaway is not possible. This corresponds to the steady state lower branch in Figure 1.1. As the scalar dissipation rate decreases, as for instance in a diesel engine after injection, heat release by chemical reactions will exceed heat loss out of the reaction zone, leading to auto-ignition. The scalar dissipation rate at auto-ignition is denoted by

$$\chi_i = \chi_{st,ign}. \qquad (3.158)$$

For ignition under diesel engine conditions this has been investigated by Pitsch and Peters (1998b). An example of auto-ignition of a n-heptane–air mixture calculated with the RIF code (cf. Paczko et al., 1999) is shown in Figure 3.18. The initial air temperature is 1,100 K and the initial fuel temperature is 400 K. Mixing of fuel and air leads to a straight line for the enthalpy in mixture fraction space, but not for the temperature $T_u(Z)$ in Figure 3.18, since the heat capacity c_p depends on temperature. A linear decrease of the scalar dissipation rate from $\chi_{st} = 30/s$ to $\chi_{st} = 10/s$ within a time interval of 0.3 ms was prescribed. It

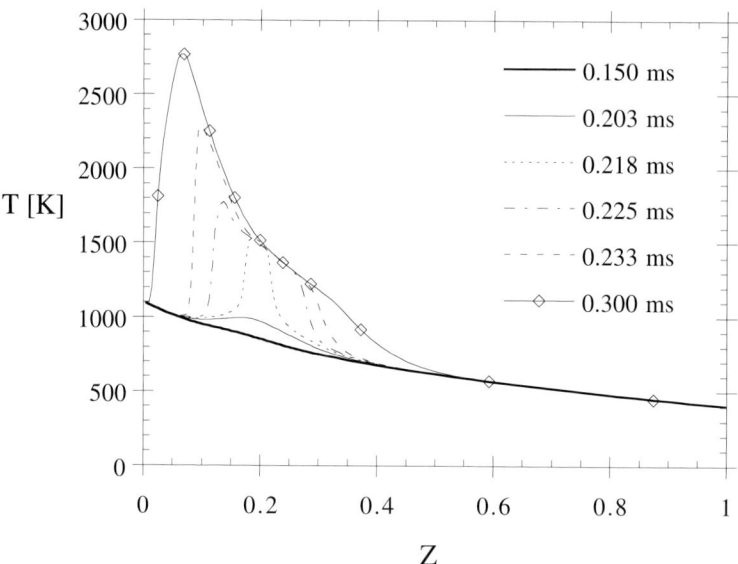

Figure 3.18. Auto-ignition of a n-heptane–air mixture calculated in mixture fraction space by solving the flamelet equations (cf. Paczko et al., 1999). (Reprinted with permission by the author.)

is seen that auto-ignition starts after 0.203 ms, when the temperature profile already shows a small increase over a broad region around $Z = 0.2$. At $t = 0.218$ ms there has been a fast thermal runaway in that region, with a peak at the adiabatic flame temperature. From thereon, the temperature profile broadens, which may be interpreted as a propagation of two fronts in mixture fraction space, one toward the lean and the other toward the rich mixture. Although the transport term in (3.154) contributes to this propagation, it should be kept in mind that the mixture is close to auto-ignition everywhere. The propagation of an ignition front in mixture fraction space therefore differs considerably from premixed flame propagation, which has been interpreted as an extinction wave in Section 3.5. At $t = 0.3$ ms most of the mixture, except for a region beyond $Z = 0.4$ in mixture fraction space, has reached the equilibrium temperature. A maximum value of $T = 2,750$ K is found close to the stoichiometric mixture.

The ignition of n-heptane mixtures under diesel engine conditions has been discussed in detail by Pitsch and Peters (1998b). There it is shown that auto-ignition under nonpremixed conditions occurs predominantly at locations in a turbulent flow field where the scalar dissipation rate is low. This cannot be captured by using a mean scalar dissipation rate as in (3.144) but suggests that the pdf of χ_{st} must be retained. A feasible way to account for a distribution of scalar dissipation rates in calculating unsteady flamelet histories is the use of multiple flamelets. This will be discussed below in the context of ignition in diesel engines.

Even though it appears convenient to use the flamelet equations to calculate ignition in nonpremixed turbulent flows, it should be kept in mind that chemistry is not fast at the early stages of auto-ignition. As can be seen from Figure 3.18 there is at $t = 0.203$ ms a broad layer of elevated temperature in mixture fraction space prior to thermal runaway. The width $\Delta Z \approx 0.1$ of that profile in mixture fraction space is of the same order as the diffusion thickness $(\Delta Z)_F$. This is consistent with the analysis of Liñán (1974) for the ignition regime, which does not exhibit a thin layer. If an analysis similar to that in Section 3.6 is performed one finds that the Karlovitz number Ka is of order unity, if ΔZ_R is of the order $(\Delta Z)_F$. This criterion happens to correspond to $Ka = 1$ for the corrugated flamelets regime in premixed combustion.

Mastorakos et al. (1997) have analyzed DNS data for auto-ignition using one-step kinetics with a large activation energy. They stress the importance of statistical fluctuations and criticize the use of a single flamelet with an ensemble-averaged scalar dissipation rate for auto-ignition problems by Zhang et al. (1995). Fedotov (1992) had shown that in a homogeneous system with random heat loss, ignition may occur at conditions that are not predicted by models that only take the average heat loss into account. Mastorakos et al. (1997) find that the auto-ignition time is the shortest in the vicinity of the "most reactive" mixture fraction $Z_{MR} = 0.13$ (cf. Liñán and Crespo, 1976), but that it remains at the same order of magnitude over a width of the order of Z_{MR} in mixture fraction space. Starting from the equations for auto-ignition in a homogeneous system they construct modeled equations containing additional transport terms for the mean temperature and for the temperature variance. These terms are derived by assuming that the temperature depends on the mixture fraction only. Therefore only transport in mixture fraction space appears in the model while transport in other directions in physical space is not taken into account. Correlation coefficients between the scalar dissipation rates and the temperature and its variance are derived from the DNS data. The modeled equations are discussed with reference to Conditional Moment Closure.

3.13 Steady versus Unsteady Flamelet Modeling

In the past the steady state version of the flamelet equations

$$\frac{\rho}{Le_i} \frac{\chi}{2} \frac{\partial^2 \psi_i}{\partial Z^2} + \omega_i = 0, \tag{3.159}$$

also called the Stationary Laminar Flamelet Model (SLFM), has sometimes been applied in engineering calculations mainly because of its simplicity of implementation. A library of steady state flamelet profiles can then be generated for arbitrary complex chemistry and the presumed shape pdf approach can be used. The library is typically parameterized in terms of values of the scalar dissipation rate χ_{st}. Because the time-dependent term has been omitted, the underlying assumption is that the imposed value of the scalar dissipation rate varies slowly enough. If the scalar dissipation rate changes rapidly, however, the unsteady term in the flamelet equations must be retained leading to a slow relaxation of the profiles. This has been noted by Haworth et al. (1988) who proposed limiting the effect of changes of χ on the flamelet profiles rather than using the local value of the scalar dissipation rate in a turbulent jet. Mell et al. (1994) compare the magnitude of the unsteady term in (3.134) with that of the reaction term and develop a criterion for the validity of the steady state assumption. Cuenot and Poinsot (1994) have used results from DNS of flame–vortex interactions to construct a phase diagram showing domains where the SLFM was found to be valid and where it was not. They distinguish three main effects that limit SLFM: unsteadiness, curvature, and quenching of flamelets.

Peters (1984) has introduced a fast time scale in the asymptotic derivation of the flamelet equations, thereby retaining the unsteady term. Unsteady flamelets were first used by Mauss et al. (1990) to simulate flamelet extinction and reignition in a steady turbulent jet diffusion flame. A Lagrangian time defined by

$$t = \int_0^x \frac{1}{\tilde{u}(x \mid \tilde{Z} = Z_{st})} dx \qquad (3.160)$$

was introduced to account for history effects in the flamelet structure. Nine different flamelet histories with random distribution of initial times have been used to show that the scatter in experimental data found in one-point Raman measurements are likely to be caused by unsteady changes of the flamelet structure. Such scatter plots were shown in Figure 3.13, for example.

Barlow and Chen (1992) imposed periodic variations as well as step changes of the scalar dissipation rate on the flamelet structure. They found that periodic oscillations have minor effects on the species concentrations, but that sudden decreases in the scalar dissipation rate lead to an overshoot of OH and CO. This is consistent with the results of Mauss et al. (1990). Darabiha (1992) analyzed strain rate variations of laminar hydrogen–air counterflow flames. He found that for high frequencies the strain rate may exceed the steady quenching limit without completely extinguishing the flame. Moreover, there is a limiting frequency at which the flame becomes insensitive to perturbations and behaves like a steady state flame.

The steady state flamelet equations had independently been derived by Peters (1980) and by Kuznetsov (1982). Buriko et al. (1994) compare SLFM model predictions of mean concentrations of fuel, CO, OH, NO, and flame temperature with experimental data from jet diffusion flames. They find a good agreement in the high temperature flame region but also note some limitations. They see these in the postflame regions where the scalar dissipation rate becomes small. They also argue that unsteady effects due to fluctuations of the scalar dissipation rate are important at extinction and that unsteady relaxation should be accounted for in the far downstream region of the jet.

Whether transient effects are important in modeling steady turbulent jet diffusion flames has also been discussed by Pitsch et al. (1998) who use the unsteady flamelet model with a Lagrangian time to account for rapid changes in flow direction. In addition, they introduce a diffusion time

$$t_\chi = \frac{(\Delta Z)^2}{\tilde{\chi}_{st}}, \tag{3.161}$$

which is the time needed to exchange mass and energy over a distance ΔZ in mixture fraction space. Here ΔZ is typically the flame thickness $(\Delta Z)_F$ in mixture fraction space. If this time is short compared to the Lagrangian time, the flamelet is able to follow changes in the scalar dissipation rate rapidly and the unsteady term in the flamelet equations can be neglected. This is found by Pitsch et al. (1998) to be valid in the region up to 30 diameters from the nozzle. Further downstream the conditional scalar dissipation rate decreases rapidly and therefore t_χ becomes larger than the Lagrangian time t. In this region the unsteady term must be retained to correctly predict the slow NO_x formation process, while the main combustion reactions are already very close to chemical equilibrium.

Figure 3.19 shows a diagram that illustrates this behavior. The conditional scalar dissipation rate is plotted as a function of the Lagrangian time or, in an unsteady situation, as a function of the real time t. Also plotted as a solid line is the conditional scalar dissipation rate as a function of t_χ according to (3.161). The region on the lower l.h.s. of the solid line corresponds to the regime of unsteady flamelets while that on the upper r.h.s. marks the steady flamelet regime. In the jet diffusion flame studied by Pitsch and Peters (1998a) and in the flames considered by Buriko et al. (1994) the conditional scalar dissipation rate decreases more rapidly than t^{-1} (which is the time dependence in (3.161)), thereby moving from the steady to the unsteady flamelet regime. If, however, the scalar dissipation rate was maintained at a constant value during an unsteady calculation as in the DNS by Swaminathan and Bilger (1999), one would start with an unsteady flamelet calculation at the initial conditions to reach the steady flamelet regime at late times.

222 3. Nonpremixed turbulent combustion

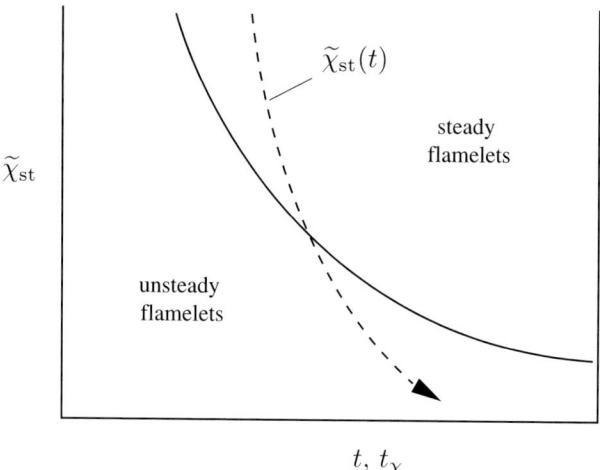

Figure 3.19. Schematic diagram of the steady versus the unsteady flamelet regimes. The solid curve represents (3.161). The dashed line illustrates the more rapid decrease of $\tilde{\chi}_{st}$ with residence time in a turbulent jet flame.

3.14 Predictions of Reactive Scalar Fields and Pollutant Formation in Turbulent Jet Diffusion Flames

In this section predictions by flamelet and pdf models will be discussed. The first attempt to use the steady laminar flamelet model to predict mean scalars in turbulent jet diffusion flames is due to Liew et al. (1984) who used the presumed shape pdf approach for Z and χ and a flamelet library. Later on Rogg et al. (1986) extended the model to account for partial premixing and Haworth et al. (1988) accounted for time-dependent flame structures by an "equivalent strain model." Fairweather et al. (1991) used a single profile to predict essentially the fluid-dynamical structure of a reacting jet in a cross flow measured by Birch et al. (1989). A library with different values of the strain rate ($a = 100/s, 200/s, 500/s$) had been generated to prescribe the instantaneous thermochemical structure of the flame. Since the mean density was found to be rather insensitive to the strain rate used, a value of $a = 100/s$ was finally chosen.

Flamelet models have also been used to predict soot formation and oxidation in jet diffusion flames. The steady laminar flamelet model has been applied to model soot formation in laminar ethylene–air diffusion flames by Balthasar et al. (1996). The chemical model consisted of 581 chemical reactions among 71 species. The flamelet solutions were stored in a library that was parameterized in terms of mixture fraction and the scalar dissipation rate. At every grid point within the two-dimensional flame the reactive scalars were interpolated from the flamelet libraries. For the soot mass fraction, however, a 2D transport equation was solved, where the chemical source term was also taken from the flamelet library. The calculated soot volume fractions compare favorably with experimental data of Gomez et al. (1987). The Stretched Laminar Flamelet Model has even been applied by Lentini and Puri (1995) to turbulent chloromethane–air jet flames using a library of seven steady state flamelets with scalar dissipation rates defined at the maximum temperature ranging from $\chi_{max} = 1.91/s$ to $\chi_{max} = 29.35/s$, the latter corresponding to extinction. More recent calculations using steady flamelet models are those in Ferreira (1996), Oevermann (1997) (discussed earlier), and in the dissertation by Heyl (2000) (cf. also Heyl and Bockhorn, 1998).

Heyl (2000) validates the laminar flamelet model first on the laminar axisymmetric diffusion flame by Feese and Turns (1998). He compares the experimental results with a full numerical solution of the governing axisymmetric equations and with two flamelet models. The first flamelet model contained a flamelet library of seven flamelets with scalar dissipation rates ranging from $\chi_{st} = 1/s$ to extinction, while the library of the second flamelet model contained 13 flamelets ranging from $\chi_{st} = 0.003/s$ to extinction. The NO_x emission index was 3.5 for the experiments, 3.2 for the full numerical solution, 3.3 for the second flamelet model, but only 1.7 for the first flamelet model (cf. Heyl and Bockhorn, 2000). This shows that even steady flamelet models are able to predict NO formation very well, provided a flamelet library containing flamelets with sufficiently small scalar dissipation rates is used. This conclusion was reconfirmed by Heyl (2000) for the diluted laminar methane–air diffusion flame of Smooke et al. (1996) and, as far as the temperature and H_2O is concerned, for the turbulent H_2–CO–N_2–air diffusion flame by Correa and Gulati (1992). Good agreement for the major species and the temperature was also obtained by comparing the steady flamelet model with the experimental data set of the so-called H_3 flame documented by Tacke et al. (1998) (cf. also Pfuderer et al., 1996). It was however less satisfactory (50% overprediction) for NO in the turbulent H_2 flame.

Pitsch et al. (2000) have presented a calculation of soot volume fractions in turbulent diffusion flames based on the unsteady flamelet model using a detailed reaction mechanism compiled by Mauss (1997). Starting from gas phase chemistry for the formation of the first aromatic rings, the mechanism describes the formation, growth, and oxidation of soot particles. Flamelet Equations for the statistical moments of the particle size distribution were also derived. The importance of gas and soot particle radiation are investigated and the model is validated using experimental temperature profiles of Kent and Honnery (1987). When soot volume fractions from these measurements were compared with the model predictions, Pitsch et al. (2000) found that differential diffusion effects should be considered in predictions of soot formation.

The same conclusion is drawn by Dederichs et al. (1999). Rather than solving the unsteady flamelet equations they constructed libraries for the source terms of soot from steady flamelet equations (cf. also Bai et al., 1998), using the detailed mechanism by Mauss (1997). They compare three different models: a differential diffusion model for all species, one that allows differential diffusion only for H and H_2, and finally a model where the Lewis numbers for all species are set equal to unity. Dederichs et al. (1999) found that the second model gives the best results, when compared with experimental data.

Masri and Pope (1990) were apparently the first to explore the potential of using the joint velocity mixture fraction pdf transport equation to calculate the velocity, turbulence, and mixture fraction fields in the piloted turbulent jet flames of Masri and Bilger (1986). They used a flamelet library from counterflow diffusion flames of methane with strain rates varying from $a = 1/s$ to $a = 360/s$. Chen et al. (1990) used the pdf transport equation model for the reactive scalars only, while the flow field was described by second-order moment closure. The chemistry is presented by a three-step reduced mechanism validated for CO–H_2–N_2–air counterflow diffusion flames. Comparisons were made with the turbulent jet diffusion flames of Masri and Dibble (1988). At high jet velocities, the model predicts bimodal joint statistics among the scalars, indicating frequent occurence of local flame extinction, while the experimental data show no extinction. The effects of the pilot flame or the neglect of differential diffusion were proposed as plausible reasons for this discrepancy.

Taing et al. (1993) based their calculations on the joint velocity-composition pdf transport equation and used a three-step reduced mechanism for the H_2–CO_2 fuel–air mixture of the piloted turbulent jet diffusion flame of Masri et al. (1992). They generated a five-dimensional look-up table for computationally efficient calculations using between 30,000 and 50,000 Monte-Carlo particles. For the lower and intermediate velocity flames the temperature and the mass fraction of stable species agree reasonably well with the experimental data. Larger discrepancies occur at high velocities when the flames are close to blow-off and finite rate chemical kinetic effects become very significant.

The piloted methane–air flames of Masri, Bilger, and Dibble documented in three consecutive papers (cf. Masri et al., 1988a,b,c) also have been subject of transported pdf simulations. Jones and Kakhi (1998) base their calculations on the joint composition pdf and solve the flow field on an Eulerian grid either with a k–ε model or a Reynolds stress model. They investigate both the lower velocity L flame and the higher velocity B flame, the latter showing local extinction events. A global hydrocarbon reaction scheme was employed to represent the chemistry. A major objective of the paper was a comparison between two mixing models: the Interchange by Exchange with the Mean (IEM) model and the

3.14 Predictions of reactive scalar fields and pollutant formation

Coalescence–Dispersion (C–D) model (cf. Section 1.10). They find that for the B flame, where local extinction was observed experimentally, the predictions with IEM showed extinction with no subsequent relighting, whereas stable burning was obtained with the C–D model. They discuss the foundations of both mixing models and point out the deficiency of using a single turbulent time scale for mixing. They conclude that a model would be needed that accounts for the fine scale mixing processes in the presence of reaction.

In contrast, Saxena and Pope (1998) use the joint velocity-frequency-composition transported pdf method to calculate the L flame of Masri, Dibble, and Bilger. Their calculations include detailed chemistry and are achieved by the novel In Situ Adaptive Tabulation (ISAT) procedure. They had tried to use the IEM model but found that the flame blew out. With the Euclidian Minimum Spanning Tree (EMST) based model this problem could be solved. They compare their results with previous calculations (Masri and Pope, 1990) where chemistry had been represented by a single flamelet profile. Although the temperature now is largely overpredicted (while it was fairly well predicted before), a considerable improvement is obtained for the minor species CO and H_2.

Turbulent combustion models have also been used to predict NO_x formation in turbulent diffusion flames. This is a problem of great practical importance, but because of the many physical aspects involved, it is also a very demanding test for any combustion model. A very knowledgeable review on the various aspects of the problem has been given by Turns (1995).

A global scaling law for NO production in turbulent jet flames has been derived by Peters and Donnerhack (1981) assuming equilibrium combustion chemistry and a thin NO reaction zone around the maximum temperature in mixture fraction space. An asymptotic solution for the mean turbulent NO production rate can be obtained by realizing that in the expression

$$\bar{\omega}_{NO} = \bar{\rho}\tilde{S}_{NO} = \bar{\rho} \int_0^1 S_{NO}(Z)\tilde{P}(Z)\,dZ \tag{3.162}$$

the function $S_{NO}(Z)$ has a very strong peak in the vicinity of the maximum temperature but decreases very rapidly to both sides. This is shown in Figure 3.20 for the case of a hydrogen flame. The NO reaction rate acts nearly like a δ-function inside the integral in (3.162). It has been shown by Peters (1978) and Janicka and Peters (1982) that an asymptotic expansion of the reaction rate around the maximum temperature leads to

$$\bar{\omega}_{NO} = \bar{\rho}\tilde{P}(Z_b)\varepsilon S_{NO}(Z_b), \tag{3.163}$$

where Z_b is the mixture fraction at the maximum temperature T_b and $S_{NO}(Z_b)$ is the maximum reaction rate. The quantity ε represents the reaction zone thickness

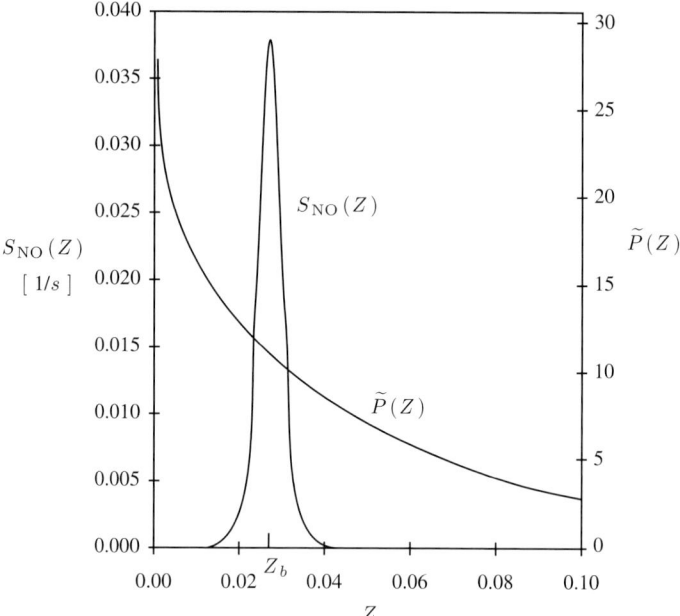

Figure 3.20. NO reaction rate and pdf for a hydrogen–air mixture. From Peters and Donnerhack (1981). (Reprinted with permission by The Combustion Institue.)

of NO production in mixture fraction space. That quantity was derived from the asymptotic theory as

$$\varepsilon = \left(\frac{-2RT_b^2}{Z_b^2 E_{NO}(d^2T/dZ^2)_{T_b}} \right)^{1/2}. \quad (3.164)$$

Here E_{NO} is the activation energy of the NO production rate. Finally, Peters and Donnerhack (1981) predicted the NO emission index EINO, which represents the total mass flow rate of NO produced per mass flow rate of fuel, as being proportional to

$$\text{EINO} \sim S_{NO}(Z_b)\varepsilon \left(\frac{L}{d} \right)^3 \frac{d}{u_0}. \quad (3.165)$$

Here L is the flame length, d the nozzle diameter, and u_0 the jet exit velocity. The normalized flame length L/d is constant for momentum dominated jets but scales with the Froude number $Fr = u_0^2/(gd)$ as $L/d \sim Fr^{1/5}$ for buoyancy dominated jets as shown in Figure 3.12. This explains, for instance, the $Fr^{3/5}$ dependence of the emission index found in the buoyancy dominated propane jet diffusion flames of Røkke et al. (1992). These data are reproduced in Figure 3.21 together with the prediction of the NO_x emission index (expressed here in

3.14 Predictions of reactive scalar fields and pollutant formation

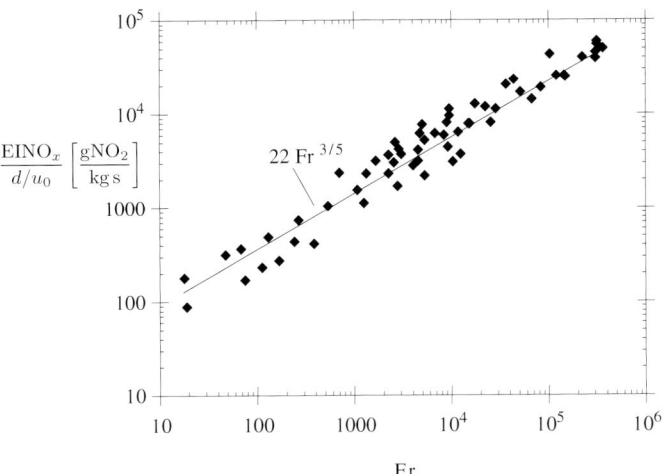

Figure 3.21. Emission index versus Froude number for buoyant jet diffusion flames of propane in air. Experimental data and prediction from Røkke et al. (1992). (Reprinted with permission by The Combustion Institue.)

terms of NO_2):

$$\frac{\text{EINO}_x}{d/u_0} = 22 Fr^{3/5} \left[\frac{\text{gNO}_2}{\text{kg fuel s}}\right]. \qquad (3.166)$$

It is interesting to note that by taking the values $S_{NO}(Z_b) = 10.8 \times 10^{-3}$/s and $\varepsilon = 0.109$ for propane from Peters and Donnerhack (1981) and using (3.113) in the buoyancy dominated limit one calculates a factor of 27.2 rather than 22 in (3.166).

Since Peters and Donnerhack (1981) had assumed equilibrium combustion chemistry, the second derivative of the temperature in (3.164) was calculated from an equilibrium temperature profile like the one shown in Figure 3.2. Therefore ε was tabulated as a constant for each fuel. If the quantity $S_{NO}(Z_b)d/u_0$ is interpreted as a Damköhler number the rescaled emission index from (3.165) is

$$\frac{\text{EINO}}{(L/d)^3} \sim \varepsilon Da. \qquad (3.167)$$

Turns (1995) has called this the leading order scaling. He argues that this scaling law must be modified to account for changes of the mean NO formation rate due to flame radiation, strain, and nonequilibrium chemistry. In hydrocarbon flames one must consider not only the dominating thermal NO formation path, but also additional chemical pathways, such as the prompt and the nitrous-oxide mechanism, as in Røkke et al. (1992).

Particularly challenging sets of experimental data are those by Chen and Driscoll (1990) and Driscoll et al. (1992) for diluted hydrogen flames. In contrast to the leading order scaling, these data show a square root dependence of the rescaled emission index on the Damköhler number. This $Da^{1/2}$ dependence had been reproduced by Chen and Kollmann (1992) using the transported pdf approach and by Smith et al. (1992) with the CMC method. An explanation for this scaling may be found by using the steady state flamelet equation for the second derivative of temperature in (3.164) rather than the equilibrium profile. This may be written as

$$\frac{d^2 T}{dZ^2} \sim \frac{\omega_T}{\chi}. \qquad (3.168)$$

Here the term on the r.h.s., evaluated at and divided by the maximum temperature, may also be interpreted as a Damköhler number. This becomes evident if one realizes that in a turbulent jet diffusion flame χ scales with u_0/d. Inserting this into (3.164) the quantity ε becomes proportional to $Da^{-1/2}$. This finally leads with (3.167) to

$$\frac{\text{EINO}}{(L/d)^3} \sim Da^{1/2}. \qquad (3.169)$$

This scaling law indicates that the experimentally observed $(d/u_0)^{1/2}$ dependence of the rescaled NO emission index is a residence time effect, modified by the temperature sensitivity of the NO reaction rate, on which the asymptotic theory by Peters and Donnerhack (1981) was built.

This result contrasts with the conclusions of Chen et al. (1995) who incorporated five reduced mechanisms for hydrogen into a transported pdf model to predict the NO emission indices measured by several authors. By increasing the number of global steps from one to five they observed a change from a linear to a square root dependence of the NO emission index on the Damköhler number.

Sanders et al. (1997) have reexamined steady state flamelet modeling using the two-variable presumed shape pdf model for the mixture fractions and either the scalar dissipation rate or the strain rate as second variable. Their study revealed that only the formulation using the scalar dissipation rate as the second variable was able to predict the $Da^{1/2}$ dependence of the data of Driscoll et al. (1992). This is in agreement with the results of Ferreira (1996) referenced above. In addition Sanders et al. (1997) examined whether there is a difference between using a lognormal pdf of χ_{st} with a variance of unity and a delta-function pdf and found that both assumptions gave similar results. Their predictions improved with increasing Damköhler number and their results also suggest that $Le_i = 1$ is the best choice for these hydrogen flames.

For a turbulent jet flame with a fuel mixture of 31% methane and 69% hydrogen Chen and Chang (1996) performed a detailed comparison between steady state flamelet and pdf modeling. They found that radiative heat loss becomes increasingly important for NO predictions further downstream in the flame. This is in agreement with the comparison of time scales by Pitsch et al. (1998) who found that radiation is too slow to be effective as far as the combustion reactions are concerned but that it affects NO levels considerably.

Recently Barlow et al. (1999) have isolated the effects of radiation in comparing predictions using the pdf transport equation model and the Conditional First Moment Closure model with experiments in undiluted and helium-diluted hydrogen flames. For the undiluted flames it was found that uncertainties in the Planck absorption coefficients of optically thin radiation models amounted to changes of the predicted NO_x emission index that were comparable to changes between the two turbulent combustion models. For the helium-diluted flame the effect of radiation was reduced to a point where comparisons between the combustion models could be made. Predictions using appropriate radiation models were within 30% of the NO measurements.

3.15 Combustion Modeling of Gas Turbines, Burners, and Direct Injection Diesel Engines

We will not report on results using the Eddy Break-Up Model or the Eddy Dissipation Model for these applications but focus on the more advanced flamelet and pdf models.

The recent review on combustion in industrial gas turbines by Correa (1998) reports, among other matters, on the state of CFD methods to simulate combustion in these engines. The pdf transport equation model and flamelet models are the most advanced combustion models being used for that purpose. Tolpadi et al. (1997) report on calculations using a Lagrangian Monte-Carlo pdf method for composition and an Eulerian method for the flow field. The model was first validated on the Raman data by Correa and Gulati (1992). Calculations for an aircraft engine combustor were based on the fast chemistry assumption. Computational time requirements are discussed and are compared with those needed by the presumed shape pdf approach.

The Eulerian Particle Flamelet Model was used by Barths et al. (1998) to simulate the pollutant formation in a gas turbine combustor. The turbulent flow and mixture fraction field in the combustor were steady and had been calculated in advance. Several different unsteady flamelets were attached to fictitious marker particles that were generated on the fuel-rich side of the mean stoichiometric mixture. The probability of finding these marker particles further downstream was then calculated by solving a convective–diffusive equation with gradient transport. In contrast to the Monte-Carlo method, where the Lagrangian motion of individual particles is calculated, the numerical solution was based entirely on Eulerian equations. The initial profiles of the flamelets are calculated using the scalar dissipation rate in the region of initialization. The flamelet time evolved from the time at which the particles were released. During that time they encountered different values of the scalar dissipation rate depending on the location within the flow field. Different flamelet histories were calculated with the time-dependent scalar dissipation obtained from a weighted average over the region where the particles were dispersed. By integration over the residence time that the particles needed to reach a particular location in space, and using the presumed shape pdf approach at that location, the weighted sums of all reactive scalars including NO_x and soot could be calculated. The residence time effect proved to be particularly important with respect to NO_x: The residence time of the particles within the high temperature region was fairly short, resulting in lower NO_x emissions than had been calculated using steady flamelets. The NO_x emission index was predicted as 29.3g NO_2/kg fuel, whereas 26g NO_2/kg fuel had been measured.

The Eulerian Particle Flamelet Model was also used by Coelho and Peters (2000) to predict combustion and pollution formation in a low-flowrate burner operating in the MILD combustion mode (cf. de Joannon et al., 1999). This combustion mode had previously been refered to as flameless oxidation by Wünning and Wünning (1997). It has also been reviewed by Katsuki and Hasegawa (1998). This new and very promising technology uses strong dilution and exhaust gas recirculation combined with strong air-preheating. Plessing et al. (1998a) have performed simultaneous laser-optical measurements of temperature and OH in the flameless oxidation burner described by Wünning and Wünning (1997). Evaluating local events they could demonstrate that because of the small local heat release the temperature difference between the highly preheated unburnt mixture and the burnt gas was so small that instead of the S-shaped curve shown in Figure 1.1 in Chapter 1, a monotonic dependence of the temperature on the Damköhler number existed. As a consequence ignition and extinction phenomena cannot occur. This explains the extremely silent operation of the burner. Plessing et al. (1998a) concluded that the flameless oxidation mode of combustion could be described by well-stirred reactors. It is tempting to conclude that this would rule out the application of flamelet models.

In spite of this expectation, the use of unsteady flamelets simulations by Coelho and Peters (2000) shows promising results. Two different codes were used to predict the velocity and the mixture fraction fields: the FLUENT code and an in-house code. Both codes provided similar results, which were in good

3.15 Combustion modeling of gas turbines, burners

agreement with velocity measurements. Residence times of real TiO_2 particles were also measured and predicted showing discrepancies close to the inlet but good agreement further downstream. There is also a fair agreement, given the remaining uncertainties in the kinetics, between the NO_x measurements and predictions in the exhaust gas: Measurements showed 4 ppm, while the predictions were at 15 ppm. This low value seems to indicate that despite the long residence times of fluid particles in the burner the generation of large scalar dissipation rates in regions of relatively high temperatures prevented excessive NO_x formation.

Unsteady flamelet calculations have also been used in a series of papers to predict auto-ignition, combustion, and the emissions of NO_x and soot from a direct-injection (DI) diesel engine. Exhaust gas emission data from a Volkswagen 1.9 liter engine were available to validate the predictions. In the first in this series of papers, by Pitsch et al. (1996b), *n*-heptane was used as a fuel for both the experiments and the simulations. The unsteady flamelet calculations were based on a detailed kinetic mechanism comprising 530 reactions. This mechanism describes the low temperature auto-ignition; combustion; NO_x formation including thermal, prompt, nitrous oxide, and reburn contributions; and soot formation and oxidation. The low temperature chemistry was based on that by Chevalier et al. (1990), enhanced and revised by later kinetic data from Baulch et al. (1992). For NO_x chemistry the mechanism of 106 reactions proposed by Hewson and Bollig (1996) was used. Soot precursor chemistry and the growth of small polycyclic aromatic hydrocarbons (PAHs) are included in the mechanism for gaseous species up to the fourth aromatic ring. This part of the mechanism follows the work by Mauss (1998) and Mauss and Bockhorn (1995) and uses data from Frenklach and Warnatz (1987) and Miller and Melius (1992). The formation, growth, and oxidation of soot particles is based on the method of statistical moments, further developed by Frenklach and Wang (1990).

The unsteady flamelet calculations were performed using a separate code that solves the system of parabolic equations in mixture fraction space quite efficiently. The interaction with the computational fluid dynamics (CFD) code is shown schematically in Figure 3.22. The concept is called RIF (Representative Interactive Flamelets). The CFD code (the KIVA code in this case) solves the three-dimensional Navier–Stokes equations, those for $\tilde{k}$ and $\tilde{\varepsilon}$, and the equations for $\tilde{Z}$, $\widetilde{Z''^2}$, and $\tilde{h}$, Equations (3.74)–(3.77), (3.79), and (3.83), respectively. From (3.79) and (3.155) the conditional mean scalar dissipation rate $\tilde{\chi}_z$ is available at each grid point of the CFD calculation. Since only the temporal evolution of a single flamelet was calculated in the paper by Pitsch et al. (1996b) an average was taken over all values of $\tilde{\chi}$ in the cylinder at each time step. The

232   3. Nonpremixed turbulent combustion

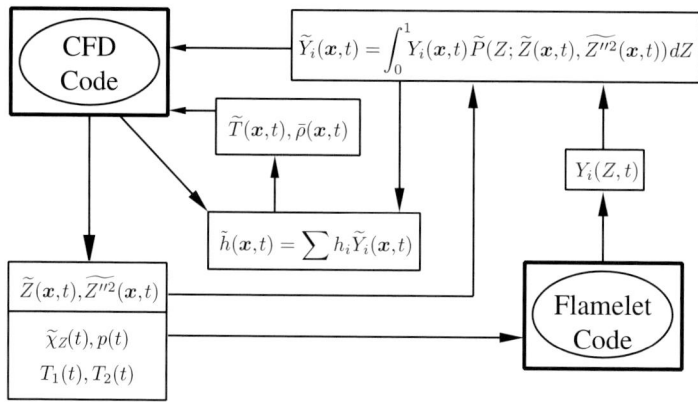

Figure 3.22. Code structure of the Representative Interactive Flamelet concept.

conditional scalar dissipation rate $\tilde{\chi}_Z$ obtained from (3.155) is then used in the flamelet equations. This profile is fed together with the pressure p, the fuel temperature $T_1(t)$, and the oxidizer temperature $T_2(t)$ into the flamelet code. This is shown by the arrow on the bottom of Figure 3.22. The time evolutions of the temperatures on the fuel and oxidizer sides of the flamelet are calculated from the evolution of the pressure using the law of adiabatic compression.

With these input parameters the unsteady evolution of the flamelet profiles $Y_i(Z, t)$ and $T(Z, t)$ are calculated. Using the presumed shape pdf approach the Favre average values of the mass fractions $\tilde{Y}_i(x, t)$ are calculated for each grid point in the cylinder. This is shown in the upper right corner of Figure 3.22. To evaluate the presumed shape pdf at each grid point, the local values of $\tilde{Z}(x, t)$ and $\widetilde{Z''^2}(x, t)$, calculated in the CFD code, are needed. The mean temperature $\tilde{T}(x, t)$ is not calculated from the flamelet solutions, but from the definition of the enthalpy, applied here to the mean values. The mean enthalpy $\tilde{h}(x, t)$ obtained from the CFD code is nonuniform, since heat losses to the wall are taken into account. This results in lower mean temperatures in the vicinity of the walls. This procedure only roughly approximates the combustion processes occurring close to the walls and needs to be improved, in particular with respect to soot formation. Finally, with mean mass fractions, the mean temperature, and the mean density available (the latter is calculated from the thermal equation of state) at each grid point in the CFD code the interaction cycle for one time step is computed and the next time step in the CFD calculation can be performed.

The only adjustable parameter that determines ignition, combustion, and pollutant formation in these calculations is the constant c_χ in the model (3.79) for the scalar dissipation rate. It has been set equal to 2.0 in the equation for the variance $\widetilde{Z''^2}$, but for the use in the flamelet code, $\tilde{\chi}_Z$ has been multiplied by an

3.15 Combustion modeling of gas turbines, burners

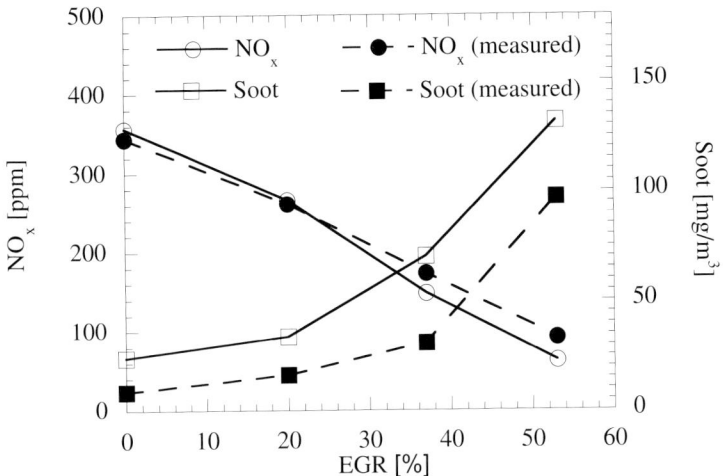

Figure 3.23. Comparison of measured and predicted emissions of NO$_x$ and soot in a direct injection diesel engine as a function of exhaust gas recirculation (EGR). (Reprinted with permission from SAE paper 962057 © 1996 Society of Automotive Engineers Inc.)

additional factor 3.0. It is not entirely clear why this artificial enhancement of scalar transport in mixture fraction space is needed to predict the experimentally observed ignition delay times correctly. A phenomenological justification has been given by Pitsch et al. (1995) who argued that unsteady fluctuations of the scalar dissipation are likely to suppress ignition. This effect can be modeled by artificially increasing the value of $\tilde{\chi}_{st}$. Another reason for needing to increase the scalar dissipation rate may be that transport in directions other than normal to the mixture fraction iso-surfaces is modeled by this enhancement of scalar transport. An evaluation of the Karlovitz number at ignition in simulations of diesel engine combustion shows that it is around $Ka=4$, thereby exceeding the limit $Ka=1$ estimated for flamelet ignition at the end of Section 3.12.

The main result of the paper by Pitsch et al. (1996b) is shown in Figure 3.23 where the emissions of soot and NO$_x$ are plotted against the exhaust gas recirculation (EGR) rate. This figure illustrates the trade-off between NO$_x$ and soot emissions: With increasing exhaust gas recirculation rates the NO$_x$ emissions decrease as desired but soot emissions increase. The agreement between predictions and measurements is quite good, which is particularly surprising for the soot predictions. As far as NO$_x$ is concerned it is noteworthy that the agreement would not be as good if only thermal NO$_x$ had been considered. Test calculations have shown that in that case the decrease of NO$_x$ with increasing EGR would be much steeper. This indicates that the additional reaction paths in the prompt, nitrous, and reburn mechanisms need to be taken into account if trends are to be well predicted.

The use of a single unsteady flamelet for the entire combustion chamber leads to a simultaneous ignition of a large partially premixed part of the mixture in the vicinity of the spray. To predict ignition locations along the spray in a pressure chamber, Wan et al. (1997) used up to ten different flamelet histories, each representing a disk-shaped volume at increasing distances from the nozzle. Experimental data for diesel spray ignition were available from the literature. As far as ignition is concerned n-heptane, having a cetane number of 56, is thought to represent commercial diesel fuel with cetane numbers around 50 sufficiently well. Calculations with 1,4,6, and 10 flamelets were performed. Although the ignition delay time changed by only 10% when four flamelets instead of one were used, the effect on ignition location was more pronounced: At least six flamelets were needed to predict the ignition location along the spray axis within 20% accuracy. It was found that ignition takes place close to the nozzle if the injected mass and duration of injection are small, whereas it occurs close to the tip if the injected mass and the injection duration are large. This is in agreement with experimental observation.

Multiple flamelets were also used by Barths et al. (1998) in a simulation of a n-decane fueled DI diesel engine using the Eulerian Particle Flamelet Model described above. Up to twenty different flamelets were initialized at the nozzle to account for different and temporally varying scalar dissipation rates along the time history of the flamelet. An Eulerian equation for the probability of finding marker particles was solved for each flamelet. Since the problem was inherently time dependent, no integration over residence times was needed as in the steady flow applications mentioned above. Problems were encountered with the prediction of soot at part load: At the end of the cycle the pdf of mixture fraction showed vanishingly small values in the mixture fraction range corresponding to rich mixture. Since all flamelets predicted soot on the rich side only and a complete burnout of soot at the surface of the stoichiometric mixture, there was little overlap between the pdf profile and the soot volume fraction profile in mixture fraction space. The soot volume fraction calculated by using (3.87) was therefore very small, much smaller than in the emission measurements for part load. For full load the agreement with measurements was satisfactory.

In a recent paper Barths et al. (1999) have performed simulations with multiple flamelets using a two-component surrogate fuel. This is a 70%/30% mixture of n-decane and α-methylnaphthalene, which is also known as the IDEA fuel. Engine tests with the 1.9l Volkswagen engine were performed to compare the surrogate fuel with diesel fuel in terms of exhaust gas emissions. The agreement was found to be satisfactory. An kinetic mechanism of 557 elementary reactions was derived for the surrogate fuel. For a better prediction of the ignition process up to nine different flamelet histories were used. The time evolution of the scalar dissipation rate is shown in Figure 3.24. Injection starts at $8°$ crank angle (CA)

3.16 Concluding remarks

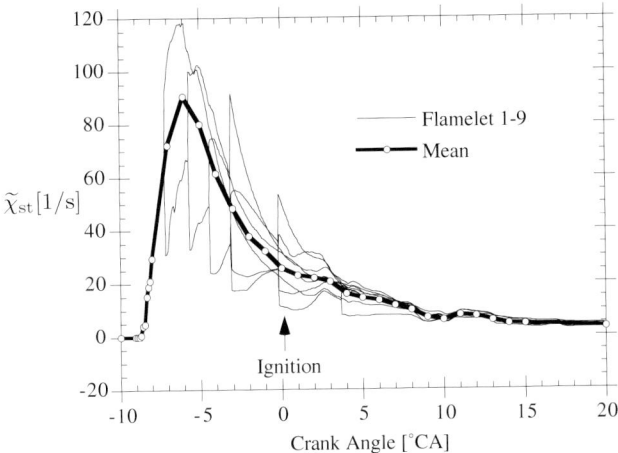

Figure 3.24. Time history of the scalar dissipation rates of up to nine flamelets in a CFD simulation of a direct injection diesel engine. Ignition starts with the flamelet that is the first to have a low scalar dissipation rate. By Barths et al. (1999). Reprinted with permission by editions technip.

before top dead center (TDC). Owing to the turbulence generated by the spray the scalar dissipation rate increases at first. At about 6° CA before TDC a maximum of the mean scalar dissipation rate is reached. From thereon it decreases because turbulence production by the spray decreases. The calculation starts with a single flamelet. At about 7° CA before TDC the first flamelet is split into two such that one of them represents the larger and the other the lower values of the scalar dissipation rate. This splitting continues, sometimes into three flamelets at a time. At TDC, one of the flamelets has a scalar dissipation rate low enough to ignite. The other flamelets ignite successively, each one contributing their part to the total heat release. The pressure curve plotted over the crank angle is much smoother and in better agreement with experimental data than in cases where only a single flamelet is used. The simulations using the IDEA surrogate fuel predict full load emissions of NO_x and soot obtained with diesel fuel very well. The use of multiple flamelets accounts for those ignition events that are triggered by small values of χ_{st}. Therefore it approximates the effect that would be modeled by the use of a distribution function $P(\chi_{st}; \boldsymbol{x}, t)$ in determining mean values of the reaction scalars, as in (3.143).

3.16 Concluding Remarks

This chapter has shown that relatively high standards have been reached in both flamelet and transported pdf methods to model nonpremixed turbulent combustion. There also seems to be a certain convergence of ideas and not so much difference in the results that are obtained by these two modeling approaches. There is agreement that for a model to be predictive in a large variety of cases, it should include

- a reliable chemistry model, either based on elementary or on validated reduced kinetic mechanisms;
- a reliable molecular transport model that accounts for differential diffusion in regions where this is necessary; and
- a radiation model, if NO_x emissions are to be predicted.

The success of any prediction strongly depends on the way the flow field and turbulence are modeled and on the way the combustion model is implemented into a numerical code. There are, for instance, contradictory statements on whether steady or unsteady flamelets need to be employed. Unsteadiness is built into the pdf transport equation model and into the unsteady flamelet model for the calculations that have been reported. With pdf transport equation models there is a debate on the best choice among the various mixing models that are available, while with the flamelet model choices are to be made between single and multiple flamelets and on the modeling of the scalar dissipation rate.

To substantiate these statements the reader is refered to the proceedings of the Third and the Fourth International Workshop on Measurement and Computation of Turbulent Nonpremixed Flames (cf. Barlow et al., 1999). Many different groups have participated in these workshops, trying to validate their favorite combustion model with a known set of experimental data. Comparisons are made by the organizers between the measurements and the different predictions. In comparing all these different calculations with the experimental data and among each other, one cannot conclude that any of the combustion models gives the best results for all cases. The differences between the models seem to result from certain approximations that have been introduced and perhaps from the way the models have been implemented.

The relative success of combustion models for nonpremixed combustion as compared to those for premixed combustion is certainly due to the controlling role of conserved scalar mixing. Even formation of pollutants such as NO_x and soot depends very much on the mixture fraction pdf, although residence time also plays a crucial role. Combustion models that are based on reliable conserved scalar mixing and that also model residence time effects adequately are certainly a good choice for predictions of nonpremixed turbulent combustion.

4

Partially premixed turbulent combustion

4.1 Introduction

In terms of mixing, there are two extremes: 1. premixed combustion where fuel and oxidizer are completely mixed prior to their entering the combustion chamber and 2. nonpremixed combustion where fuel and oxidizer enter separately. We have treated these two cases in the preceding two chapters. In technical applications, however, the optimum often lies somewhere between the extremes, trying to profit from advantageous features of both while avoiding their adverse effects. If fuel and oxidizer enter separately, but partially mix by turbulence, combustion takes place in a stratified medium, once the mixture is ignited. Such a mode of combustion has traditionally been called partially premixed combustion.

Turbulent flame propagation in a stratified mixture occurs, for instance, in aircraft gas turbines. Liquid kerosene is fed by an air-blast injector into the gas turbine combustion chamber, where it is mixed with the compressed air. There is typically a pilot injector for idling and a main injector for part load and full load operations. When the main injector is started, an inhomogeneous ignitable mixture is formed at its inlet. When this comes into contact with the hot exhaust gases from the pilot injector, flame propagation takes place from the pilot burner to the main burner through a stratified mixture. This process is difficult to control and modeling of the mixing and of partially premixed combustion poses a main challenge to CFD simulations of gas turbine combustion.

Another example is partially premixed flame propagation in direct injection gasoline engines, where a spray of liquid gasoline is injected directly into the cylinder rather than into the intake manifold as in conventional homogeneous charge spark-ignition engines. The injection occurs early enough during the cycle to allow for vaporization of the fuel and its partial mixing with the intake air. The key problem of this technology is generating an ignitable mixture at the spark plug at the right time. Once a flame kernel is initiated and has grown, a turbulent flame propagates through a stratified mixture. The advantage of the technology lies in the improved efficiency at part load, but the increased HC and soot emissions are disadvantgeous.

With respect to diesel engines, where auto-ignition is the controlling process prior to combustion, it was argued in Section 3.12 that ignition of an inhomogeneous mixture is followed by the propagation of two fronts in mixture fraction space. The burnout of the partially premixed part of the charge in a direct injection diesel engine therefore occurs essentially by flame propagation normal to the iso-mixture fraction surfaces. In Section 4.3 below we will present DNS results by Domingo and Vervisch (1996) that suggest that triple flame propagation through a stratified mixture is an important mechanism in the early phase of combustion in a diesel engine. This phase is often refered to as the premixed combustion phase.

Another important manifestation of partially premixed combustion is the lift-off and stabilization at the lift-off height in turbulent jet diffusion flames. This is the canonical problem for studying partially premixed combustion. It commonly occurs in flames of large industrial boilers, where lifting the flame base off the burner has the advantage of avoiding a thermal contact between the flame and the rim, which would lead to erosion of the burner material. The disadvantage of this flame stabilization technique is that lifted flames blow off more easily than attached flames and therefore must continuously be controlled.

4.2 Lifted Turbulent Jet Diffusion Flames

At a sufficiently low exit velocity u_0 a turbulent jet diffusion flame is attached to the nozzle. By increasing the exit velocity, the diffusion flame sheet will be stretched and finally disrupted. As a consequence the flame lifts off and stabilizes itself further downstream within the jet. The velocity at which this occurs is called the lift-off velocity and the distance between the burner rim and the base of the lifted flame is called the lift-off height H. A lifted jet diffusion flame is shown schematically in Figure 4.1. The lift-off height increases with increasing jet exit velocities but cannot exceed a critical value $H_{\max}$ at which the flame completely blows out. If one reduces the jet exit velocity of a lifted flame again, there is a critical velocity at which the flame will reattach to the burner. This velocity is different from the lift-off velocity and hence there is a hysteresis phenomenon between lift-off and reattachment. Since the problem of turbulence–chemistry interaction at the lift-off height of diffusion flames

4.2 Lifted turbulent jet diffusion flames

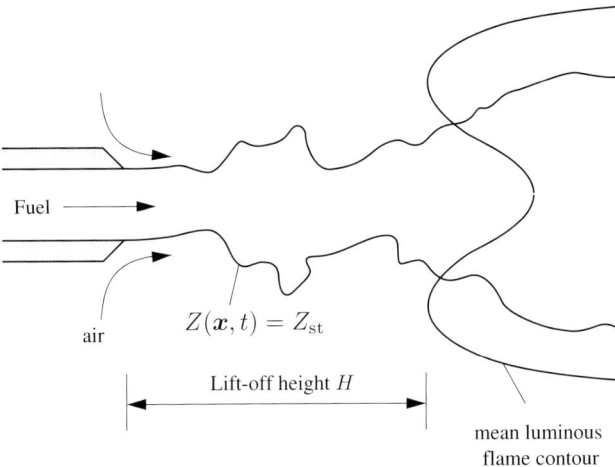

Figure 4.1. Schematic representation of a lifted jet diffusion flame.

is representative for partially premixed turbulent combustion in general and since explanations of the stabilization mechanism have been quite controversial, we will at first present an extended literature review. This will guide us later in developing an expression for the turbulent burning velocity for partially premixed combustion.

A first analysis of the mechanism responsible for lift-off of nonpremixed jet flames apparently is due to Wohl et al. (1949), presented at the Third Symposium on Combustion. They proposed that a diffusion flame will lift when the mean velocity gradient at the burner rim exceeds a certain critical value, and that it will stabilize itself at the position where the burning velocity is equal to the mean flow velocity. Later Vanquickenborne and van Tiggelen (1966) proposed, based on a large amount of measurements of lift-off heights in turbulent jets of methane, that fuel and air are fully premixed at the base of a lifted diffusion flame and that stabilization occurs at the position where the mean flow velocity at the contour of mean stoichiometric mixture is equal to the burning velocity of a stoichiometric premixed turbulent flame. Along the same line Eickhoff et al. (1984) reported on time-averaged temperature, concentration, and velocity measurements at positions just upstream and downstream of the stabilization region of lifted flames. They found that upstream of the stabilization region there is a substantial amount of air entrainment, which generates premixed conditions at the flame base. They concluded that stabilization is governed by premixed turbulent flame propagation. They found this to be supported by a more intense burnout at the flame base than at any position in an attached diffusion flame.

In contrast to the premixed flame propagation model Peters and Williams (1983) proposed that diffusion flamelet extinction is responsible for flame stabilization. They argued that there is insufficient residence time below the flame base to achieve spatial and temporal uniformity of the mixture. According to this model, since the governing parameter of flamelet extinction is the scalar dissipation rate χ_q, lift-off is expected to occur, if the probability of burning flamelets, calculated from the integral over the pdf of χ_{st} from $\chi_{st} = 0$ to $\chi_{st} = \chi_q$, is smaller than a certain threshold. Peters and Williams (1983) introduce notions from percolation theory and make use of the analogy between the threshold behavior of flame propagation at the flame base and the threshold of electrical conductivity that is found when holes are randomly punched into electrically conducting sheets. This analogy was later used by Masri and Bilger (1986) to classify piloted jet diffusion flames as electrically intermittent, when the electrical current between the nozzle and a simple wire placed downstream within the flame starts to show spikes of nonconductivity.

Whether quenching of diffusion flamelets is the relevant mechanism explaining flame stabilization at the lift-off height must be questioned, as will be shown below. There is little doubt, however, that diffusion flame quenching is responsible for the lift-off of an initially attached flame. Scholefield and Garside (1949) had already observed that lift-off of a turbulent flame is due to local extinction at the point where transition from laminar to turbulent flow occurs in the jet. In fact, as reported by Takeno and Kotani (1975) it is possible to have blowout of the main turbulent flame, while retaining a stable residual laminar flame attached at the rim.

Local extinction and the subsequent lift-off of diffusion flames has also been observed by Takahashi and Goss (1992). They emphasized the role of high-frequency jet-fluid vortices from the core region of the jet. The presence of a quasi-laminar diffusion flame near the nozzle inlet and the influence of the density ratio was more recently confirmed by a Laser-Induced Fluorescence (LIF) imaging performed by Clemens and Paul (1995), who showed that the effect of heat release is to laminarize the local turbulence and that extinction occurs when large three-dimensional vortical structures appear further downstream. Similar conclusions about local and occasional flame extinction at the transition point of jet diffusion flames were drawn by Yamashita et al. (1996) from their two-dimensional unsteady simulations of turbulent jet diffusion flames. They found occasional extinction when large values of the local scalar dissipation rate, exceeding a certain threshold, produce a rupture of the reaction zone. Hence, it seems evident that the initial lift-off of the turbulent flame results primarily from local extinction of the attached quasi-laminar diffusion flame.

4.2 Lifted turbulent jet diffusion flames

Janicka and Peters (1982) had applied the laminar flamelet concept for diffusion flames to the lift-off problem and derived a model for the conditional scalar dissipation rate by using a pdf transport equation. They compared the results of their predictions for the conditional and the unconditional scalar dissipation rates with data from the dissertation of Horch (1978) and found good agreement for small lift-off heights but discrepancies for large ones. They attributed those discrepancies to the fact that in turbulent shear layers large coherent structures are formed that lead to instantaneous peaks of the scalar dissipation rate, and they argued that local extinction is likely to occur during these events. They concluded that it is not the mean scalar dissipation rate that controls local extinction but rather some isolated events.

More recent measurements by Everest et al. (1996), who examined images of strained flammable layers at the base of a lifted turbulent jet diffusion flame, confirm the occurrence of strong instantaneous peaks of the scalar dissipation rate. They found local values of χ that exceed the predicted extinction value by a factor of 60. They attribute this to the thinning of the scalar layer by extensive strain similar to the predictions by Ashurst and Williams (1990) who had studied the effect of a vortex on a diffusion flame.

The classical set of experimental data on lift-off heights is by Kalghatgi (1984). He investigated jets of hydrogen, propane, methane, and ethylene for a large range of exit velocities and used nozzle diameters ranging between 1 and 10 mm. He found that the lift-off height was essentially a linear function of the jet exit velocity u_0; it is nearly independent of the nozzle diameter and inversely proportional to the square of the maximum laminar burning velocity. He explained these results by the premixed flame stabilization model of Vanquickenborne and van Tiggelen. He also calculated the burning velocity ratio and found it to be proportional to the square root of the turbulent Reynolds number:

$$\frac{s_T}{s_L} = b_4 \left(\frac{v'\ell}{\nu}\right)^{1/2}. \tag{4.1}$$

This corresponds to Damköhler's scaling for small scale turbulence, with the important difference that the constant of proportionality b_4 depends on the fuel and is significantly smaller than the value $b_3 = 1.0$ used in (2.191). A similar analysis using some additional data on lift-off heights has recently been presented by Burgess and Lawn (1999).

Donnerhack and Peters (1984) have measured lift-off heights of methane flames for various dilutions, nozzle diameters, and exit velocities. Normalized lift-off heights H/d of undiluted methane flames are plotted as a function of the exit velocity u_0 in Figure 4.2. Except for the nozzle diameter of 3 mm

242   4. Partially premixed turbulent combustion

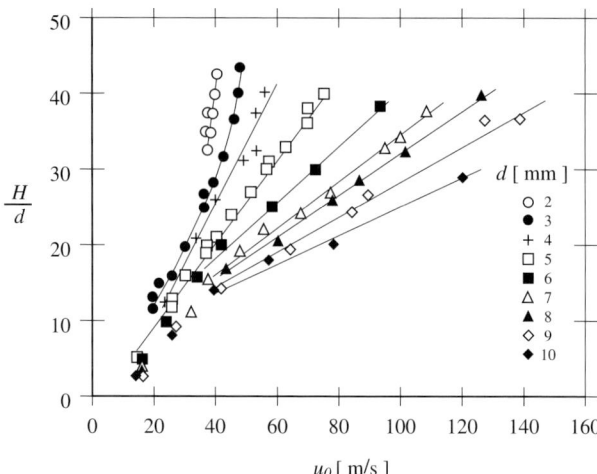

Figure 4.2. Nondimensional lift-off heights for methane–air flames. From Donnerhack and Peters (1984). (Reprinted with permission from Gordon and Breach.)

there is a linear dependence of H on u_0. The maximum value of H/d was found to be around 40. Beyond this value blow-off of the entire flame occurs. Another set of data on lift-off heights is due to Miake-Lye and Hammer (1988). Studying methane, ethylene, and natural gas flames that were partially diluted with air, they also reported a linear dependence of the lift-off height H on the jet exit velocity. For small dilutions they found that the lift-off height was inversely proportional to the fuel mass fraction in the mixture issuing from the nozzle, while there were systematic deviations from that dependence for larger dilutions. They measured the r.m.s. fluctuations of the lift-off height, which were around 8% of the mean, and attributed these fluctuations to large scale fluid-dynamic motions within the turbulent jet.

A very comprehensive data base on lift-off heights of partially premixed propane–air flames is provided by Røkke et al. (1994). Fuel mass fractions at the nozzle exit were varied between 0.15 and 1.0 and six different nozzle diameters, ranging from 3.2 to 29.5 mm, were used. The jet exit velocity was varied from 1 to 130 m/s. The paper is complemented by a table containing the original data base.

The idea that large scale turbulent structures control flame stabilization has also been discussed by Broadwell et al. (1984) who proposed a model where flame stabilization is due to a mechanism in which hot reaction products are supposed to be carried by large vortices to the edges of the jet from where they reenter into it together with fresh air to ignite the combustible mixture. In their view, both lift-off and blow-out occur when the reentrained products are mixed so rapidly with unburnt jet fluid that there is insufficient time to initiate the reaction before the temperature and the radical species drop below some critical value.

In his review of several of the previously mentioned theories Pitts (1988) assessed their capabilities to predict experimentally observed lift-off and blow-out behavior. His own concentration measurements have shown that at the base of the lifted flame the composition of the mixture is highly intermittent (i.e., for a short period of time one finds mixture within the flammability limits, while during the remaining time there is only either air or fuel present). This means that during long periods of time it is impossible for a premixed flame front to propagate against the upstream velocity. Hence, the assumption of premixed flame propagation is not fully in line with these experimental observations, because it does not include the observed intermittency. Pitts therefore concluded that the premixed flame stabilization model does not incorporate the true physical mechanism that controls partially premixed combustion at the lift-off height. The diffusion flamelet quenching model is not thought to be in line with the experiments either, because Pitts's as well as Eickhoff's measurements clearly show that a considerable amount of premixing has occured upstream of the flame base. With respect to the model of Broadwell et al. (1984), which incorporates the influence of large structures, Pitts argued that when the model is applied to lift-off it leads to predictions that are inconsistent with experimental findings. Hence, Pitts's main conclusion at that time was that none of the existing theories was satisfactory.

In recent years laser diagnostics have been used extensively to provide insight into the local phenomena occuring in the stabilization region. Schefer et al. (1988) reported on instantaneous planar images of CH_4 concentrations. Their comparison of a lifted flame with a nonreacting jet revealed a lower centerline decay rate of CH_4 in the lifted flame, steeper instantaneous gradients, and higher concentration fluctuations. Similar measurements on lifted jet flames and bluff-body flames by Namazian et al. (1988) included the CH radical, which was used as a marker for the reaction zone, and confirmed that this zone is located at the edge of the jet and does not penetrate into its core. Namazian et al. showed that large scale structures create CH_4 gradients steep enough to lead to extinction. Later on, Schefer et al. (1994a), performing planar imaging measurements of CH_4, CH, and temperature, found that the flame zone, identified as a region of high CH concentrations, forms in a narrow band along the outer boundary of the central fuel jet, and its location and curvature varies in response to the dynamics of the central fuel jet. Fluctuations of the flame lift-off heights and of the radial position of flame stabilization were comparable to the local jet width and the size of the large structures formed along the outer jet boundaries, respectively. Hence, at increased lift-off heights the flame zone was more convoluted owing to the larger turbulent length scales downstream. Their measurements showed that at the instantaneous point of flame stabilization, fuel and air were premixed within the flammability limits. Furthermore, the magnitude of the measured scalar dissipation rate was found to be considerably below the critical value for extinction of a corresponding laminar counterflow diffusion flame. Schefer et al. (1994a) therefore concluded that both three-dimensional flame propagation and large scale turbulence motion control flame stabilization.

In a further analysis Schefer et al. (1994b) emphasized the time-dependent character of the stabilization mechanism and reported on two-pulse planar LIF images of combined CH/CH_4 signals in a lifted turbulent CH_4 jet flame. The results showed that the lift-off height varied from shot to shot and in general was not the same on both sides of the jet. As seen before, the flame zone was located around large scale vortical structures formed along the fuel jet boundary. Interactions between those vortical structures and the flame may cause excessive stretch leading to local flame extinction. Once the flame is extinguished the instantaneous point of flame stabilization can be carried over significant distances downstream. The pictures also showed that during subsequent time steps the instantaneous flame front may move upstream, mainly into regions where the jet shows large scale structures.

One-dimensional profiles of temperature, mixture fraction, and major species upstream of the stabilization region in a lifted hydrogen diffusion flame have been reported by Brockhinke et al. (1966) using Rayleigh–Raman scattering along a line. Using two consecutive pulses of varying delay temporal information was obtained and, in specific cases, the local flame speed could be determined. Measurements of axial and radial components of the scalar dissipation rates showed that the maximum values in the high temperature region are well below 10/s. Brockhinke et al. conclude that the scalar dissipation rate is not the dominant parameter in the flame-stabilization process.

The stabilization region in undiluted and in several diluted lifted turbulent hydrogen flames has been investigated by Tacke et al. (1998) by a combined line-Raman–Rayleigh–OH-PLIF technique. The instantaneous point of flame stabilization was defined as the leading edge in the planar OH image. Several scalar fields, conditioned with respect to that point, were presented as contour plots. It is found from these conditional scalar fields that the flame stabilizes on the lean side of the stoichiometric mixture and that this location in mixture fraction space is independent of dilution.

Muñiz and Mungal (1997) used particle image velocimetry (PIV) to measure instantaneous two-dimensional velocity fields in the region of the lifted flame base of methane and ethylene flames with Reynolds numbers ranging from 3,800 to 22,000. They found that at the instantaneous leading edge of the flame front the local axial velocity does not exceed three times the maximum laminar burning velocity while the most probable local velocity was 1.5 times the laminar burning velocity. They compare these finding with the theoretical prediction of Ruetsch et al. (1995) that the overall velocity of a triple flame is up to 2.6 times the laminar burning velocity. They also found that, when a co-flow was added to the jet, the methane flames blew out immediately, if the co-flow velocity exceeded three times the laminar burning velocity. A detailed evaluation of velocities at the two stabilization regions on each side of the centerline showed that the local velocities at the flame front position are close to the laminar burning velocity. The velocity profiles resemble very much those of a laminar triple flame to be discussed in the next section. The paper by Muñiz and Mungal (1997) contains some beautiful color photographs of lifted turbulent diffusion flames. The conclusion of the paper is that the structure of the leading-edge flame front exhibits similarities to triple flames.

In a subsequent study Hasselbrink and Mungal (1998a) presented histograms of the flow velocity, the burning velocity, and the resulting flame propagation velocity. They also show histograms of strain rate normalized by the flame time, which demonstrate that the resulting Karlovitz numbers are one order of magnitude lower than the values required for premixed flame extinction. They conclude that the concept of premixed flame extinction by large scale vortical structures appears dubious. In another study, Hasselbrink and Mungal (1988b) combine measurements of visible CH chemiluminescence, planar OH-PLIF, and PIV to examine the stabilization region of methane jet flames in a transverse jet.

4.3 Triple Flames as a Key Element of Partially Premixed Combustion

When the laminar burning velocity is plotted as a function of mixture fraction there is a maximum burning velocity that lies close to the stoichiometric mixture (cf. Figure 4.7). If a range of the mixture fractions $Z_{min} < Z < Z_{max}$ exists in a partially premixed field, and the stoichiometric mixture fraction lies between these limits, flame propagation generates a flame structure called a triple flame. The leading edge of the flame, called the triple point, propagates along a surface that is in the vicinity of stoichiometric mixture. On the lean side of that surface there is a lean premixed flame branch and on the rich side there is a rich premixed flame branch, both propagating with a lower burning velocity. Behind the triple point, a diffusion flame develops into which unburnt intermediates such as H_2 and CO diffuse from the rich premixed flame branch and the left over oxygen diffuses from the lean premixed flame branch. A more detailed description of the different chemical and transport processes occuring in a triple flame structure

246   4. Partially premixed turbulent combustion

can be found in a paper by Echekki and Chen (1997) reporting on a DNS of methanol triple flames. In that paper effects of differential diffusion are also discussed.

Triple flames have attracted much interest because of the crucial role they may play in many partially mixed combustion situations. Liñán (1994) and Veyante et al. (1994b) have shown theoretically that in a laminar flow lifted flames are stabilized by a triple flame configuration. A model, proposed by Müller et al. (1994) assumes that the leading edge of a partially premixed turbulent flame is composed of an ensemble of laminar triple flamelets. This model was used to simulate and to explain the flame stabilization mechanism in lifted turbulent jet diffusion flames.

Numerical simulations by Domingo and Vervisch (1996), performed for auto-ignition in a nonuniform mixture, illustrate the dynamics of triple flame structures very convincingly. Two-dimensional simulations have been performed using a one-step reaction with a Zeldovich number $Ze = 8$ and a gas expansion parameter $\gamma = 0.8$. A result of their simulation showing reaction rate iso-contours is presented in Figure 4.3. To simulate the effect of compression

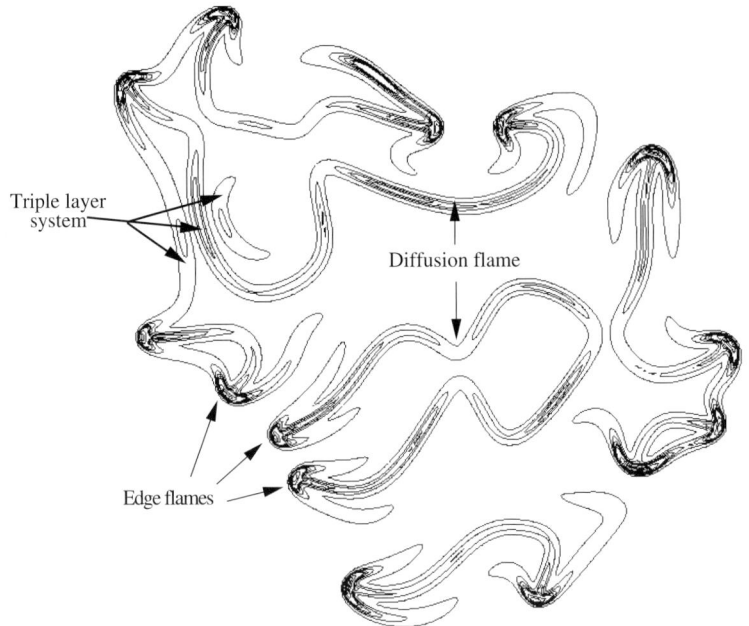

Figure 4.3. Triple flame propagation through a nonhomogeneous mixture field kindly provided by L. Vervisch. (Reprinted with permission by L. Vervisch.)

4.3 Triple flames as a key element

in a diesel engine, homogeneous source terms were added to all governing equations, thereby increasing the temperature to conditions close to auto-ignition. The different stages of ignition and flame propagation may be described as follows:

- Ignition occured in the vicinity of the stoichiometric line in regions where the scalar dissipation rate was low.
- Two premixed flame fronts containing lean and rich branches propagate in opposite directions along the stoichiometric lines. They have the shapes of arrowheads as shown in Figure 4.3.
- A diffusion flame develops on the stoichiometric line between the premixed flames. The tails of the premixed flames are lying nearly parallel to the diffusion flame and are propagating into the lean and rich mixture. As they depart from the diffusion flame they become weaker and finally disappear.
- When premixed flame fronts try to propagate into regions of very high scalar dissipation rates, local extinction is likely to occur.
- The dissipation rate and the heat release rate are inversely correlated. Maximum values of the dissipation rate correspond to minimum levels of heat release and vice versa.

It was also shown by Domingo and Vervisch (1996), that a triple flame is able to survive strong interaction with vortices by adjusting its structure to a new transient environment. Even if one premixed wing is extinguished the other may sustain combustion. Thus, triple flames are more robust than pure diffusion flames. This has been substantiated by Favier and Vervisch (1998) by a DNS study of triple flames that are squeezed through two counterrotating vortices.

An early paper demonstrating the structure of a laminar triple flame is due to Phillips (1965). Also, pictures of Savaş and Gollahalli (1986) show that the flame front consists of a highly curved premixed front with a lean and a rich premixed wing and a trailing diffusion flame. Kioni et al. (1993) have shown experimentally and theoretically that in a laminar mixing layer the flame stabilizes as a triple flame configuration.

The structure of triple flames, also called tribrachial flames or edge flames, has been analyzed theoretically by large activation energy asymptotics. Buckmaster and Matalon's (1988) analysis considers the case that the Lewis number of the fuel is larger than unity, which leads them to a solution where the fuel-rich branch points upstream rather than downstream. Dold (1989) calculates the shapes of the two premixed branches using the constant-density assumption based on the theory of slowly varying flames, valid for asymptotically small mixture fraction gradients. This work is significantly extended by Hartley and Dold (1991) and Dold et al. (1991) for cases of larger mixture fraction gradients. Treating the elliptic equation in the preheat zone by a Green's function approach, they derive a nonlinear integral equation for the shape of the premixed flames as a function of the mixture fraction gradient. While the analysis of Hartley and Dold (1991) treats the flame front as a free boundary, Ghosal and Vervisch (2000) approximate it by a parabolic profile and apply asymptotic expansions in parabolic-cylinder coordinates, thereby obtaining closed form solutions for the flame curvature as well as the velocity and temperature fields. The paper by Ghosal and Vervisch also extends the analysis to include the effect of small but finite gas expansion.

Buckmaster (1996) has proposed a one-dimensional model for edge flames. Buckmaster and Weber (1996) apply this model to the case where the edge flame is fixed by diffusive interaction with a burner rim. The stability of the steady state solution is analyzed and conditions for blow-off are derived. In a subsequent paper Buckmaster (1997) analyzes unbounded flames for Lewis numbers equal to and different from unity. For the latter case it is shown that propagating unbounded edge flames may not exist. For unity Lewis numbers an analytic expression for the velocity of the triple flame structure is derived, which may be positive or negative, depending on the Damköhler number. Buckmaster (1997) also points out the similarity to the strain rate correction used in the turbulent burning velocity expression by Müller et al. (1994).

Daou and Liñán (1998) formulate the problem of triple flame propagation as a distinguished asymptotic limit by relating the Zeldovich number Ze to the length scale ratio Λ/ℓ_F (cf. (2.246)), both assumed to be large. They define a parameter

$$\varepsilon = \frac{\ell_F}{\Lambda} Ze, \qquad (4.2)$$

which for the general case is of order unity. The analysis allows for small derivations from unity for the Lewis number of the fuel and for that of oxygen. By considering small values of the parameter ε Daou and Liñán find that if either Lewis number is larger than unity extinction of the triple flame occurs. This is in agreement with the anomalous Lewis number effects reported by Buckmaster and Matalon (1988).

A methane–air triple flame stabilized in a laminar round jet was studied experimentally and numerically by Plessing et al. (1998). A layered mixture was generated by three concentric flows, where methane diluted by N_2 was issued from the central nozzle, which was surrounded by a lean methane–air mixture, which again was surrounded by an air co-flow. These mixtures had interdiffused at the stabilization height to form a layered mixture.

The concentration fields of several stable species were obtained from line-Raman measurements. The temperature field, obtained from 2D Rayleigh measurements, is shown in the picture on the left of Figure 4.4, where it is compared with direct numerical simulations. The numerical simulations are based on a 10-step reduced mechanism for methane oxidation that had been derived from

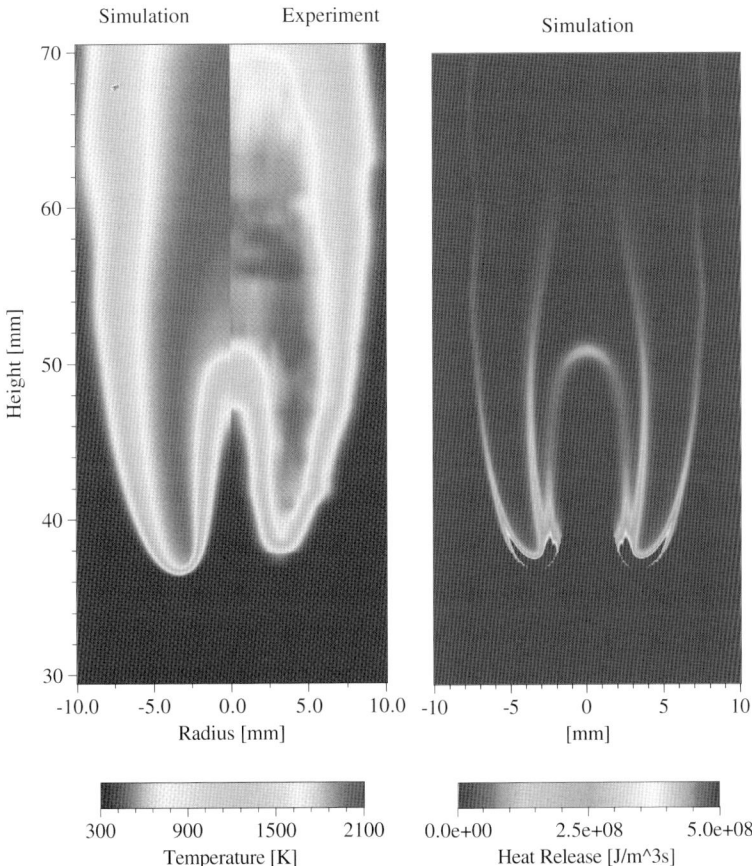

Figure 4.4 (see Color Plate II, Bottom). The laminar triple flame investigated by Plessing et al. (1998). The picture on the l.h.s. shows a comparison between the calculated (left) and the measured (right) temperature field. The picture on the r.h.s. shows the three branches on both sides of the axisymmetric flame in terms of calculated heat release rates. (Reprinted by permission of Elsevier Science from "An experimental and numerical study of a laminar triple flame," by T. Plessing, P. Terhoeven, M. S. Mansour and N. Peters, Combustion & Flame, Vol. 115, pp. 335–353, Copyright 1998 by The Combustion Institute.)

a 61-step elementary mechanism. A 7-step reduced mechanism that had been used in earlier studies for diffusion flames turned out to be insufficient because it did not predict the burning velocities of rich mixtures (beyond the rich flammability of planar flames) accurately enough. In the triple flame premixed burning of such rich mixtures is possible because it is supported by heat transfer from the diffusion flame. In the concentric triple flame the very rich mixture on the centerline is therefore able to burn as a premixed flame.

The numerically calculated heat release rate in the picture on the right of Figure 4.4 clearly shows the triple flame structure. It is interesting to note that the triple point lies on the lean premixed flame branch where also the strongest heat release occurs. This is in agreement with the findings of Tacke et al. (1998) in lifted turbulent flames. Since in both flames there was a central fuel rich jet with a higher flow velocity than in the air stream, the flame tends to stabilize in the lower velocity lean mixture.

Triple flames are always curved at the triple point, as has been analyzed in detail by Kioni et al. (1993) and by Ruetsch et al. (1995). The curving occurs simply because the local burning velocity, by which the premixed wings of the triple flame propagate in the direction normal to the premixed front, depends on the local mixture fraction. It decreases as one moves from the triple point downstream on the lean or the rich branch. The triple point therefore propagates the fastest while the rich and lean premixed flame branches stay behind. Since on these branches the flames are inclined, the gradient σ introduced by (2.61) increases. If the flow velocity u at the premixed flame front of a stationary triple flame was uniform, assuming that it was not modified by gas expansion, a simple kinematic balance like that by Damköhler (1940) for premixed flames states that

$$s_L(Z)\sigma(Z) = u. \tag{4.3}$$

Here the burning velocity $s_L(Z)$ is that at the local mixture fraction and the gradient $\sigma(Z)$ depends on Z as a consequence of this balance. Equation (4.3) indicates that, in the constant-density approximation, the decreasing burning velocity of the premixed branches in a triple flame is compensated by a larger flame surface area ratio.

In reality, gas expansion in the premixed flame front influences the flow field ahead of the flame by generating a normal velocity from the front into the unburnt mixture. Since the front is curved, this leads to a diverging flow field and a lower oncoming velocity directly ahead of the triple point. This is illustrated in Figure 4.5, taken from Plessing et al. (1998b), where the axial velocity through the triple point, obtained from PIV measurements, is compared with a numerical simulation. The oncoming velocity decreases from the upstream value of 90

4.4 Modeling turbulent flame propagation

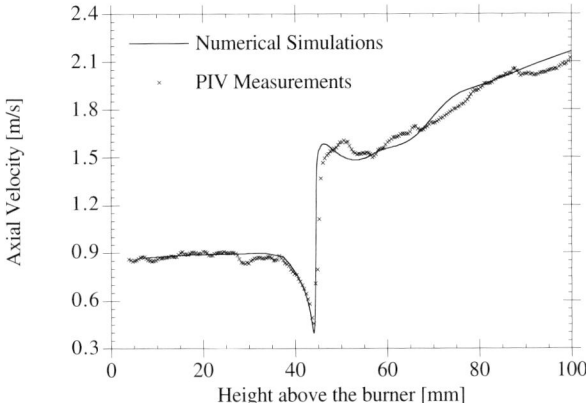

Figure 4.5. Axial velocity profiles at $r = 3.2$ mm through the triple point of the laminar triple flame investigated by Plessing et al. (1998). (Reprinted by permission of Elsevier Science from "An experimental and numerical study of a laminar triple flame," by T. Plessing, P. Terhoeven, M. S. Mansour and N. Peters, Combustion & Flame, Vol. 115, pp. 335–353, Copyright 1998 by The Combustion Institute.)

to 40 cm/s, the latter corresponding to the laminar burning velocity at the triple point.

Recently the propagation speed of unsteady tribrachial flames has been measured by Ko and Chung (1999) in laminar methane jets by high speed shadowgraphy. Propagation speeds vary between 65 and 90 cm/s and depend on curvature, strain, and mixture fraction gradient. When plotted in terms of these local quantities, the propagation speed becomes Reynolds number independent.

4.4 Modeling Turbulent Flame Propagation in Partially Premixed Systems

In view of the many different aspects that are potentially important for a physically correct description of flame stabilization at the lift-off height, a turbulent combustion model is not easily formulated. Bradley et al. (1990) proposed a flamelet model where combustion proceeds essentially as premixed turbulent flame propagation with the flammability limits imposed as a constraint. Differently from Bradley et al. (1994) discussed in Section 2.16, the mean heat rate was evaluated by integrating over the pdfs of both the progress variable and the mixture fraction. The lift-off height is determined as the location where a substantial increase of the mean heat release rate occurs. The predicted lift-off heights compare favorably with the experimental data of Kalghatgi (1984). Later on Bradley et al. (1998a, 1998b) improved this model by allowing for premixed flame extinction at both positive and negative strain rates. In addition, as in Bradley et al. (1994) a Reynolds stress model and second-order closure for the temperature are used. The model not only predicts the lift-off heights of Donnerhack and Peters (1984) but also the blow-off velocity as a function of nozzle diameter reported by Kalghatgi (1981), McCaffrey and Evans (1986), and Birch et al. (1988).

Sanders and Lamers (1994) formulated a model based on diffusion flamelet extinction. Instead of the scalar dissipation rate they used the strain rate imposed by Kolmogorov eddies as the parameter to describe flame stretch. A detailed discussion is given to motivate this choice. In addition, to model the influence of large scale structures, they use a pdf of spatial fluctuations to determine the mean reactive scalars. This model reproduces approximately the correct slope in the linear dependence of lift-off heights on exit velocities.

Two-dimensional DNS have been performed by Kaplan et al. (1994) to study the stabilization mechanism in lifted jet diffusion flames. The computations, using one-step chemistry, show that the flame is stabilized on a vortical Kelvin–Helmholtz structure at stoichiometry conditions. The simulations agree favorably with the turbulent burning velocities reproduced from Kalghatgi's (1984) correlation. In a subsequent paper from the same group, Montgomery et al. (1998) used the four-step reduced mechanism (1.92) for methane–air combustion and reproduced the previous predictions as well as Kalghatgi's experimental results for a 10 mm nozzle diameter. In addition, they predicted the effect of co-flow velocity on the lift-off height. They proposed an effective velocity, which is a linear combination of the jet velocity and the co-flow velocity, by which the lift-off data collapse into a single curve.

Three-dimensional direct numerical simulations of partially premixed combustion related to applications in stratified spark-ignition engines have been preformed by Hélie and Trouvé (1998). Studying different mixture fraction variances around a mean stoichiometric mixture they find that the overall reaction rate is progressively reduced by partial premixing as the mixture fraction variance increases compared to the fully premixed case.

Müller et al. (1994) have developed a model for partial premixed turbulent combustion that they validate with lift-off heights in turbulent jet diffusion flames. Their model is based on the two scalar fields $G(x, t)$ and $Z(x, t)$ and turbulent modeling is introduced for each of these two quantities. While the mixture fraction Z determines the local equivalence ratio and thereby the value of the laminar burning velocity as a function of mixture fraction, the scalar G determines the location of the premixed flame front. The features of the model are the following: For the flame propagation process a G-equation similar to

4.4 Modeling turbulent flame propagation

(2.151) is formulated, but now the laminar burning velocity is a function of the mixture fraction:

$$\rho \frac{\partial G}{\partial t} + \rho v \cdot \nabla G = (\rho s_{L,p}) - (\rho D)\kappa \sigma. \tag{4.4}$$

Here $s_{L,p} = s_L(Z)$ stands for the laminar burning velocity in a partially premixed system. Modeling this G-equation is the same as in Chapter 2. Therefore the turbulent counterpart of (4.4), corresponding to (2.151), is

$$\bar{\rho} \frac{\partial \tilde{G}}{\partial t} + \bar{\rho} \tilde{v} \cdot \nabla \tilde{G} = (\bar{\rho} s_{T,p}) |\nabla \tilde{G}| - \bar{\rho} D_t \tilde{\kappa} |\nabla \tilde{G}|. \tag{4.5}$$

In (4.5) the turbulent burning velocity $s_{T,p}$ for partially premixed combustion was introduced. Modeling this quantity for the purpose of calculating lifted flames, for example, requires a choice between the two mechanisms that have been discussed in Section 4.1 – premixed flame propagation or diffusion flamelet extinction. Müller et al. (1994) claim to have resolved that controversy by formulating a model for the turbulent burning velocity $s_{T,p}$ that contains three terms: a term modeling premixed flame propagation, a term accounting for partial premixing, and a flamelet quenching term. Their model has been applied to predict the upstream propagation within a partially premixed jet and the stabilization of the turbulent flame at the lift-off height. For the calculation of the unsteady upstream propagation process the model predicts that the premixed flame propagation term dominates. As the stabilization region is approached, however, the modeling of the flamelet quenching term controls the lift-off height.

Figure 4.6 shows as a result from the paper of Müller et al. (1994) the blowup of the stabilization region in a lifted turbulent jet flame of methane with a diameter $d = 4$ mm and a fuel exit velocity of 20 m/s. The shaded area indicates the burnt gas region $\bar{G} > G_0$ and the solid lines represent iso-contours of the mixture fraction. The expansion at the flame front deflects the streamlines and thereby the mixture fraction iso-contours at the flame base. Also shown is the point $\tilde{\chi} = \chi_q$ where the flame would stabilize according to the diffusion flamelet extinction model.

A closer inspection of the flamelet quenching term in the model of Müller et al. (1994) shows that the model for the scalar dissipation rate, which does not take mixture fraction fluctuations into account, makes this term proportional to the inverse of the Damköhler number. Therefore there is no fundamental difference between this functional dependence and the scaling of the burning velocity for premixed turbulent combustion shown in Figure 2.24. This suggests that while a dependence of $s_{T,p}$ on the mixture fraction and its fluctuations

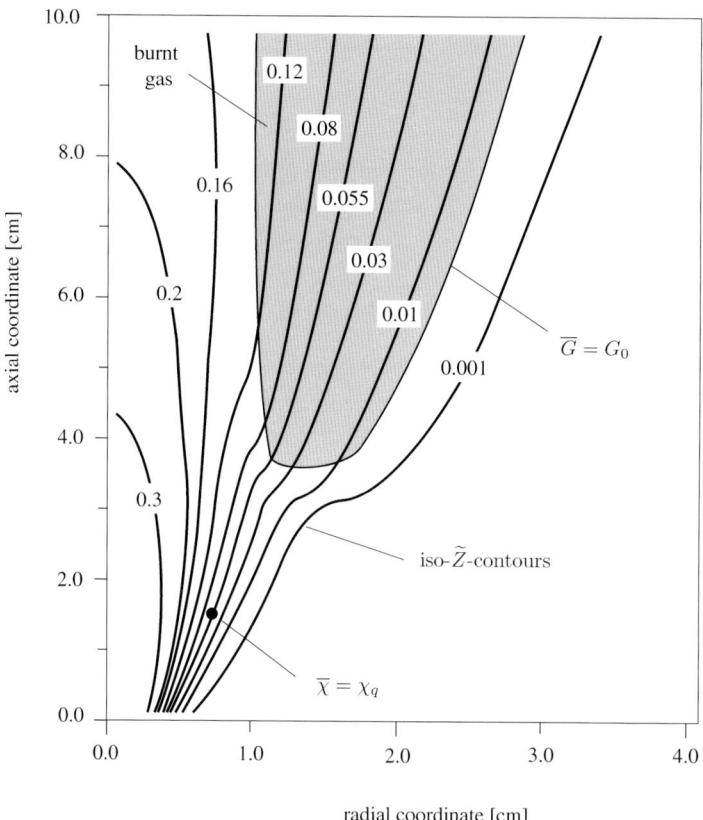

Figure 4.6. Stabilization region of a lifted turbulent jet diffusion flame. The flame stabilizes in the vicinity of the surface of mean stoichiometric mixture $\tilde{Z} = Z_{st} = 0.055$. The shaded region corresponds to the burnt gas ($\bar{G} > G_0$). From Müller et al. (1994). (Reprinted with permission by The Combustion Institute.)

remains important, diffusion flamelet extinction is not the mechanism that has been modeled by Müller et al. (1994).

Since that model has been successful in predicting lift-off heights in turbulent diffusion flames as well as in predicting partially premixed flame propagation in gas turbine combustion chambers using the TASC-Flow code (C. Müller, private communication), we will modify the formulation and base it entirely on the premixed flame propagation mechanism. In doing so we give due credit to Vanquickenborne and van Tiggelen (1966), Eickhoff et al. (1984), and Kalghatgi (1984), among many others. Since the simple balance in (4.3) shows that each part of a triple flame front, parameterized in terms of mixture fraction, contributes to the propagation velocity of the structure in nearly the same way,

Herrmann (2000) has suggested that we consider them separately and introduce a conditional turbulent burning velocity as

$$s_T(Z) = s_L(Z) + v' f\{Da(Z)\}, \tag{4.6}$$

where $Da(Z)$ is a conditional Damköhler number

$$Da(Z) = \frac{s_L(Z)\ell}{v'\ell_F(Z)} = \frac{s_L^2(Z)\ell}{v'D} \tag{4.7}$$

and $f\{Da\} = \Delta s/v'$ is the correlation given by (2.196). In the second equality in (4.7) following the definitions (2.5) and (2.6), $\ell_F(Z)$ has been replaced by $D/s_L(Z)$, where D has been assumed mixture fraction independent.

The mean turbulent burning velocity $s_{T,p}$ is then calculated from

$$(\bar{\rho}s_{T,p}) = \int_0^1 \rho(Z)s_T(Z)\tilde{P}(Z)\,dZ. \tag{4.8}$$

This is similar to (3.155) for the conditional scalar dissipation rate. In Equation (4.8) $\tilde{P}(Z)$ is obtained by using the presumed shape pdf approach described in Section 3.8. Since $s_T(Z)$ is defined with respect to the unburnt mixture, $\rho(Z)$ is to be evaluated there.

If partially premixed flame propagation is assumed to proceed by an ensemble of triple flamelets, the burning velocity $s_L(Z)$ should be the velocity normal to the premixed flame surface of a triple flame. This velocity has been evaluated by Plessing et al. (1998b), for instance, as a function of Z. When this was compared with the burning velocity of a planar premixed flame, both plotted as a function of the mixture fraction, a significant difference was noted. This difference is due to the multidimensional nature of triple flames, which not only introduces curvature and strain effects (cf. Ko and Chung, 1999) but also heat transfer between the trailing diffusion flame and the premixed wings.

4.5 Numerical Simulation of Lift-Off Heights in Turbulent Jet Flames

The turbulent burning velocity model described by (4.6)–(4.8) has been used by Chen (2000) (cf. also Chen et al., 2000) to calculate the stabilization at the lift-off height in turbulent jet flames. For this purpose the mean $\tilde{G}$-equation and those for $\tilde{Z}$ and $\widetilde{Z''^2}$ were implemented into the FLUENT code. The $\tilde{G}$-field was solved using the turbulent burning velocity $s_{T,p}$. After each time step the $\tilde{G}$-field outside of the flame surface $\tilde{G}(\mathbf{x}, t)$ was reinitialized as in the numerical example in Section 2.17. In addition, an equation for the mean total enthalpy $\tilde{h}$

was solved:

$$\frac{\partial \bar{\rho}\tilde{h}}{\partial t} + \nabla \cdot (\bar{\rho}\tilde{v}\tilde{h}) = \frac{\partial \bar{p}}{\partial t} + \tilde{v} \cdot \nabla \bar{p} + \nabla \cdot \left(\frac{\bar{\rho}v_t}{\Pr_t}\nabla \tilde{h}\right). \quad (4.9)$$

This equation replaces the original energy equation of the FLUENT code.

Combustion was initiated at a downstream position. Upstream unsteady flame propagation then takes place until the flame front stabilizes at the lift-off height. To describe the scalar fields a flame sheet model was adopted. This model does not resolve the premixed flame structure but replaces it by a jump at $G = G_0$. The dependence of the scalar fields on the mixture fraction, however, was taken into account by calculating the diffusion flamelet structure. There are two possible states for a diffusion flamelet: burning or nonburning. For the burning flamelets the mass fractions of the chemical species were determined by using a steady state flamelet library with the scalar dissipation rate as a parameter. If a computational cell is in the burnt gas region the enthalpy and mass fractions are calculated from

$$\tilde{Y}_{i,b}(\tilde{Z}, \widetilde{Z''^2}, \tilde{\chi}_{st}) = \int_0^1 Y_i(Z, \chi_{st}) \tilde{P}(Z)\, dZ. \quad (4.10)$$

Here $Y_i(Z, \chi_{st})$ is determined by interpolation from a library of burning flamelets, using the function dependence $\chi(Z)$ given by (3.47) and setting the scalar dissipation rate χ_{st} of the flamelets equal to the conditional mean scalar dissipation rate $\tilde{\chi}_{st}$ calculated from (3.157). In (4.10) a beta function pdf was used. The integration was performed in advance and values for $\tilde{Y}_{i,b}$ were tabulated as functions of $\tilde{Z}$, $\widetilde{Z''^2}$, and $\tilde{\chi}_{st}$.

If a computational cell is in the region of the unburnt mixture, all reactive species mass fractions are zero except those of fuel and oxygen. Those mass fractions, being linear in mixture fraction space, are evaluated from (3.7) and (3.8), respectively, using the local mean mixture fraction $\tilde{Z}$:

$$\tilde{Y}_{i,u} = Y_{i,u}(\tilde{Z}). \quad (4.11)$$

Finally, if the computational cell is located within the flame brush thickness, the weighted sum

$$\tilde{Y}_i = p_b \tilde{Y}_{i,b} + (1 - p_b)\tilde{Y}_{i,u} \quad (4.12)$$

is used. The probability p_b of burning is calculated from

$$p_b = \int_{G=G_0}^{\infty} \frac{1}{\sqrt{2\pi \widetilde{G''^2}}} \exp\left\{-\frac{(G - \tilde{G})^2}{2\widetilde{G''^2}}\right\} dG, \quad (4.13)$$

where it was assumed that the pdf of G is a Gaussian distribution. The conditional variance $\widetilde{G''^2}$ was assumed to have reached its steady state value equal to

4.5 Lift-off heights in turbulent jet flames

$(b_2\ell)^2$, which was calculated using the constants in Table 2.1. The temperature $\tilde{T}$ in the cell was calculated from the mean enthalpy by using the local mean mass fractions,

$$\sum_{i=1}^{n} \tilde{Y}_i h_i(\tilde{T}) = \tilde{h}(\boldsymbol{x}, t), \quad (4.14)$$

where the specific enthalpies are taken from NASA polynomials.

The quantities needed to calculate $s_{T,p}$ are plotted as a function of the mixture fraction in Figure 4.7. Rather than taking the laminar burning velocity from measurements or calculations of triple flames, which were not always available with a sufficient accuracy, calculated laminar burning velocities of one-dimensional planar flames were used. For methane $s_L(Z)$ was taken from the solid curve in Figure 2.2, extended to the rich side by the values plotted for the ten-step mechanism in Figure 5 of Plessing et al. (1998b). The conditional Damköhler number was evaluated with a diffusivity D of 7×10^{-5} m^2/s calculated from (2.6). The conditional Damköhler number $Da(Z)$ and the conditional turbulent burning velocity $s_T(Z)$ are plotted in Figure 4.7 as a function of mixture fraction using (4.6) and (4.7).

Using the formulation (4.6)–(4.8) for the turbulent burning velocity, Chen (2000) has calculated lift-off heights in turbulent methane–air diffusion flames

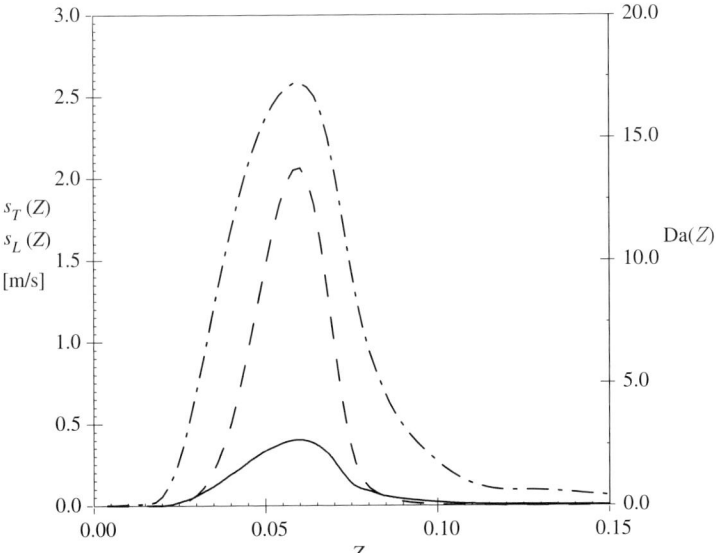

Figure 4.7. Quantities at the lift-off height. The laminar burning velocity (———), the conditional Damköhler number (- - - -), and the conditional turbulent burning velocity (— · — · — · —), evaluated for $v' = 1.4$ m/s and $\ell = 0.85$ mm as a function of the mixture fraction. (Reprinted with permission by M. Chen.)

258   4. Partially premixed turbulent combustion

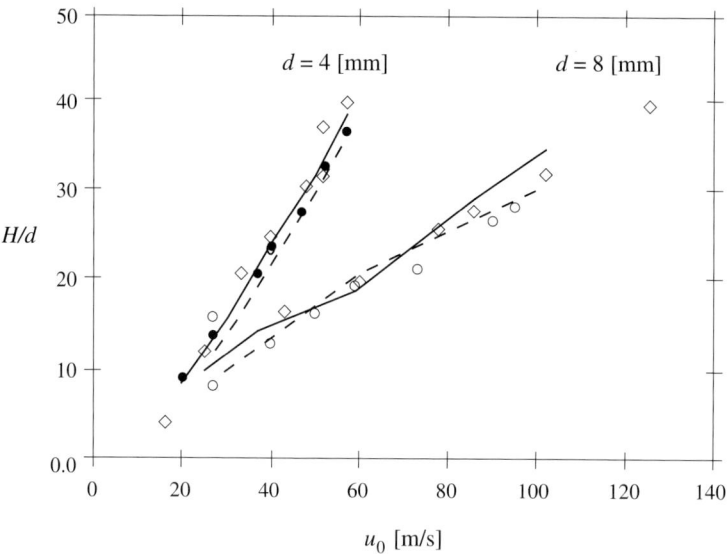

Figure 4.8. Normalized lift-off heights H/d of methane–air jet diffusion flames. Comparison of predictions by Müller et al. (1994) (———), and Chen (2000) (- - - -), with experimental data of Donnerhack and Peters (1984) ($\diamond$), Kalghatgi (1984)($\circ$) and Miake-Lye and Hammer (1988)($\bullet$). (Reprinted with permission by The Combustion Institute and by M. Chen.)

as a function of the jet exit velocity for the nozzle diameters $d = 4$ mm and $d = 8$ mm. These results are compared in Figure 4.8 with measurements of Donnerhack and Peters (1984), Kalghatgi (1984), and Miake-Lye and Hammer (1988) and with the previous calculations of Müller et al. (1994). We see that the newly calculated values using (4.6)–(4.8) agree well with the previous calculations of Müller et al. (1994) and with the experiments.

In conclusion, stabilization of a partially premixed flames at the lift-off height in turbulent jet diffusion flames can be well predicted by using the expression for the turbulent burning velocity (4.5), which is based on the premixed flame propagation model.

4.6 Scaling of the Lift-Off Height

The linear dependence of the lift-off height on the exit velocity in turbulent jet diffusion flames that has been found for all fuels and nozzle diameters by many authors requires a more fundamental explanation. We start from the kinematic balance

$$\tilde{v}(x, r) = \boldsymbol{n} s_{T,p} \qquad (4.15)$$

4.6 Scaling of the lift-off height

at the stabilization point in a two-dimensional steady turbulent jet and assume that stabilization occurs on the iso-surface of mean stoichiometric mixture. The mean velocity $\tilde{v}$ vector is assumed to point downstream and the flame normal vector n is assumed to point upstream and to be aligned with the mean stoichiometric iso-mixture fraction surface. An analysis of the approximate solutions (3.101)–(3.104) for the axial velocity and the mixture fraction field in a turbulent jet flame shows that for $Sc_t = 1$ both fields become similar, that is,

$$\frac{\tilde{u}(x,r)}{u_0} = \tilde{Z}(x,r). \tag{4.16}$$

Therefore, using this as a first approximation, the mean axial velocity along the surface of mean stoichiometric mixture is constant:

$$\tilde{u}(x, r_{st}) = u_0 Z_{st}. \tag{4.17}$$

Neglecting the contribution of the much smaller radial velocity component we obtain the balance (4.15) in flame normal direction:

$$s_{T,p} = \tilde{u}(x, r_{st}). \tag{4.18}$$

We now want to analyze the two limits of $\Delta s/v' = f\{Da\}$ in (2.196) separately, namely

$$\frac{s_T - s_L}{v'} = f\{Da\} = \begin{cases} b_1, & \text{for } Da \to \infty, \\ a_4 b_3^2 Da^{1/2}, & \text{for } Da \to 0. \end{cases} \tag{4.19}$$

In the limit $Da \to \infty$, integration of (4.8) over Z leads to

$$s_{T,p} = s_L(Z_{st})\tilde{P}(Z_{st})(\Delta Z)_{s_L} + b_1 v'. \tag{4.20}$$

Here the integral over the laminar burning velocity was replaced by

$$\int_0^1 s_L(Z)\tilde{P}(Z)\,dZ = s_L(Z_{st})\tilde{P}(Z_{st})(\Delta Z)_{s_L}, \tag{4.21}$$

which defines the width in mixture fraction space $(\Delta Z)_{s_L}$ over which the laminar burning velocity is significantly larger than zero.

Neglecting the first term on the r.h.s. in (4.20) as being small compared to the second term, we obtain in the limit $Da \to \infty$ for the turbulent burning velocity

$$s_{T,p} = b_1 v', \tag{4.22}$$

which is the same as in a homogeneous mixture. In the nonreacting jet upstream of the lift-off height, it may be assumed that the r.m.s. velocity fluctuation v' is approximately proportional to the mean velocity. At the surface of mean stoichiometric mixture this leads to

$$v' = b_5 \tilde{u}(x, r_{st}) = b_5 u_0 Z_{st}, \tag{4.23}$$

where b_5 is of the order of 0.1. Introducing this into (4.22) we obtain

$$s_{T,p} = b_1 b_5 \tilde{u}(x, r_{st}). \tag{4.24}$$

This leads to a burning velocity that is typically smaller than $\tilde{u}$. Comparing this with (4.18) we see that there is no balance possible between the flow velocity and the burning velocity such that the premixed turbulent will always been blown away by the flow in the limit $Da \to \infty$.

In the small Damköhler number limit, however, evaluation of (4.8) leads to

$$s_{T,p} = s_L(Z_{st}) \tilde{P}(Z_{st})(\Delta Z)_{s_L} \left[1 + \left(\frac{v'\ell}{D} \right)^{1/2} \right]. \tag{4.25}$$

If the first term in the square brackets is neglected and this is compared with (4.1), one obtains for $\nu = D$ Kalghatgi's result. It is interesting to note that the coefficient b_4 in (4.1) represents $\tilde{P}(Z_{st})(\Delta Z)_{s_L}$ and thereby the influence of partial premixing on the turbulent burning velocity. This explains why b_4 was found to be fuel dependent and smaller than unity, since $\tilde{P}(Z_{st})(\Delta Z)_{s_L}$ is typically smaller than unity.

If (4.25), with the first term in the square brackets neglected, and (4.23) are inserted into (4.18) one obtains

$$s_L(Z_{st}) \tilde{P}(Z_{st})(\Delta Z)_{s_L} b_5 \left(\frac{v'\ell}{D} \right)^{1/2} = v'. \tag{4.26}$$

This can be rearranged to show that the conditional Damköhler number at stoichiometric mixture is a constant at the lift-off height:

$$Da(Z_{st}) = \frac{s_L^2(Z_{st}) \ell}{v' D} = \text{const}. \tag{4.27}$$

Replacing now v' by $b_5 u_0 Z_{st}$ according to (4.23) and using the fact that the integral length scale in a round jet increases linearly with x,

$$\ell \sim x, \tag{4.28}$$

one obtains at the lift-off height $H = x$ the relation

$$H \sim u_0 Z_{st} \frac{D}{s_L^2(Z_{st})}. \tag{4.29}$$

This shows the linear dependence of the lift-off height on the exit velocity and also the dependence on the inverse of the square of the laminar burning velocity reported by Kalghatgi (1984). It also explains why the slope of the plots of H over u_0 in Kalghatgi (1984) and in Miake-Lye and Hammer (1988), for instance, depends on the fuel and on dilution.

4.7 Concluding Remarks

As a combination of the premixed and the nonpremixed mode of combustion, partially premixed combustion shows features of both. For the example of lifted turbulent diffusion flames, it was shown that, contrary to a previous theory to which the author has contributed, stabilization at the lift-off height occurs by premixed flame propagation and not by diffusion flamelet quenching. It is remarkable that the approximation (2.196) for the turbulent burning velocity that was validated only for homogeneous premixtures can be used without modifications or additional constants for the prediction of flame stabilization at the lift-off height in turbulent jet diffusion flames. Predicting the lift-off height is probably the most severe test for any model for partially premixed combustion, because turbulence is very intense owing to the relatively high shear in a jet. Since the premixed flame propagation model is found to be valid under those conditions, it certainly can be expected to be valid for partially premixed flame propagation in layered mixtures of lower turbulence intensity, such as in a direct injection gasoline engine.

Epilogue

Models of turbulent combustion require input from two areas:

- theory and modeling of turbulence and
- asymptotic flame theory.

In both of these areas mathematical procedures have been developed that are appropriate for the respective subject, but are quite incompatible with each other.

Turbulence modeling is a very powerful tool as long as it is recognized that it is strictly valid only in the large Reynolds number limit where all turbulent quantities become Reynolds number independent. As a consequence molecular quantities such as the viscosity ν and the diffusivity D must not appear in the modeled equations for any turbulent mean quantity. In the chapter on premixed turbulent combustion similar reasoning has been used to show that the laminar burning velocity s_L must disappear from the model in the limit of large values v'/s_L. The resulting equations for mean quantities derived in that chapter are only strictly valid in the limit of $v'/s_L \to \infty$. Since in practical applications v' and s_L are often of the same order of magnitude, more work is needed to extend the theory to finite values of v'/s_L.

The modeled equations for turbulent quantities used in this book are those for the mean and the variance of the scalar G, the equation for the flame surface area ratio $\bar{\sigma}$, and the equations for the mean and the variance of the mixture fraction Z. Molecular quantities do not appear in the equations for the means and the variances of G and Z. As far as the $\bar{\sigma}$-equation is concerned, the appearance of s_L^0 and D may at first seem to contradict the above statements. It should be recalled, however, that the $\bar{\sigma}$-equation is the sum of two equations: One was derived for the corrugated flamelets regime and contains only the laminar burning velocity, whereas the other was derived for the thin reaction zones regime and contains only the diffusivity. If the former is multiplied with s_L^0, it is seen that the quantity $s_L^0 \bar{\sigma}$ obeys an equation that contains no molecular quantities. If the latter is multiplied with $D\bar{\sigma}$, the same is true for the scalar dissipation $D\bar{\sigma}^2$.

Asymptotic flame theory is an equally powerful tool, as long as appropriate choices are made for the order of magnitude of the relevant terms. In asymptotic theory, the problem is not defined in advance, but the choice of the leading order terms defines the problem. By introducing a fast time variable in the flamelet equations, for instance, these equations are defined as being unsteady. Similarly, other quantities that would originally appear in higher order terms, such as strain and curvature, can be assumed to be large enough to appear in the leading order equation. Whether such an assumption is appropriate and useful depends on the problem to be solved.

Since the methodology in the two areas mentioned above is so different, only leading order results can be used when the input from both areas is to be combined. The basis of such a combination is *scale separation* between chemistry occuring on small scales and the turbulent scales in the inertial range. Because of the self-accelerating nature of combustion, the assumption of scale separation seems to be valid in nearly all practical combustion applications. Using the established methods in both areas mentioned above appropriately, one has a good chance of formulating consistent and predictive models for turbulent combustion.

Glossary

BML: Bray–Moss–Libby
CA: Crank Angle
C–D: Coalescence–Dispersion
CFD: Computational Fluid Dynamics
CFM: Coherent Flame Model
CFMC: Conditional First Moment Closure
CMC: Conditional Moment Closure
CSEM: Conserved Scalar Equilibrium Model
CSP: Computational Singular Perturbations
DNS: Direct Numerical Simulation
EBU: Eddy-Break-Up
ECFM: Extended Coherent Flame Model
EMST: Euclidean Minimum Spanning Tree
FDP: Filtered Density Function
IEM: Interaction by Exchange with the Mean
ILDM: Intrinsic Low Dimensional Manifold
ISAT: In Situ Adaptive Tabalation
LEPDF: Large Eddy Probability Density Function
LEM: Linear Eddy Model
LES: Large Eddy Simulation
l.h.s.: left hand side
LIF: Laser Induced Fluorescence
LMST: Linear Mean Square Estimation
pdf: probability density function
PIV: Particle Image Velocimetry
PLIF: Planar Laser Induced Fluorescence
ODT: One-Dimensional Turbulence

QSSA: Quasi-Steady-State Assumption
RANS: Reynolds Averaged Navier–Stokes
r.h.s.: right hand side
RIF: Representative Interactive Flamelet
r.m.s.: root mean square
SDRL: Strained Dissipation and Reaction Layer
SLFM: Strained Laminar Flamelet Model
TDC: Top Dead Center

Bibliography

Abdel-Gayed, R. G. and Bradley, D. (1977). Dependence of turbulent burning velocity on turbulent Reynolds number and ratio of laminar burning velocity to r.m.s. turbulent velocity, *Sixteenth Symposium (International) on Combustion*, pages 1725–1735, The Combustion Institute, Pittsburgh.

(1981). A two-eddy theory of premixed turbulent flame propagation, *Phil. Trans. Roy. Soc. Lond.* **301**, 1–25.

(1989). Combustion regimes and the straining of turbulent premixed flames, *Combust. Flame* **76**, 213–218.

Abdel-Gayed, R. G., Bradley, D., Hamid, M. N., and Lawes, M. (1984). Lewis number effects on turbulent burning velocity, *Twentieth Symposium (International) on Combustion*, pages 505–512, The Combustion Institute, Pittsburgh.

Abdel-Gayed, R. G., Bradley, D., and Lawes, M. (1987). Turbulent burning velocities: A general correlation in terms of strain rates, *Proc. Roy. Soc. Lond.* **A414**, 389–413.

Adalsteinsson, D. and Sethian, J. A. (1999). The fast construction of extension velocities in level-set methods, *J. Comput. Phys.* **148**, 2–22.

Aldredge, R. C., Vaezi, V., and Ronney, P. D. (1998). Premixed-flame propagation in turbulent Taylor–Couette flow, *Combust. Flame* **115**, 395–405.

Andrews, G. E., Bradley, D., and Lwakabamba, S. B. (1975). Turbulence and turbulent flame propagation – A critical appraisal, *Combust. Flame* **24**, 285–304.

Ashurst, W. T. (1994). Modelling turbulent flame propagation, *Twenty-Fifth Symposium (International) on Combustion*, pages 1075–1089, The Combustion Institute, Pittsburgh.

(1995). Turbulent flame motion in a pancake chamber via a Lagrangian, two-dimensional vortex dynamics simulation, *Combust. Sci. and Tech.* **109**, 227–253.

(1996). Flame propagation along a vortex: The baroclinic push, *Combust. Sci. and Tech.* **112**, 175–185.

(1997). Darrieus–Landau instability, growing cycloids and expanding flame acceleration, *Combust. Theory Modelling* **1**, 405–428.

Ashurst, W. T., Kerstein, A. R., Kerr, R. M., and Gibson, C. H. (1987a). Alignment of vorticity and scalar gradient with strain rate in simulated Navier–Stokes turbulence, *Phys. Fluids* **30**, 2343–2353.

Ashurst, W. T., Peters, N., and Smooke, M. D. (1987b). Numerical simulation of turbulent flame structure with non-unity Lewis number, *Combust. Sci. and Tech.* **53**, 339–375.

Ashurst, W. T. and Shepherd, I. G. (1997). Flame front curvature distributions in a turbulent premixed flame zone, *Combust. Sci. and Tech.* **124**, 115–144.

Ashurst, W. T. and Williams, F. A. (1990). Vortex modification of diffusion flamelets, *Twenty-Third Symposium (International) on Combustion*, pages 543–550, The Combustion Institute, Pittsburgh.

Aung, K. T., Hassan, M. I., and Faeth, G. M. (1998). Effects of pressure and nitrogen dilution on flame/stretch interactions of laminar premixed $H_2/O_2/N_2$ flames, *Combust. Flame* **112**, 1–15.

Bai, X.-S., Balthasar, M., Mauss, F., and Fuchs, L. (1998). Detailed soot modeling in turbulent jet diffusion flames, *Twenty-Seventh Symposium (International) on Combustion*, pages 1623–1630, The Combustion Institute, Pittsburgh.

Balthasar, M., Heyl, A., Mauss, F., Schmitt, F., and Bockhorn, H. (1996). Flamelet modeling of soot formation in laminar ethylene/air diffusion flames, *Twenty-Sixth Symposium (International) on Combustion*, pages 2369–2377, The Combustion Institute, Pittsburgh.

Baritaud, T. A., Duclos, J. M., and Fusco, A. (1996). Modeling turbulent combustion and pollutant formation in stratified charge SI engines, *Twenty-Sixth Symposium (International) on Combustion*, pages 2627–2635, The Combustion Institute, Pittsburgh.

Barlow, R. S. and Carter, C. D. (1994). Raman/Rayleigh/LIF measurements of nitric oxide formation in turbulent hydrogen jet flames, *Combust. Flame* **97**, 261–280.

Barlow, R. S. and Chen, J.-Y. (1992). On transient flamelets and their relationship to turbulent methane–air jet flames, *Twenty-Fourth Symposium (International) on Combustion*, pages 231–237, The Combustion Institute, Pittsburgh.

Barlow, R. S., Chen, J.-Y., Bilger, R. W., Hassel, E. P., Janicka, J., Masri, A., Peeters, T. W. J., Peters, N., and Pope, S. (1999). *Third and Fourth International Workshop on Measurement and Computation of Turbulent Nonpremixed Flames*, http://www.ca.sandia.gov/tdf/Workshop.html.

Barlow, R. S., Dibble, R. W., Chen, J.-Y., and Lucht, R. P. (1990). Effect of the Damköhler number on super-equilibrium OH concentration in turbulent nonpremixed jet flames, *Combust. Flame* **82**, 235–251.

Barlow, R. S. and Frank, J. H. (1998). Effects of turbulence on species mass fractions in methane/air jet flames, *Twenty-Seventh Symposium (International) on Combustion*, pages 1087–1095, The Combustion Institute, Pittsburgh.

Barlow, R. S., Smith, N. S. A., Chen, J.-Y., and Bilger, R. W. (1999). Nitric oxide formation in diluted hydrogen jet flames: Isolation of the effects of radiation and turbulence–chemistry submodels, *Combust. Flame* **117**, 4–31.

Barths, H., Antoni, C., and Peters, N. (1998). Three-dimensional simulation of pollutant formation in a DI-Diesel engine using multiple interactive flamelets, SAE paper 982459.

Barths, H., Peters, N., Brehm, N., Mack, A., Pfitzner, M., and Smiljanovski, V. (1998). Simulation of pollutant formation in a gas turbine combustor using unsteady flamelets, *Twenty-Seventh Symposium (International) on Combustion*, pages 1841–1847, The Combustion Institute, Pittsburgh.

Barths, H., Pitsch, H., and Peters, N. (1999). 3D simulation of DI Diesel combustion and pollutant formation using a two-component reference fuel, *Oil & Gas Sci. and Tech. Rev. IFP* **54**, 233–244.

Batchelor, G. K. (1952). The effect of homogeneous turbulence on material lines and surfaces, *Proc. Roy. Soc.* **A213**, 349–366.

Baulch, D. L., Cobos, C. J., Cox, R. A., Frank, P., Hayman, G., Just, Th., Kerr, J. A., Murrells, T., Pilling, M. J., Troe, J., Walker, R. W., and Warnatz, J. (1992).

Evaluated kinetic data for combustion modelling, *J. Phys. Chem. Ref. Data* **21**, 411–429.

(1994). Summary table of evaluated kinetic data for combustion modelling: Supplement 1, *Combust. Flame* **98**, 59–79.

Bédat, B. and Cheng, R. K. (1995). Experimental study of premixed flames in intense isotropic turbulence, *Combust. Flame* **100**, 485–494.

Bergmann, V., Meier, W., Wolff, D., and Stricker, W. (1998). Application of spontaneous Raman and Rayleigh scattering and 2D LIF for the characterization of a turbulent $CH_4/H_2/N_2$ jet diffusion flame, *Appl. Phys.* **B66**, 489–502.

Bilger, R. W. (1976). The structure of diffusion flames, *Combust. Sci. and Tech.* **13**, 155–170.

(1980). Turbulent flows with nonpremixed reactants, in P. A. Libby and F. A. Williams, editors, *Turbulent Reacting Flows*, pages 65–113, Springer-Verlag, Berlin.

(1988). The structure of turbulent nonpremixed flames, *Twenty-Second Symposium (International) on Combustion*, pages 475–488, The Combustion Institute, Pittsburgh.

(1993). Conditional moment closure for turbulent reacting flows, *Phys. Fluids* **A5**, 436–444.

Birch, A. D., Brown, D. R., Cook, D. K., and Hargrave, G. K. (1988). Flame stability in underexpanded natural gas jets, *Combust. Sci. and Tech.* **58**, 267–280.

Birch, A. D., Brown, D. R., Fairweather, M., and Hargrave, G. K. (1989). An experimental study of a turbulent natural gas jet in a cross-flow, *Combust. Sci. and Tech.* **66**, 217–232.

Bish, E. S. and Dahm, W. J.(1995). Strained dissipation and reaction layer analysis of nonequilibrium chemistry in turbulent reacting flows, *Combust. Flame* **100**, 457–464.

Boger, M., Veynante, D., Boughanem, H., and Trouvé, A. (1998). Direct numerical simulation analysis of flame surface density concept for large eddy simulation of turbulent premixed combustion, *Twenty-Seventh Symposium (International) on Combustion*, pages 917–925, The Combustion Institute, Pittsburgh.

Borghi, R. (1985). On the structure and morphology of turbulent premixed flames, in C. Casci, editor, *Recent Advances in the Aerospace Science*, pages 117–138, Plenum, New York.

Borghi, R. (1988). Turbulent combustion modeling, *Prog. Energy Combust. Sci.* **14**, 245–292.

Borghi, R. and Gonzalez, M. (1986). Application of Lagrangian models to turbulent combustion, *Combust. Flame* **63**, 239–250.

Boudier, P., Henriot, S., Poinsot, T., and Baritaud, T. A. (1992). A model for turbulent flame ignition and propagation in spark ignition engines, *Twenty-Fourth Symposium (International) on Combustion*, pages 503–510, The Combustion Institute, Pittsburgh.

Bowman, C. T., Hanson, R. K., Davidson, D. F., Gardiner, W. C., Jr., Lissianski, V., Smith, G. P., Golden, D. M., Frenklach, M., and Goldenberg, M. (1999). http://www.me.berkeley.edu/gri_mech/.

Bradley, D. (1992). How fast can we burn?, *Twenty-Fourth Symposium (International) on Combustion*, pages 247–262, The Combustion Institute, Pittsburgh.

Bradley, D., Gaskell, P. H., and Gu, X. J. (1994). Application of a Reynolds stress, stretched flamelet, mathematical model to computations of turbulent burning velocities and comparison with experiments, *Combust. Flame* **96**, 221–248.

(1996). Burning velocities, Markstein lengths and flame quenching for spherical methane–air flames: A computational study, *Combust. Flame* **104**, 176–198.

(1998a). The modeling of aerodynamic strain rate and flame curvature effects in premixed turbulent combustion, *Twenty-Seventh Symposium (International) on Combustion*, pages 849–856, The Combustion Institute, Pittsburgh.

(1998b). The mathematical modeling of liftoff and blowoff of turbulent non-premixed methane jet flames at high strain rates, *Twenty-Seventh Symposium (International) on Combustion*, pages 915–922, The Combustion Institute, Pittsburgh.

Bradley, D., Gaskell, P. H., and Lau, A. K. C. (1990). A mixedness-reactedness flamelet model for turbulent diffusion flames, *Twenty-Third Symposium (International) on Combustion*, pages 685–692, The Combustion Institute, Pittsburgh.

Bradley, D., Lau, A. K. C., and Lawes, M. (1992). Flame stretch as a determinant of turbulent burning velocity, *Phil. Trans. Soc. Lond.* **338**, 359–387.

Branley, N. and Jones, W. P. (1997). Large eddy simulation of turbulent non-premixed flame, *Proceedings of the Eleventh Symposium on Turbulent Shear Flows*, pages 4.1–4.6, Grenoble, France.

(1999). Large eddy simulation of a nonpremixed turbulent swirling flame, in W. Rodi and D. Laurence, editors, *Engineering Turbulence Modelling and Experiments* **4**, 861–870, Elsevier Science, Amsterdam.

Bray, K. N. C. (1980). Turbulent flows with premixed reactants, in P. A. Libby and F. A. Williams, editors, *Turbulent Reacting Flows*, pages 113–183, Springer-Verlag, Berlin.

(1990). Studies of the turbulent burning velocity, *Proc. Roy. Soc. Lond.* **A431**, 315–335.

(1995). Turbulent transport in flames, *Proc. Roy. Soc. Lond.* **A451**, 231–256.

(1996). The challenge of turbulent combustion, *Twenty-Sixth Symposium (International) on Combustion*, pages 1–26, The Combustion Institute, Pittsburgh.

Bray, K. N. C., Champion, M., and Libby, P. A. (1991). Premixed flames in stagnating turbulence: Part I. The general formulation for counterflowing streams and gradient models for turbulent transport, *Combust. Flame* **84**, 391–410.

(1992). Premixed flames in stagnating turbulence: Part III. – The $\tilde{k}$-$\tilde{\varepsilon}$-theory for reactants impinging on a wall, *Combust. Flame* **91**, 165–186.

(1998). Premixed flames in stagnating turbulence: Part II. – The mean velocities and pressure and the Damköhler number, *Combust. Flame* **112**, 635–654.

Bray, K. N. C. and Libby, P. A. (1986). Passage time and flamelet crossing frequencies in premixed turbulent combustion, *Combust. Sci. and Tech.* **47**, 253–274.

(1994). Recent developments in the BML model of premixed turbulent combustion, in P. A. Libby and F. A. Williams, editors, *Turbulent Reacting Flows*, pages 115–151, Academic Press, London.

Bray, K. N. C., Libby, P. A., Masuya, G., and Moss, J. B. (1981). Turbulence production in premixed turbulent flames, *Combust. Sci. and Tech.* **25**, 127–140.

Bray, K. N. C., Libby, P. A., and Moss, J. B. (1984a). Flamelet crossing frequencies and mean reaction rates in premixed turbulent combustion, *Combust. Sci. and Tech.* **41**, 143–172.

(1984b). Scalar length variations in premixed turbulent flames, *Twentieth Symposium (International) on Combustion*, pages 421–427, The Combustion Institute, Pittsburgh.

Bray, K. N. C. and Moss, J. B. (1977). A unified statistical model of the premixed turbulent flame, *Acta Astronautica* **4**, 291–319.

Bray, K. N. C. and Peters, N. (1994). Laminar flamelets in turbulent flames, in P. A. Libby, and F. A. Williams, editors, *Turbulent Reacting Flows*, pages 63–113, Academic Press, London.

Broadwell, J. E., Dahm, W. J. A., and Mungal, M. G. (1984). Blowout of turbulent diffusion flames, *Twentieth Symposium (International) on Combustion*, pages 303–310, The Combustion Institute, Pittsburgh.

Brockhinke, A., Andresen, P., and Kohse-Höinghaus, K. (1966). Contribution to the analysis of temporal and spatial structures near the lift-off region of a turbulent hydrogen diffusion flame, *Twenty-Sixth Symposium (International) on Combustion*, pages 153–159, The Combustion Institute, Pittsburgh.

Brown, M. J., McLean, I. C., Smith, D. B., and Taylor, S. C. (1996). Markstein lengths of CO/H_2/air flames, using expanding spherical flames, *Twenty-Sixth Symposium (International) on Combustion*, pages 875–881, The Combustion Institute, Pittsburgh.

Buckmaster, J. (1996). Edge-flames and their stability, *Combust. Sci. and Tech.* **115**, 41–68.

(1997). Edge Flames, *J. Eng. Math.* **31**, 269–284.

Buckmaster, J. D. and Ludford, G. S. S. (1982). Theory of laminar flames, *Cambridge Monographs on Mechanics and Applied Mathematics*, Cambridge University Press, Cambridge.

Buckmaster, J. and Matalon, M. (1988). Anomalous Lewis number effects in tribrachial flames, *Twenty-Second Symposium (International) on Combustion*, pages 1527–1533, The Combustion Institute, Pittsburgh.

Buckmaster, J. and Weber, R. (1996). Edge-flame-holding, *Twenty-Sixth Symposium (International) on Combustion*, pages 1143–1149, The Combustion Institute, Pittsburgh.

Bui-Pham, M., Seshadri, K., and Williams, F. A. (1992). The asymptotic structure of premixed methane–air flames with slow CO oxidation, *Combust. Flame* **89**, 343–362.

Burcat, A. (1984). Thermochemical data for combustion calculations, in W. C. Gardiner, Jr., editor, *Combustion Chemistry*, pages 455–504, Springer-Verlag, New York.

Burgess, C. P. and Lawn, C. J. (1999). The premixture model of turbulent burning to describe lifted jet flames, *Combust. Flame* **119**, 95–108.

Buriko, Yu. Ya., Kuznetsov, V. R., Volkov, D. V., Zaitsev, S. A., and Uryvsky, A. F. (1994). Test of a flamelet model for turbulent non-premixed combustion, *Combust. Flame* **96**, 104–120.

Burke, S. P. and Schumann, T. E. W. (1928) Diffusion flames, *First Symposium (International) on Combustion*, pages 2–11, The Combustion Institute, Pittsburgh.

Buschmann, A., Dinkelacker, F., Schäfer, T., and Wolfrum, J. (1996). Measurement of the instantaneous detailed flame structure in turbulent premixed combustion, *Twenty-Sixth Symposium (International) on Combustion*, pages 437–445, The Combustion Institute, Pittsburgh.

Candel, S. and Poinsot, T. J. (1990). Flame stretch and the balance equation for the flame area, *Combust. Sci. and Tech.* **70**, 1–15.

Candel, S. M., Veynante, D., Lacas, F., Maistret, E., Darabiha, N., and Poinsot, T. J. (1990). Coherent flamelet model: Applications and recent extensions, in B. Larrouturou, editor, *Recent Advances in Combustion Modelling*, Series on Advances in Mathematics for Applied Sciences **6**, 19–64, World Scientific, Singapore.

Cant, R. S. and Bray, K. N. C. (1988). Strained laminar flamelet calculations of premixed turbulent combustion in a closed vessel, *Twenty-Second Symposium (International) on Combustion*, pages 791–799, The Combustion Institute, Pittsburgh.

Cant, R. S., Bray, K. N. C., Kostiuk, L. W., and Rogg, B. (1994). Flow divergence effects in strained laminar flamelets for premixed turbulent combustion, *Combust. Sci. and Tech.* **95**, 261–276.

Cant, R. S., Pope, S. B., and Bray, K. N. C. (1990). Modelling of flamelet surface-to-volume ratio in turbulent premixed combustion, *Twenty-Third Symposium (International) on Combustion*, pages 809–815, The Combustion Institute, Pittsburgh.

Cheatham, S. and Matalon, M. (2000). A general asymptoic theory of diffusion flames with application to cellular instability, submitted to *J. Fluid Mech.*

Chelliah, H. K., Seshadri, K., and Law, C. K. (1993). Reduced kinetic mechanisms for counterflow methane–air diffusion flames, in N. Peters and B. Rogg, editors, *Reduced Kinetic Mechanisms for Applications in Combustion Systems*, 224–240, Springer-Verlag, Berlin.

Chen, H., Chen, S., and Kraichnan, R. H. (1989). Probability distribution of a stochastically advected scalar field, *Phys. Rev. Lett.* **63**, 2657–2660.

Chen, J. H., Echekki, T., and Kollman, W. (1999). The mechanism of two-dimensional pocket formation in lean premixed methane–air flames with implications to turbulent combustion, *Combust. Flame* **116**, 15–48.

Chen, J. H. and Im, H. G. (1998). Correlation of flame speed with stretch in turbulent premixed methane/air flames, *Twenty-Seventh Symposium (International) on Combustion*, pages 819–826, The Combustion Institute, Pittsburgh.

Chen, J.-Y. and Chang, W.-C. (1996). Flamelet and pdf modeling of CO and NO_x emissions from a turbulent, methane hydrogen jet nonpremixed flame, *Twenty-Sixth Symposium (International) on Combustion*, pages 2207–2214, The Combustion Institute, Pittsburgh.

Chen, J.-Y., Chang, W.-C., and Koszykowski, M. (1995). Numerical simulation and scaling of NO_x emissions from turbulent jet flames with various amounts of helium dilution, *Combust. Sci. and Tech.* **110–111**, 505–529.

Chen, J.-Y., Dibble, R. W., and Bilger, R. W. (1990). Pdf modeling of turbulent nonpremixed $CO/H_2/N_2$ jet flames with reduced mechanisms, *Twenty-Third Symposium (International) on Combustion*, pages 775–780, The Combustion Institute, Pittsburgh.

Chen, J.-Y. and Kollmann, W. (1992). Pdf modeling and analysis of thermal NO formation in turbulent nonpremixed hydrogen–air jet flames, *Combust. Flame* **88**, 397–412.

Chen, M. (2000). Simulation of flame stabilization in lifted turbulent jet diffusion flames, Dissertation, RWTH Aachen.

Chen, M., Herrmann, M., and Peters, N. (2000). Flamelet modeling of lifted turbulent methane/air and propane/air jet diffusion flames, to appear in the *Twenty-Eighth Symposium (International) on Combustion*.

Chen, R.-H. and Driscoll, J. F. (1990). Nitric oxide levels of jet diffusion flames: Effects of coaxial air and other mixing parameters, *Twenty-Third Symposium (International) on Combustion*, pages 281–288, The Combustion Institute, Pittsburgh.

Chen, Y.-C. and Mansour, M. (1997). Simultaneous Rayleigh scattering and laser-induced CH fluorescence for reaction zone imaging in premixed hydrocarbon flames, *Appl. Phys.* **B64**, 599–609.

Chen, Y.-C., Peters, N., Schneemann, G. A., Wruck, N., Renz, U., and Mansour, M. S. (1996). The detailed flame structure of highly stretched turbulent premixed methane–air flames, *Combust. Flame* **107**, 223–244.

Cheng, R. K. (1995). Velocity characteristics of premixed turbulent flames stabilized by weak swirl, *Combust. Flame* **101**, 1–14.

Cheng, R. K. and Shepherd, I. G. (1991). The influence of burner geometry on premixed turbulent flame propagation, *Combust. Flame* **85**, 7–26.

Chevalier, C., Louessard, P., Müller, U. C., and Warnatz, J. (1990). A detailed low-temperature reaction mechanism of n-heptane auto-ignition, *Int. Symposium on Diagnostics and Modeling of Combustion in Internal Engines COMODIA 90*, Kyoto, The Japan Society of Mechanical Engineers.

Choi, C. R. and Huh, K. Y. (1998). Development of a coherent flamelet model for a spark-ignited premixed flame in a closed vessel, *Combust. Flame* **114**, 336–348.

Chung, S. H. and Law, C. K. (1983). Structure and extinction of convective diffusion flames with general Lewis numbers, *Combust. Flame* **52**, 59–79.

Clavin, P. (1985). Dynamic behaviour of premixed flame fronts in laminar and turbulent flows, *Prog. Energy Combust. Sci.* **11**, 1–59.

Clavin, P. and Garcia, P. (1983). The influence of the temperature dependence of diffusivities on the dynamics of flame fronts, *J. Méc. Théor. Appl.* **2**, 245–263.

Clavin, P. and Nicoli, C. (1985). Effect of heat losses on the limits of stability of premixed flames propagating downwards, *Combust. Flame* **60**, 1–14.

Clavin, P. and Williams, F. A. (1982). Effects of molecular diffusion and of thermal expansion on the structure and dynamics of premixed flames in turbulent flows of large scale and low intensity, *J. Fluid Mech.* **116**, 251–282.

Clemens, N. T. and Paul, P. H. (1995). Effects of heat release on the near field flow structure of hydrogen jet diffusion flames, *Combust. Flame* **102**, 271–284.

Coelho, P. J. and Peters, N. (2000). Numerical simulation of a MILD combustion burner, submitted to *Combust. Flame*.

Colucci, P. J., Jaberi, F. A., Givi, P., and Pope, S. B. (1998). Filtered density function for large eddy simulation of turbulent reacting flows, *Phys. Fluids* **10**, 499–515.

Cook, A. W. (1997). Determination of the constant coefficient in scale similarity models of turbulence, *Phys. Fluids* **9**, 1485–1487.

Cook, A. W. and Bushe, W. K. (1999). A subgrid-scale model for the scalar dissipation rate in nonpremixed combustion, *Phys. Fluids* **11**, 746–748.

Cook, A. W. and Riley, J. J. (1994). A subgrid model for equilibrium chemistry in turbulent flows, *Phys. Fluids* **6**, 2868–2870.

 (1998). Subgrid-scale modeling for turbulent reacting flows, *Combust. Flame* **112**, 593–606.

Cook, A. W., Riley, J. J., and Kosály, G. (1997). A laminar flamelet approach to subgrid-scale chemistry in turbulent flows, *Combust. Flame* **109**, 332–341.

Correa, S. M. (1998). Power generation and aeropropulsion gas turbines: From combustion science to combustion technology, *Twenty-Seventh Symposium (International) on Combustion*, pages 1793–1807, The Combustion Institute, Pittsburgh.

Correa, S. M. and Gulati, A. (1992). Measurements and modeling of a bluff body stabilized flame, *Combust. Flame* **89**, 195–213.

Correa, S. M. and Pope, S. B. (1992). Comparison of a Monte Carlo pdf/finite volume mean flow model with bluff-body Raman data, *Twenty-Fourth Symposium (International) on Combustion*, pages 279–285, The Combustion Institute, Pittsburgh.

Corrsin, S. (1961). The reactant concentration spectrum in turbulent mixing with a first order reaction, *J. Fluid Mech.* **11**, 407–416.

Cuenot, B. and Poinsot, T. J. (1994). Effects of curvature and unsteadiness in diffusion flames. Implications for turbulent diffusion combustion, *Twenty-Fifth Symposium (International) on Combustion*, pages 1383–1390, The Combustion Institute, Pittsburgh.

 (1996). Asymptotic and numerical study of diffusion flames with variable Lewis number and finite rate chemistry, *Combust. Flame* **104**, 111–137.

Curl, R. L. (1963). Dispersed phase mixing: I Theory and effects in single reactors, *AICHE J.* **9**, 175–181.

Damköhler, G. (1940). Der Einfluß der Turbulenz auf die Flammengeschwindigkeit in Gasgemischen, *Z. Elektrochem.* **46**, 601–652, 1947, English translation *NASA Tech. Mem.* 1112.

Dandekar, A. and Collins, L. R. (1995). Effect of Lewis number on premixed flame propagation through isotropic turbulence, *Combust. Flame* **101**, 428–440.

Daou, J. and Liñán, A. (1998). Triple flames in mixing layers with nonunity Lewis numbers, *Twenty-Seventh Symposium (International) on Combustion*, pages 667–674, The Combustion Institute, Pittsburgh.

Darabiha, N. (1992). Transient behaviour of laminar counterflow hydrogen–air diffusion flames with complex chemistry, *Combust. Sci. and Tech.* **86**, 163–181.

Darabiha, N., Giovangigli, V., Trouvé, A., Candel, S. M., and Esposito, E. (1987). Coherent flame description of turbulent premixed ducted flames, in R. Borghi and S. N. B. Murthy, editors, *Turbulent Reactive Flows, Lecture Notes in Engineering* **40**, 591–637, Springer-Verlag, Berlin.

de Bruyn Kops, S. M., Riley, J. J., Kosály, G., and Cook, A. W. (1998). Investigation of modeling for non-premixed turbulent combustion, *Flow, Turbulence and Combustion* **60**, 105–122.

Dederichs, A. S., Balthasar, M., Mauss, F., and Bai, X.-S. (1999). Pollution formation in non-premixed combustion using different flamelet models, presented at the *Seventeenth International Colloquium on the Dynamics of Explosions and Reactive Systems*, Heidelberg.

de Goey, L. P. H., Mallens, R. M. M., and ten Thije Boonkkamp, J. H. M. (1997). An evaluation of different contributions to flame stretch for stationary premixed flames, *Combust. Flame* **110**, 54–66.

de Goey, L. P. H. and ten Thije Boonkkamp, J. H. M. (1997). A mass based definition of flame stretch for flames with finite thickness, *Combust. Sci. and Tech.* **122**, 399–405.

(1999). A flamelet description of premixed flames and the relation to flame stretch, *Combust. Flame* **119**, 253–271.

de Joannon, M., Langella, G., Beretta, F., Cavaliere, A., and Noviello, C. (1999). Mild combustion: Process features and technological constraints, *Proceedings of the Mediterranean Combustion Symposium*, pages 347–360.

Dekena, M. (1998). Numerische Simulation der turbulenten Flammenausbreitung in einem direkteinspritzenden Benzinmotor mit einem Flamelet-Modell, Dissertation, RWTH Aachen.

Dekena, M. and Peters, N. (1999). Combustion modelling with the G-equation, *Oil & Gas Sci. and Tech. Rev. IFP* **54**, 1–6.

Deschamps, B. M., Boukhalfa, A., Chauveau, C., Gokälp, I., Shepherd, I. G., and Cheng, R. K. (1992). An experimental estimation of flame surface density and mean reaction rate in turbulent premixed flames, *Twenty-Fourth Symposium (International) on Combustion*, pages 469–475, The Combustion Institute, Pittsburgh.

Deschamps, B. M., Smallwood, G. J., Prieur, J., Snelling, D. R., and Gülder, Ö. L. (1996). Surface density measurements of turbulent premixed turbulent flames in a spark-ignition engine and a Bunsen-type burner using planar laser induced fluorescence, *Twenty-Sixth Symposium (International) on Combustion*, pages 427–435, The Combustion Institute, Pittsburgh.

Deshaies, B. and Cambray, P. (1990). The velocity of a premixed flame as a function of the flame stretch: An experimental study, *Combust. Flame* **82**, 361–375.

DesJardin, P. E. and Frankel, S. H. (1998). Large eddy simulation of a nonpremixed reacting jet: Application and assessment of subgrid-scale combustion models, *Phys. Fluids* **10**, 2298–2314.

Dixon-Lewis, G., David, T., Gaskell, P. H., Fukutani, S., Jinno, H., Miller, J. A., Kee, R. J., Smooke, M. D., Peters, N., Effelsberg, E., Warnatz, J., and Behrendt, F. (1985). Calculation of the structure and extinction limit of a methane–air counterflow diffusion flame in the forward stagnation region of a porous cylinder, *Twentieth Symposium (International) on Combustion*, pages 1893–1904, The Combustion Institute, Pittsburgh.

Dold, J. W. (1989). Flame propagation in a nonuniform mixture: Analyses of a slowly varying triple flame, *Combust. Flame* **76**, 71–88.

Dold, J. W., Hartley, L. J., and Green, D. (1991). Dynamics of laminar triple-flamelet structures in non-premixed turbulent combustion, in P. C. Fife, A. Liñán, and F. A. Williams, editors, *Dynamical Issues in Combustion Theory*, IMA Volumes in Mathematics and Its Application, **35**, 35–105, Springer-Verlag, Berlin.

Domingo, P. and Vervisch, L. (1996). Triple flames and partially premixed combustion in autoignition of non-premixed turbulent mixtures, *Twenty-Sixth Symposium (International) on Combustion*, pages 233–240, The Combustion Institute, Pittsburgh.

Donnerhack, S. and Peters, N. (1984). Stabilization heights in lifted methane–air jet diffusion flames diluted with nitrogen, *Combust. Sci. and Tech.* **41**, 101–108.

Dopazo, C. (1975). Probability density function approach for a turbulent axisymmetric heated jet. Centerline evolution, *Phys. Fluids* **18**, 397–404.

(1994). Recent developments in pdf methods, in P. A. Libby and F. A. Williams, editors, *Turbulent Reacting Flows*, 375–474, Academic Press, London.

Dowdy, D. R., Smith, D. B., and Taylor, S. C. (1990). The use of expanding spherical flames to determine burning velocities and stretch effects in hydrogen/air mixtures, *Twenty-Third Symposium (International) on Combustion*, pages 325–332, The Combustion Institute, Pittsburgh.

Driscoll, J. F., Chen, R.-H., and Yoon, Y. (1992). Nitric oxide levels of turbulent jet diffusion flames: Effects of residence time and Damköhler number, *Combust. Flame* **88**, 37–49.

Driscoll, J. F., Sutkus, D. J., Roberts, W. L., Post, M. E., and Goss, L. P. (1994). The strain exerted by a vortex on a flame determined from velocity field images, *Combust. Sci. and Tech.* **96**, 213–229.

Duclos, J. M., Bruneaux, G., and Baritaud, T. A. (1996). 3D modeling of combustion and pollutants in a 4-valve SI engine; Effect of fuel and residuals distribution and spark location, SAE paper 961964.

Duclos, J. M., Veynante, D., and Poinsot, T. J. (1993). A comparison of flamelet models for premixed turbulent combustion, *Combust. Flame* **95**, 101–117.

Duclos, J.-M., Zolver, M., and Baritaud, T. A. (1999). 3D modeling of combustion for DI-SI engines, *Oil & Gas Sci. and Tech. Rev. IFP* **54**, 259–264.

Echekki, T. (1997). A quasi-one-dimensional premixed flame model with cross-stream effects, *Combust. Flame* **110**, 335–350.

Echekki, T. and Chen, J. H. (1996). Unsteady strain rate and curvature effects in turbulent premixed methane/air flames, *Combust. Flame* **106**, 184–202.

(1997). Structure and propagation of methanol–air triple flames, *Combust. Flame* **114**, 231–245.

(1999). Analysis of the contribution of curvature to premixed flame propagation, *Combust. Flame* **118**, 308–311.

Echekki, T. and Mungal, M. G. (1990). Flame speed measurements at the tip of a slot burner: Effects of flame curvature and hydrodynamic stretch. *Twenty-Third Symposium (International) on Combustion*, pages 455–461, The Combustion Institute, Pittsburgh.

Effelsberg, E. and Peters, N. (1983). A composite model for the conserved scalar pdf, *Combust. Flame* **50**, 351–360.

(1988). Scalar dissipation rates in turbulent jets and jet diffusion flames, *Twenty-Second Symposium (International) on Combustion*, pages 693–700, The Combustion Institute, Pittsburgh.

Egolfopoulos, F. N. and Law, C. K. (1990b). An experimental and computational study of the burning rates of ultra-lean to moderately-rich $H_2/O_2/N_2$ laminar flames with pressure variations, *Twenty-Third Symposium (International) on Combustion*, pages 333–340, The Combustion Institute, Pittsburgh.

Egolfopoulos, F. N., Zhu, D. L., and Law, C. K. (1990a). Experimental and numerical determination of laminar flame speeds: Mixtures of C_2-hydrocarbons with oxygen and nitrogen, *Twenty-Third Symposium (International) on Combustion*, pages 471–478, The Combustion Institute, Pittsburgh.

Eickhoff, H., Lenze, B., and Leuckel, W. (1984). Experimental investigation on the stabilization mechanism of jet diffusion flames, *Twentieth Symposium (International) on Combustion*, pages 311–318, The Combustion Institute, Pittsburgh.

Elperin, T., Kleeorin, N., and Rogachevskii, I. (1998). Effect of chemical reactions and phase transitions on turbulent transport of particles and gases, *Phys. Rev. Lett.* **80**, 69–72.

Erard, V., Boukhalfa, A., Puechberty, D., and Trinité, M. (1996). A statistical study on surface properties of freely propagating premixed turbulent flames, *Combust. Sci. Tech.* **113–114**, 313–327.

Everest, D. A., Feikema, D. A., and Driscoll, J. A. (1996). Images of the strained flammable layer used to study the lift-off of turbulent jet flames, *Twenty-Sixth Symposium (International) on Combustion*, pages 129–136, The Combustion Institute, Pittsburgh.

Fairweather, M., Jones, W. P., Lindstedt, R. P., and Marquis, A. J. (1991). Predictions of a turbulent reacting jet in a cross-flow, *Combust. Flame* **84**, 361–375.

Favier, V. and Vervisch, L. (1998). Investigating the effects of edge flames in liftoff in non-premixed turbulent combustion, *Twenty-Seventh Symposium (International) on Combustion*, pages 1239–1245, The Combustion Institute, Pittsburgh.

Fedotov, S. P. (1992). Statistical model of the thermal ignition of a distributed system, *Combust. Flame* **91**, 65–70.

Feese, J. J. and Turns, S. R. (1998). Nitric oxide emissions from laminar diffusion flames: Effect of air-side versus fuel-side dilutant addition, *Combust. Flame* **113**, 66–78.

Ferreira, J. C. (1996). Flamelet modelling of stabilization in turbulent non-premixed combustion, Dissertation, ETH Zürich, No. 11984.

Forkel, H. and Janicka, J. (1999). Large eddy simulation of a turbulent hydrogen diffusion flame, in S. Banerjee and J. K. Eaton, editors, *Turbulence and Shear Flow Phenomena – 1*, pages 65–70, Begell House Inc., New York.

Frankel, M. L. (1990). An equation of surface dynamics modeling flame fronts as density discontinuities in potential flows, *Phys. Fluids* **A2**, 1879–1883.

Frenklach, M. and Wang, H. (1990). Detailed modelling of soot particle nucleation and growth, *Twenty-Third Symposium (International) on Combustion*, pages 1559–1566, The Combustion Institute, Pittsburgh.

Frenklach, M. and Warnatz, J. (1987). Detailed modelling of PAH profiles in a sooting low-pressure acetylene flame, *Combust. Sci. and Tech.* **51**, 265–283.

Gagnepain, L., Chauveau, C., and Gökalp, I. (1998). A comparison between dynamic and scalar timescale in lean premixed turbulent flames, *Twenty-Seventh Symposium (International) on Combustion*, pages 775–787, The Combustion Institute, Pittsburgh.

Gao, F. (1991a). An analytic solution for the scalar probability density function in homogeneous turbulence, *Phys. Fluids*, **A3**, 511–513.

(1991b). Mapping closure and non-Gaussianity of the scalar probability density functions in isotropic turbulence, *Phys. Fluids*, **A3**, 2438–2444.

Germano, M., Maffio, A., Sello, S., and Mariotti, G. (1997). On the extension of the dynamic modeling procedure to turbulent reacting flows, in J. P. Collet, P. R. Voke, and L. Kleiser, editors, *Direct and Large Eddy Simulation II*, pages 291–300, Kluwer Academic Publishers, Amsterdam.

Germano, M., Piomelli, U., Moin, P., and Cabot, W. H. (1991). A dynamic subgrid scale eddy viscosity model, *Phys. Fluids* **A3**, 1760–1765.

Ghosal, S. (1999). Mathematical and physical constraints on large-eddy simulation of turbulence, *AIAA J.* **37**, 425–433.

Ghosal, S. and Moin, P. (1995). The basic equations for the large eddy simulation of turbulent flows in complex geometry, *J. Comput. Phys.* **118**, 24–37.

Ghosal, S. and Vervisch, L. (2000). Asymptotic theory of triple flames, submitted to *J. Fluid Mech.*

Gibson, C. H., Ashurst, W. T., and Kerstein, A. R. (1988). Mixing of strongly diffusive passive scalars like temperature by turbulence, *J. Fluid Mech.* **194**, 261–293.

Giovangigli, V. and Smooke, M. D. (1987). Extinction of strained premixed laminar flames with complex chemistry, *Combust. Sci. and Tech.* **53**, 23–49.

Girimaji, S. S. (1992). On the modelling of scalar diffusion in isotropic turbulence, *Phys. Fluids* **A4**, 2529–2537.

(1993). A study of multi-scalar mixing, *Phys. Fluids* **A5**, 1802–1809.

Girimaji, S. S. and Zhou, Y. (1996). Analysis and modeling of subgrid scalar mixing using numerical data, *Phys. Fluids* **A8**, 1224–1236.

Gomez, A., Littman, M. G., and Glassman, I. (1987). Comparative study of soot formation on the centerline of axisymmetric laminar diffusion flames: Fuel and temperature effects, *Combust. Flame* **70**, 225–241.

Göttgens, J., Mauss, F., and Peters, N. (1992). Analytic approximations of burning velocities and flames thicknesses of lean hydrogen, methane, ethane, ethylene, acetylene and propane flames, *Twenty-Fourth Symposium (International) on Combustion*, pages 129–135, The Combustion Institute, Pittsburgh.

Göttgens, J., Peters, N., and Seshadri, K. (2000). The influence of fluctuating flame front curvature on the burning velocity of premixed flamelets, submitted to *Combust. Theory Modelling*.

Gouldin, F. C. (1987). An application of fractals to modeling premixed turbulent flames, *Combust. Flame* **68**, 249–266.

Gu, X. J., Haq, M. Z., Lawes, M., and Woolley, R. (1999). Laminar burning velocity and Markstein lengths of methane–air mixtures, *Combust. Flame* **121**, 41–58.

Gulati, A. and Driscoll, J. F. (1986). Velocity–density correlations and Favre averages measured in a premixed turbulent flame, *Combust. Sci. and Tech.* **48**, 285–307.

Gülder, Ö. L. (1990a). Turbulent premixed combustion modelling using fractal geometry, *Twenty-Third Symposium (International) on Combustion*, pages 835–842, The Combustion Institute, Pittsburgh.

(1990b). Turbulent premixed flame propagation models for different combustion regimes, *Twenty-Third Symposium (International) on Combustion*, pages 743–750, The Combustion Institute, Pittsburgh.

(1999). Fractal characteristics and surface density of flame fronts in turbulent premixed combustion, *Mediterranean Combustion Symposium–99*, pages 130–154, Antalya, Turkey.

Gülder, Ö. L. and Smallwood, G. J. (1995). Inner cut-off scale of flame surface wrinkling in turbulent premixed flames, *Combust. Flame* **103**, 107–114.

Hartley, L. J. and Dold, J. W. (1991). Flame propagation in a nonuniform mixture: Analysis of a propagating triple-flame, *Combust. Sci. and Tech.* **80**, 23–46.

Hasselbrink E. F., Jr. and Mungal, M. G. (1998a). Characteristics of the velocity field near the instantaneous base of lifted non-premixed turbulent jet flames, *Twenty-Seventh Symposium (International) on Combustion*, pages 867–873, The Combustion Institute, Pittsburgh.

(1998b). Observations on the stabilization region of lifted non-premixed methane transverse jet flames, *Twenty-Seventh Symposium (International) on Combustion*, pages 1167–1173, The Combustion Institute, Pittsburgh.

Haworth, D. C., Drake, M. C., Pope, S. B., and Blint, R. J. (1988). The importance of time dependent flame structures in stretched laminar flamelet models for turbulent jet diffusion flames, *Twenty-Second Symposium (International) on Combustion*, pages 589–597, The Combustion Institute, Pittsburgh.

Haworth, D. C. and Poinsot, T. J. (1992). Numerical simulations of Lewis number effects in turbulent flames, *J. Fluid Mech.* **224**, 405–436.

Hawthorne, W. R., Weddell, D. S., and Hottel, H. C. (1949). Mixing and combustion of turbulent gas jets, *Third Symposium on Combustion, Flame and Explosion Phenomena*, pages 266–288, Williams and Wilkins, Baltimore.

Hélic, J. and Trouvé, A. (1998). Turbulent flame propagation in partially premixed combustion, *Twenty-Seventh Symposium (International) on Combustion*, pages 891–898, The Combustion Institute, Pittsburgh.

Herrmann, M. (2000). Numerische Simulation vorgemischter und teilweise vorgemischter turbulenter Flammen, Dissertation, RWTH Aachen.

Hewson, J. C. and Bollig, M. (1996). Reduced mechanism for NO_x emissions from hydrocarbon diffusion flames, *Twenty-Sixth Symposium (International) on Combustion*, pages 2171–2180, The Combustion Institute, Pittsburgh.

Heyl, A. (2000). Modellierung komplexer chemischer Reaktionen in turbulenten Strömungen, Dissertation, Karlsruhe.

Heyl, A. and Bockhorn, H. (1998). Modelling of pollutant formation in complex geometries under consideration of detailed chemical mechanisms, *Eccomas 98*, pages 808–813, Wiley, Chichester, U.K.

Hilka, M., Baum, M., Poinsot, T. J., and Veynante, D. (1996). in T. Baritaud, T,. Poinsot, and M. Baum, editors, Direct numerical simulation of turbulent flames with complex chemical kinetics, in *Direct Numerical Simulation for Turbulent Reacting Flows*, Edition Technip., 201–224.

Horch, K. (1978). Zur Stabilität von Freistrahl-Diffusionsflammen, Dissertation, Universität Fridericana, Karlsruhe.

Huang, Z., Bechtold, J. K., and Matalon, M. (1998). Weakly stretched premixed flames in oscillating flows, *Combust. Theory Modelling* **2**, 115–133.

Hulek, T. and Lindstedt, R. P. (1996a). Modelling of unclosed nonlinear terms in a pdf closure for turbulent flames, *Mathl. Comput. Modelling* **24**, 137–147.

(1996b). Computations of steady-state and transient premixed turbulent flames using pdf methods, *Combust. Flame* **104**, 481–504.

(1998). Joint scalar-velocity pdf modelling of finite rate chemistry in a scalar mixing layer, *Combust. Sci. and Tech.* **136**, 303–331.

Im, H. G., Lund, T. S., and Ferziger, J. H. (1997). Large eddy simulation of turbulent front propagation with dynamic subgrid models, *Phys. Fluids* **A9**, 3826–3833.

Janicka, J., Kolbe, W., and Kollmann, W. (1979). Closure of the transport equation for the probability density function of turbulent scalar fields, *J. Non-Equilib. Thermodyn.* **4**, 47–66.

Janicka, J. and Peters, N. (1982). Prediction of turbulent jet diffusion flame lift-off using a pdf transport equation, *Nineteenth Symposium (International) on Combustion*, pages 367–374, The Combustion Institute, Pittsburgh.

Jiménez, J., Liñán, A., Rogers, M. M., and Higuera, F. J. (1997). A priori testing of subgrid models for chemically reacting non-premixed turbulent shear flows, *J. Fluid Mech.* **349**, 149–171.

Jones, W. P. (1994). Turbulence modeling and numerical solution methods for variable density and combusting flows, in P. A. Libby and F. A. Williams, editors, *Turbulent Reacting Flows*, 309–374, Academic Press, London.

Jones, W. P. and Kakhi, M. (1996). Mathematical modelling of turbulent flames, in F. Culick, M. V. Heitor, and J. H. Whitelaw, editors, *Unsteady Combustion*, NATO ASI Series E, **306**, Kluwer Academic Publishers, Dordrecht.

  (1998). PDF modeling of finite-rate chemistry effects in turbulent non-premixed jet flames, *Combust. Flame* **115**, 210–229.

Jones, W. P. and Musonge, P. (1988). Closure of the Reynolds stress and scalar flux equations, *Phys. Fluids* **31**, 3589–3604.

Jones, W. P. and Prasetyo, Y. (1996). Probability density function modeling of premixed turbulent opposed jet flames, *Twenty-Sixth Symposium (International) on Combustion*, pages 275–282, The Combustion Institute, Pittsburgh.

Joulin, G. (1994). On the response of premixed flames to time-dependent stretch and curvature, *Combust. Sci. and Tech.* **97**, 219–229.

Juneja, A. and Pope, S. B. (1996). A DNS study of turbulent mixing of two passive scalars, *Phys. Fluids* **8**, 2177–2184.

Kalghatgi, G. T. (1981). Blow-out stability of gaseous jet diffusion flames. Part I: In still air, *Combust. Sci. and Tech.* **26**, 233–239.

  (1984). Lift-off heights and visible lengths of vertical turbulent jet diffusion flames in still air, *Combust. Sci. and Tech.* **41**, 17–29.

Kalt, P. A. M., Frank, J. H., and Bilger, R. W. (1998). Laser imaging of conditional velocities in premixed propane–air flames by simultaneous OH PLIF and PIV, *Twenty-Seventh Symposium (International) on Combustion*, pages 751–758, The Combustion Institute, Pittsburgh.

Kaplan, C. R., Oran, E. S., and Baek, S. W. (1994). Stabilization mechanism of lifted jet diffusion flames, *Twenty-Fifth Symposium (International) on Combustion*, pages 1183–1189, The Combustion Institute, Pittsburgh.

Karpov, V. P, Lipatnikov, A. N., and Wolanski, P. (1997). Finding the Markstein number using the measurements of expanding spherical laminar flames, *Combust. Flame* **109**, 436–448.

Karpov, V. P, Lipatnikov, A. N., and Zimont, V. L. (1996). Flame curvature as a determinant of preferential diffusion effects in premixed turbulent combustion, *Progr. Astronautics and Aeronautics* **173**, 235–250.

Katsuki, M. and Hasegawa, T. (1998). The science and technology of combustion in highly preheated air, *Twenty-Seventh Symposium (International) on Combustion*, pages 3135–3146, The Combustion Institute, Pittsburgh.

Keller, D. (1993). Herleitung und Lösung einer skalaren Feldgleichung zur Beschreibung von Flammenfronten, Dissertation, RWTH Aachen.

Keller, D. and Peters, N. (1994). Transient pressure effects in the evolution equation for premixed turbulent flames, *Theoret. Comput. Fluid Dynamics* **6**, 141–159.

Keller-Sornig, P. (1996). Berechnung der turbulenten Flammenausbreitung bei der ottomotorischen Verbrennung mit einem Flamelet-Modell, Dissertation, RWTH Aachen.

Kennel, C., Göttgens, J., and Peters, N. (1990). The basic structure of lean propane flames, *Twenty-Third Symposium (International) on Combustion*, pages 479–485, The Combustion Institute, Pittsburgh.

Kent, J. H. and Honnery, D. (1987). Soot and mixture fraction in turbulent diffusion flames, *Combust. Sci. and Tech.* **54**, 383–397.

Kent, J. H. and Wagner, H. G. (1984). Why do diffusion flames emit smoke?, *Combust. Sci. and Tech.* **41**, 245–269.

Kerstein, A. R. (1988a). Fractal dimension of turbulent premixed flames, *Combust. Sci. and Tech.* **60**, 441–445.

(1988b). Linear-eddy model of turbulent transport and mixing, *Combust. Sci. and Tech.* **60**, 391–421.

(1989). Linear-eddy modeling of turbulent transport. II: Application to shear layer mixing, *Combust. Flame* **75**, 397–413.

(1990). Linear-eddy modelling of turbulent transport. Part 3: Mixing and differential molecular diffusion in round jets, *J. Fluid Mech.* **216**, 411–435.

(1991). Linear-eddy modelling of turbulent transport, Part 6. Microstructure of diffusive scalar mixing fields, *J. Fluid Mech.* **231**, 361–394.

(1992a). Linear-eddy modelling of turbulent transport. Part 4. Structure of diffusion flames, *Combust. Sci. and Tech.* **81**, 75–96.

(1992b). Linear-eddy modelling of turbulent transport. Part 7. Finite-rate chemistry and multi-stream mixing, *J. Fluid Mech.* **240**, 289–313.

(1999). One-dimensional turbulence: Model formulation and application to homogeneous turbulence, shear flows, and buoyant statistical flows, *J. Fluid Mech.* **392**, 277–334.

Kerstein, A. R., Ashurst, W. T., and Williams, F. A. (1988). Field equation for interface propagation in an unsteady homogeneous flow field, *Phys. Rev.* **A37**, 2728–2731.

Kerstein, A. R., Cremer, M. A., and McMurtry, P. A. (1995). Scaling properties of differential diffusion effects in turbulence, *Phys. Fluids* **7**, 1999–2007.

Khoklov, A. M., Oran, E. S., and Wheeler, J. C. (1996), Scaling of buoyancy-driven turbulent premixed flames, *Combust. Flame* **105**, 28–34.

Kim, J. S. and Williams, F. A. (1997). Extinction of diffusion flames with nonunity Lewis numbers, *J. Engng. Math.* **31**, 101–118.

Kioni, P. N., Rogg, B., Bray, K. N. C., and Liñán, A. (1993). Flame spread in laminar mixing layers: The triple flame, *Combust. Flame* **95**, 276–290.

Klein, R. (1999). Numerics in combustion, in L. Vervisch, D. Veynante, and D. Olivari, editors, *Introduction to Turbulent Combustion*, Lecture Series 99–04, The von Karman Institute, Rhode Saint Genese, Belgium.

Klimenko, A. Y. (1990). Multicomponent diffusion of various scalars in turbulent flows, *Fluid Dyn.* **25**, 327–334.

(1995). Note on the conditional moment closure in turbulent shear flows, *Phys. Fluids* **7**, 446–448.

(1998). Examing the cascade hypothesis for turbulent premixed combustion, *Combust. Sci. and Tech.* **139**, 15–40.

Klimenko, A. Y. and Bilger, R. W. (2000). Conditional moment closure for turbulent combustion, to appear in *Prog. Energy Combust. Sci.*

Ko, Y. S. and Chung, S. H. (1999). Propagation of unsteady tribrachial flames in laminar non-premixed jets, *Combust. Flame* **117**, 151–163.

Kobayashi, H., Kawabata, Y., and Maruta, K. (1998). Experimental study on general correlation of turbulent burning velocity at high pressure, *Twenty-Seventh Symposium (International) on Combustion*, pages 941–948, The Combustion Institute, Pittsburgh.

Kobayashi, H., Nakashima, T., Tamura, T., Maruta, K., and Niioka, T. (1997). Turbulence measurements and observations of turbulent premixed flames at elevated pressures up to 3.0 MPa, *Combust. Flame* **108**, 104–117.

Kobayashi, H., Tamura, T., Maruta, K., Niioka, T., and Williams, F. A. (1996). Burning velocity of turbulent premixed flames in a high pressure environment, *Twenty-Sixth Symposium (International) on Combustion*, pages 389–396, The Combustion Institute, Pittsburgh.

Kollmann, W. and Chen, J. H. (1998). Pocket formation and the flame surface density equation, *Twenty-Seventh Symposium (International) on Combustion*, pages 927–934, The Combustion Institute, Pittsburgh.

Kostiuk, L. W. (1991). Premixed turbulent combustion in counterflow systems, PhD thesis, University of Cambridge, Cambridge, U.K.

Kostiuk, L. W. and Bray, K. N. C. (1994). Mean effects on stretch on laminar flamelets in a premixed turbulent flame, *Combust. Sci. and Tech.* **95**, 193–212.

Kostiuk, L. W., Bray, K. N. C., and Cheng, R. K. (1993a). Experimental study of premixed turbulent combustion in opposed streams. Part I: Nonreacting flow field, *Combust. Flame* **92**, 377–395.

(1993b). Experimental study of premixed turbulent combustion in opposed streams. Part II: Reacting flow field and extinction, *Combust. Flame* **92**, 396–409.

Kravchenko, A. G. and Moin, P. (1997). On the effects of numerical errors in Large Eddy Simulation of turbulent flows, *J. Comput. Phys.* **131**, 310–322.

Kronenburg, A. and Bilger, R. W. (1997). Modelling of differential diffusion effects in non-premixed nonreacting turbulent flow, *Phys. Fluids* **9**, 1435–1447.

Kronenburg, A., Bilger, R. W., and Kent, J. H. (1998). Second order conditional moment closure for turbulent jet diffusion flames, *Twenty-Seventh Symposium (International) on Combustion*, pages 1097–1104, The Combustion Institute, Pittsburgh.

Kuznetsov, V. R. (1982). Effect of turbulence on the formation of large superequilibrium concentration of atoms and free radicals in diffusion flames, *Mehan. Zhidkosti Gasa* **6**, 3–9.

Kwon, S., Tseng, L.-K., and Faeth, G. M. (1992). Laminar burning velocities and transition to unstable flames in $H_2/O_2/N_2$ and $C_3H_8/O_2/N_2$ mixtures, *Combust. Flame* **90**, 230–246.

Lam, S. H. and Goussis, D. A. (1988). Understanding complex chemical kinetics with computational singular perturbation, *Twenty-Second Symposium (International) on Combustion*, pages 931–941, The Combustion Institute, Pittsburgh.

Law, C. K. (1993). A compilation of recent experimental data of premixed laminar flames, in N. Peters and B. Rogg, editors, *Reduced Kinetic Mechanisms for Applications in Combustion Systems, Lecture Notes in Physics*, **15**, 19–30, Springer-Verlag, Berlin.

Lee, J. G., Lee, T. W., Nye, D. A., and Santavicca, D. A. (1995). Lewis number effects on premixed flames interacting with turbulent Kármán vortex streets, *Combust. Flame* **100**, 161–168.

Lee, J. G., North, G. L., and Santavicca, D. A. (1993). Surface properties of turbulent premixed propane/air flames at various Lewis numbers, *Combust. Flame* **93**, 445–456.

Lentini, D. (1994). Assessment of the stretched laminar flamelet approach for non-premixed turbulent combustion, *Combust. Sci. and Tech.* **100**, 95–122.

Lentini, D. and Puri, I. K. (1995). Stretched laminar flamelet modeling of turbulent chloromethane–air nonpremixed jet flames, *Combust. Flame* **103**, 328–338.

Lesieur, M. and Métais, O. (1996). New trends in large-eddy simulations of turbulence, *Annu. Rev. Fluid Mech.* **28**, 45–82.

Leung, K. M. and Lindstedt, R. P. (1995). Detailed kinetic modeling of C_1–C_3 alkane diffusion flames, *Combust. Flame* **102**, 129–160.

Libby, P. A. and Bray, K. N. C. (1981). Countergradient diffusion in premixed turbulent flames, *AIAA J.* **19**, 205–213.

Libby, P. A. and Williams, F. A. (1994). Fundamental aspects and a review, in P. A. Libby and F. A. Williams, editors, *Turbulent Reacting Flow*, pages 1–61, Academic Press, London.

Liew, S. K., Bray, K. N. C., and Moss, J. B. (1981). A flamelet model of turbulent non-premixed combustion, *Combust. Sci. and Tech.* **27**, 69–73.

(1984). A stretched laminar flamelet model of turbulent nonpremixed combustion, *Combust. Flame* **56**, 199–213.

Liñán, A. (1974). The asymptotic structure of counterflow diffusion flames for large activation energies, *Acta Astronautica* **1**, 1007–1039.

(1994). Ignition and flame spread in laminar mixing layers, in J. Buckmaster, T. L. Jackson, and A. Kumar, editors, *Combustion in High-Speed Flows*, pages 461–476, Kluwer Academic, Dordrecht.

Liñán, A. and Crespo, A. (1976). An asymptotic analysis of unsteady diffusion flames for large activation energies, *Combust. Sci. and Tech.* **14**, 95–117.

Liñán, A. and Williams, F. A. (1993). *Fundamental Aspects of Combustion*, Oxford University Press, New York.

Lindstedt, P. (1998). Modeling of the chemical complexities of flames, *Twenty-Seventh Symposium (International) on Combustion*, pages 269–285, The Combustion Institute, Pittsburgh.

Lindstedt, R. P. and Váos, E. M. (1998). Second moment modeling of premixed turbulent flames stabilized in impinging jet geometries, *Twenty-Seventh Symposium (International) on Combustion*, pages 957–962, The Combustion Institute, Pittsburgh.

(1999). Modelling of premixed turbulent flames with second moment methods, *Combust. Flame* **116**, 461–485.

Lundgren, T. S. (1967). Distribution function in the statistical theory of turbulence, *Phys. Fluids* **10**, 969–975.

Maas, U. and Pope, S. B. (1992a). Simplifying chemical kinetics: Intrinsic low-dimensional manifolds in composition space, *Combust. Flame* **88**, 239–264.

(1992b). Implementation of simplified chemical kinetics based on low-dimensional manifolds, *Twenty-Fourth Symposium (International) on Combustion*, pages 103–112, The Combustion Institute, Pittsburgh.

Magnussen, B. F. and Hjertager, B. H. (1977). On mathematical models of turbulent combustion with special emphasis on soot formation and combustion, *Sixteenth Symposium (International) on Combustion*, pages 719–729, The Combustion Institute, Pittsburgh.

Majda, A. and Sethian, J. (1985). The derivation and numerical solution of the equations for zero Mach number combustion, *Combust. Sci. and Tech.* **42**, 185–205.

Mansour, M. S. (1999). Turbulent premixed and partially premixed combustion diagnostics based on advanced laser techniques, *Mediterranean Combustion Symposium*, pages 40–69, Antalya, Turkey.

Mansour, M. S., Chen, Y.-C., and Peters, N. (1992). The reaction zone structure of turbulent premixed methane–helium–air flames near extinction, *Twenty-Fourth Symposium (International) on Combustion*, pages 461–468, The Combustion Institute, Pittsburgh.

Mansour, M. S., Peters, N., and Chen, Y.-C. (1998). Investigation of scalar mixing in the thin reaction zones regime using a simultaneous CH-LIF/Rayleigh laser technique, *Twenty-Seventh Symposium (International) on Combustion*, pages 767–773, The Combustion Institute, Pittsburgh.

Mantel, T. and Borghi, R.(1994). A new model of premixed wrinkled flame propagation based on a scalar dissipation equation, *Combust. Flame* **96**, 443–457.

Marble, F. E. and Broadwell, J. E. (1977). The coherent flame model for turbulent chemical reactions, Project Squid, *Tech. Rep.* TRW-9-PU.

Marracino, B. and Lentini, D. (1997). Radiation modelling in non-luminous non-premixed turbulent flames, *Combust. Sci. and Tech.* **125**, 23–48.

Marti, A. C., Sagués, F., and Sancho, J. M. (1997). Front dynamics in turbulent media, *Phys. Fluids* **9**, 3851–3857.

Masri, A. R. and Bilger, R. W. (1986). Turbulent non-premixed flames of hydrocarbon fuels near extinction: Mean structure from probe measurements, *Twenty-First Symposium (International) on Combustion*, pages 1511–1520, The Combustion Institute, Pittsburgh.

Masri, A. R., Bilger, R. W., and Dibble, R. W. (1988a). Conditional probability density function measured in turbulent nonpremixed flames of methane near extinction, *Combust. Flame* **74**, 267–284.

(1988b). Turbulent nonpremixed flames of methane near extinction: Mean structure from Raman measurements, *Combust. Flame* **71**, 245–266.

(1988c). Turbulent nonpremixed flames of methane near extinction: Probability density functions, *Combust. Flame* **73**, 261–285.

Masri, A. R. and Dibble, R. W. (1988). Spontaneous Raman measurements in turbulent $CO/H_2/N_2$ flames near extinction, *Twenty-Second Symposium (International) on Combustion*, pages 607–618, The Combustion Institute, Pittsburgh.

Masri, A. R., Dibble, R. W., and Barlow, R. S. (1992). Chemical kinetic effects in nonpremixed flames of H_2/CO_2 fuel, *Combust. Flame* **91**, 285–309.

(1996). The structure of turbulent nonpremixed flames revealed by Raman-Rayleigh-LIF measurements, *Progr. Energy Combust. Sci.* **22**, 307–362.

Masri, A. R. and Pope, S. B. (1990). Pdf calculations of piloted turbulent nonpremixed flames of methane, *Combust. Flame* **81**, 13–29.

Mastorakos, E. (1993). Turbulent combustion in opposed jet flows, PhD thesis, University of London, U.K.

Mastorakos, E., Pires da Cruz, A., Baritaud, T. A., and Poinsot, T. J. (1997). A model for the effects of mixing on the autoignitiion of turbulent flows, *Combust. Sci. and Tech.* **125**, 243–282.

Mastorakos, E., Taylor, A. M. K. P., and Whitelaw, J. H. (1994). Mixing in turbulent opposed jet flows, in *Turbulent Shear Flows 9*, Selected papers, 147–164, Springer-Verlag, Berlin.

Matalon, M. (1983). On flame stretch, *Combust. Sci. and Tech.* **31**, 169–181.

Matalon, M. and Matkowsky, B. J. (1982). Flames as gasdynamic discontinuities, *J. Fluid Mech.* **124**, 239–259.

Mauss, F. (1998). Entwicklung eines kinetischen Modells der Rußbildung mit schneller Polymerisation, Dissertation, RWTH Aachen, Cuvillier Verlag, Göttingen.

Mauss, F. and Bockhorn, H.(1995). Soot formation in premixed hydrocarbon flames: Prediction of temperature and pressure dependence, *Z. Phys. Chem.* **188**, 45–60.

Mauss, F., Keller, D., and Peters, N. (1990). A Lagrangian simulation of flamelet extinction and re-ignition in turbulent jet diffusion flames, *Twenty-Third Symposium (International) on Combustion*, pages 693–698, The Combustion Institute, Pittsburgh.

Mauss, F. and Peters, N., (1993). Reduced kinetic mechanisms for premixed methane–air flames, in N. Peters and B. Rogg, editors, *Reduced Kinetic Mechanisms for Applications in Combustion Systems, Lecture Notes in Physics*, **15**, 58–75, Springer-Verlag, Berlin.

McCaffrey, B. J. and Evans, D. D. (1986). Very large methane jet diffusion flames, *Twenty-First Symposium (International) on Combustion*, pages 25–31, The Combustion Institute, Pittsburgh.

McMurtry, P. A., Menon, S., and Kerstein, A. R. (1992). A linear eddy subgrid model for turbulent reacting flows: Application to hydrogen–air combustion, *Twenty-Fourth Symposium (International) on Combustion*, pages 271–278, The Combustion Institute, Pittsburgh.

Mell, W. E., Nilsen, V., Kosály, G., and Riley, J. J. (1994). Investigation of closure models for non-premixed turbulent reacting flows, *Phys. Fluids* **6**, 1331–1356.

Meneveau, C., Lund, T. S., and Cabot, W. H. (1996). A Lagrangian dynamic subgrid-scale model of turbulence, *J. Fluid Mech.* **319**, 353–385.

Meneveau, C. and Poinsot, T. J. (1991). Stretching and quenching of flamelets in premixed turbulent combustion, *Combust. Flame* **86**, 311–332.

Menon, S. and Calhoon, W. H., Jr. (1996). Subgrid mixing and molecular transport modeling in a reacting shear layer, *Twenty-Sixth Symposium (International) on Combustion*, pages 59–66, The Combustion Institute, Pittsburgh.

Menon, S. and Kerstein, A. R. (1992). Stochastic simulation of the structure and propagation rate of turbulent premixed flames, *Twenty-Fourth Symposium (International) on Combustion*, pages 443–450, The Combustion Institute, Pittsburgh.

Miake-Lye, R. C. and Hammer, J. A. (1988). Lifted turbulent jet flames: A stability criterion based on the jet large-scale structure, *Twenty-Second Symposium (International) on Combustion*, pages 817–824, The Combustion Institute, Pittsburgh.

Miller, J. A. and Melius, C. F. (1992). Kinetic and thermodynamic issues in the formation of aromatic compounds in flames of aliphatic fuels, *Combust. Flame* **91**, 21–39.

Mishra, D. P., Paul, P. J., and Mukunda, H. S. (1994). Stretch effects extracted from propagating spherical premixed flames with detailed chemistry, *Combust. Flame* **99**, 379–386.

Moin, P. (1997). Progress in large eddy simulation of turbulent flows, AIAA paper 97–0749.

Moin, P., Squires, K., Cabot, W. H., and Lee, S. (1991). A dynamic subgrid-scale model for compressible turbulence and scalar transport, *Phys. Fluids* **A3**, 2746–2757.

Monin, A. S. and Yaglom, A. M. (1975). *Statistical Fluid Mechanics: Mechanics of Turbulence*, Vol. 2, MIT Press, Cambridge, MA.

Montgomery, C. J., Kaplan, C. R., and Oran, E. S. (1998). The effect of coflow velocity on a lifted methane–air jet diffusion flame, *Twenty-Seventh Symposium (International) on Combustion*, pages 1175–1182, The Combustion Institute, Pittsburgh.

Mueller, C. J., Driscoll, J. F., Reuss, D. L., and Drake, M. C. (1996). Effect of unsteady stretch on the length of a freely propagating flame wrinkled by a vortex, *Twenty-Sixth Symposium (International) on Combustion*, pages 347–355, The Combustion Institute, Pittsburgh.

Mueller, C. J., Driscoll, J. F., Reuss, D. L., Drake, M. C., and Rosalik, M. E. (1998). Vorticity generation and attenuation as vortices convert through a premixed flame, *Combust. Flame* **112**, 342–358.

Müller, C. M., Breitbach, H., and Peters, N. (1994). Partially premixed turbulent flame propagation in jet flames, *Twenty-Fifth Symposium (International) on Combustion*, pages 1099–1106, The Combustion Institute, Pittsburgh.

Müller, U. C., Bollig, M., and Peters, N. (1997). Approximations of burning velocities and Markstein numbers for lean hydrocarbon and methanol flames, *Combust. Flame* **108**, 349–356.

Muñiz, L. and Mungal, M. G. (1997). Instantaneous flame-stabilization velocities in lifted-jet diffusion flames, *Combust. Flame* **111**, 16–31.

Najm, H. N. and Wyckoff, P. S. (1997). Premixed flame response to unsteady strain rate and curvature, *Combust. Flame* **110**, 92–112.

Namazian, M., Kelly, J. T., and Schefer, R. W. (1988). Near-field instantaneous flame and fuel concentration structures, *Twenty-Second Symposium (International) on Combustion*, pages 627–634, The Combustion Institute, Pittsburgh.

Nandula, S. P., Brown T. M., and Pitz, R. W. (1994). Measurements of scalar dissipation in the reaction zones of turbulent nonpremixed H_2–air flames, *Combust. Flame* **99**, 775–783.

Nilsen, V. and Kosály, G. (1997). Differentially diffusing scalars in turbulence, *Phys. Fluids* **9**, 3386–3397.

Nooren, P. (1998). Stochastic modelling of turbulent natural-gas flames, Dissertation, TU Delft.

Norris, A. T. and Pope, S. B. (1991). Turbulent mixing model based on ordered pairing, *Combust. Flame* **83**, 27–42.

Oberlack, M. (1997). Non-isotropic dissipation in non-homogeneous turbulence, *J. Fluid Mech.* **350**, 351–374.

Oberlack, M., Arlitt, R., and Peters, N. (2000a). On stochastic Damköhler number variations in a homogeneous flow reactor, submitted to *Combust. Theory Modelling*.

Oberlack, M., Peters, N., and Wenzel, H. (2000b). On symmetries, invariant solutions and averaging of the G-equation for premixed combustion, submitted to *Combust. Theory and Modelling*.

O'Brien, E. E. (1980). The probability density function (pdf) approach to reacting turbulent flows, in P. A. Libby and F. A. Williams, editors, *Turbulent Reacting Flows*, pages 185–218, Springer-Verlag, Berlin.

Oevermann, M. (1997). Ein Finite-Volumen-Verfahren auf unstrukturierten Dreiecksgittern zur Berechnung turbulenter Diffusionsflammen in kompressiblen Strmungsfeldern, Dissertation, RWTH Aachen.

Overholt, M. R. and Pope, S. B. (1996). Direct numerical simulation of a passive scalar with imposed mean gradient in isotropic turbulence, *Phys. Fluids* **8**, 3128–3148.

Paczko, G., Barths, H., Mikulić, I., and Peters, N. (1999). Demo-RIF user Guide, Version 1.0, http://www.flamelets.com/Rifug.pdf.

Paul, R. N. and Bray, K. N. C. (1996). Study of premixed turbulent combustion including Landua-Darrieus instability effects, *Twenty-Sixth Symposium (International) on Combustion*, pages 259–266, The Combustion Institute, Pittsburgh.

Pelce, P. and Clavin, P. (1982). Influence of hydrodynamics and diffusion upon the stability limits of laminar premixed flames, *J. Fluid Mech.* **124**, 219–237.

Peters, N. (1978). An asymptotic analysis of nitric oxide formation in turbulent diffusion flames, *Combust. Sci. and Tech.* **19**, 39–49.

(1980). Local quenching of diffusion flamelets and non-premixed turbulent combustion, *Western States Section of the Combustion Institute*, paper WSS 80-4, Spring Meeting, Irvine, CA.

(1983). Local quenching due to flame stretch and non-premixed turbulent combustion, *Combust. Sci. and Tech.* **30**, 1–17.

(1984). Laminar diffusion flamelet models in non-premixed turbulent combustion, *Prog. Energy Combust. Sci.* **10**, 319–339.

(1985). Numerical and asymptotic analysis of systematically reduced reaction schemes for hydrocarbon flames, in R. Glowinsky, B. Larrouturou, and R. Temum, editors, *Numerical Simulation of Combustion Phenomena, Lecture Notes in Physics*, **241**, 90–109, Springer-Verlag, Berlin.

(1986). Laminar flamelet concepts in turbulent combustion, *Twenty-First Symposium (International) on Combustion*, pages 1231–1250, The Combustion Institute, Pittsburgh.

(1988). Systematic reduction of flame kinetics: Principles and details, in A. L. Kuhl, J. R. Bowen, J. L. Leyer, and A. Borisov, editors, *Dynamics of Reactive Systems Part I: Flames, Progr. Astronautics and Aeronautics* **113**, 67–86, AIAA, Washington, DC.

(1991). Length scales in laminar and turbulent flames, in E. S. Oran and J. A. Boris, editors, *Numerical Approaches to Combustion Modeling, Prog. Astronautics and Aeronautics* **135**, 155–182, AIAA, Washington, DC.

(1992). A spectral closure for premixed turbulent combustion in the flamelet regime, *J. Fluid Mech.* **242**, 611–629.

(1993). Premixed, non-premixed and partially premixed turbulent combustion with fast chemistry, *Proceedings of the Anglo-German Combustion Symposium*, pages 26–33, The British Section of the Combustion Institute.

(1997). Kinetic foundation of thermal flame theory, in W. A. Sirignano, A. G. Merzhanov, and L. de Luca, editors, *Advances in Combustion Science: In Honor of Ya. B. Zel'dovich, Prog. Astronautics and Aeronautics* **173**, 73–91.

(1999). The turbulent burning velocity for large scale and small scale turbulence, *J. Fluid Mech.*, **384**, 107–132.

Peters, N. and Donnerhack, S. (1981). Structure and similarity of nitric oxide production in turbulent diffusion flames, *Eighteenth Symposium (International) on Combustion*, pages 33–42, The Combustion Institute, Pittsburgh.

Peters, N. and Göttgens, J. (1991). Scaling of buoyant turbulent jet diffusion flames, *Combust. Flame* **85**, 206–244.

Peters, N. and Kee, R. J. (1987). The computation of stretched laminar methane–air diffusion flames using a reduced four-step mechanism, *Combust. Flame* **68**, 17–29.

Peters, N. and Rogg, N., Eds. (1993). *Reduced kinetic mechanisms for applications in combustion systems, Lecture Notes in Physics*, **15**, Springer-Verlag, Heidelberg.

Peters, N., Terhoeven, P., Chen, J. H., and Echekki, T. (1998). Statistics of flame displacement speeds from computations of 2-D unsteady methane–air flames, *Twenty-Seventh Symposium (International) on Combustion*, pages 833–839, The Combustion Institute, Pittsburgh.

Peters, N., Wenzel, H., and Williams, F. A. (2000). Modification of the turbulent burning velocity by gas expansion effects, to appear in the *Twenty-Eighth Symposium (International) on Combustion*.

Peters, N. and Williams, F. A. (1983). Lift-off characteristics of turbulent jet diffusion flames, *AIAA J.* **21** (3), 423–429.

(1987). The asymptotic structure of stoichiometric methane–air flames, *Combust. Flame* **68**, 185–207.

Pfuderer, D. G., Neuber, A. A., Früchtel, G., Hassel, E. P., and Janicka, J. (1996). Turbulence modulation in jet diffusion flames: Modelling and experiments, *Combust. Flame* **106**, 301–317.

Phillips, H. (1965). Flame in a buoyant methane layer, *Tenth Symposium (International) on Combustion*, pages 1277–1283, The Combustion Institute, Pittsburgh.

Pierce, C. D. and Moin, P. (1998a). A dynamic model for subgrid-scale variance and dissipation rate of a conserved scalar, *Phys. Fluids* **10**, 3041–3044.

(1998b). Large eddy simulation of a confined coaxial jet with swirl and heat release, *AIAA paper* 98-2892, 1–11.

Piomelli, U., Cabot, W. H., Moin, P., and Lee, L. (1991). Subgrid-scale backscatter in turbulent and transitional flows, *Phys. Fluids* **A3**, 1766–1771.

Piomelli, U. and Liu, J. (1995). Large eddy simulation of rotating channel flows using a localized dynamic model, *Phys. Fluids* **A7**, 839–848.

Pitsch, H. (1998). Modellierung der Zündung und Schadstoffbildung bei der dieselmotorischen Verbrennung mit Hilfe eines interaktiven Flamelet-Modells, Dissertation, RWTH Aachen.

(1999). Unsteady flamelet modelling of differential diffusion in turbulent jet diffusion flames, submitted to *Combust. Flame*.

Pitsch, H., Barths, H., and Peters, N. (1996b). Three-dimensional modelling of NO_x and soot formation in DI-Diesel engines using detailed chemistry based on the interactive flamelet approach, SAE Paper 962057, 103–117.

Pitsch, H., Chen, M., and Peters, N. (1998). Unsteady flamelet modeling of turbulent hydrogen–air diffusion flames, *Twenty-Seventh Symposium (International) on Combustion*, pages 1057–1064, The Combustion Institute, Pittsburgh.

Pitsch, H. and Peters, N. (1998a). A consistent flamelet formulation for non-premixed combustion considering differential diffusion effects, *Combust. Flame* **114**, 26–40.

(1998b). Investigation of the ignition process of sprays under Diesel engine conditions using reduced *n*-heptane chemistry, SAE Paper 982464.

Pitsch, H., Peters, N., and Seshadri, K. (1996a). Numerical and asymptotic studies of the structure of premixed *iso*-octane flames, *Twenty-Sixth Symposium (International) on Combustion*, pages 763–771, The Combustion Institute, Pittsburgh.

Pitsch, H., Riesmeier, E., and Peters, N. (2000). Unsteady flamelet modeling of soot formation in turbulent diffusion flames, submitted to *Combust. Sci. and Tech.*

Pitsch, H. and Steiner, H. (2000). Large-eddy simulation of a turbulent piloted methane air diffusion flame (Sandia Flame D), submitted to *Phys. Fluids*.

Pitsch, H., Wan, Y. P., and Peters, N. (1995). Numerical investigations of soot formation and oxidation under Diesel engine conditions, SAE Paper 952357.

Pitts, W. M. (1988). Assessment of theories for the bahaviour and blowout of lifted turbulent jet diffusion flames, *Twenty-Second Symposium (International) on Combustion*, pages 809–816, The Combustion Institute, Pittsburgh.

Plessing, T., Kortschik, C., Mansour, M. S., and Peters, N. (2000). Measurements of the turbulent burning velocity and the flame structure of premixed flames on a low swirl burner, to appear in the *Twenty-Eighth Symposium (International) on Combustion*.

Plessing, T., Mansour, M. S., Peters, N., and Cheng, R. K. (1999). Ein neuartiger Niedrigdrallbrenner zur Untersuchung turbulenter Vormischflammen, *VDI-Berichte* **1492**, 457–462.

Plessing, T., Peters, N., and Wünning, J. G. (1998a). Laseroptical investigation of highly preheated combustion with strong exhaust gas recirculation, *Twenty-Seventh Symposium (International) on Combustion*, pages 3197–3204, The Combustion Institute, Pittsburgh.

Plessing, T., Terhoeven, P., Mansour, M. S., and Peters, N. (1998b). An experimental and numerical study of a laminar triple flame, *Combust. Flame* **115**, 335–353.

Pocheau, A. (1992). Front propagation in a turbulent medium, *Europhys. Lett.* **20**, 401–406.

Poinsot, T. J., Candel, S., and Trouvé, A. (1996). Applications of direct numerical simulation to premixed turbulent combustion, *Progr. Energy. Combust.* **21**, 531–576.

Poinsot, T. J., Haworth, D., and Bruneaux, G. (1993). Direct simulation and modelling of flame–wall interaction for premixed turbulent combustion, *Combust. Flame* **95**, 118–133.

Poinsot, T. J., Veynante, D., and Candel, S. (1990). Diagrams of premixed turbulent combustion based on direct simulation, *Twenty-Third Symposium (International) on Combustion*, pages 613–619, The Combustion Institute, Pittsburgh.
  (1991). Quenching process and premixed turbulent combustion diagrams, *J. Fluid Mech.* **228**, 561–606.
Pope, S. B. (1981). A Monte Carlo method for the pdf equations of turbulent reactive flow, *Combust. Sci. and Tech.* **25**, 159–174.
  (1982). An improved turbulent mixing model, *Combust. Sci. and Tech.* **28**, 131–145.
  (1983b). Consistent modeling of scalars in turbulent flows, *Phys. Fluids* **26**, 404–408.
  (1985). Pdf methods for turbulent reactive flows, *Prog. Energy Combust. Sci.* **11**, 119–192.
  (1988). The evolution of surfaces in turbulence, *Int. J. Engng. Sci.* **26**, 445–469.
  (1990). Computations of turbulent combustion: Progress and challenges, *Twenty-Third Symposium (International) on Combustion*, pages 591–612, The Combustion Institute, Pittsburgh.
  (1991). Mapping closures for turbulent mixing and reaction, *Theoret. Comput. Fluid Dynamics* **2**, 255–270.
  (1994a). Lagrangian pdf methods for turbulent flows, *Ann. Rev. Fluid Mech.* **26**, 23–63.
  (1994b). On the relationship between stochastic Lagrangian models of turbulence and second order closures, *Phys. Fluids* **A28**, 973–985.
  (1997). Computationally efficient implementation of combustion chemistry using in situ adaptive tabulation, *Combust. Theory Modelling* **1**, 41–63.
  (2000). *Turbulent Flows*, Cambridge University Press, Cambridge, U.K., to appear.
Pope, S. B. and Anand, M. S. (1984). Flamelet and distributed combustion in premixed turbulent flames, *Twentieth Symposium (International) on Combustion*, pages 403–410, The Combustion Institute, Pittsburgh.
Prasad, R. O. S. and Gore, J. P. (1999). An evaluation of flame surface density models for turbulent jet flames, *Combust. Flame* **116**, 1–14.
Réveillon, J. and Vervisch, L. (1997). Dynamic subgrid pdf modelling for nonpremixed turbulent combustion, in J. P., Chollet, P. R. Voke, and L. Kleiser, editors, *Direct and Large Eddy Simulation II*, pages 311–320, Kluwer Academic Publishers, Amsterdam.
  (1998). Subgrid mixing modeling: A dynamic approach, *AIAA J.* **36**, 336–341.
Rodi, W. and Spalding, D. B. (1970). A two-parameter model of turbulence and its application to free jets, *Wärme- und Stoffübertragung* **3**, 85–95.
Rogg, B. (1988). Response and flamelet structure of stretched premixed methane–air flames, *Combust. Flame* **73**, 45–65.
Rogg, B., Behrendt, F., and Warnatz, J. (1986). Turbulent non-premixed combustion in partially premixed diffusion flamelets with detailed chemistry, *Twenty-First Symposium (International) on Combustion*, pages 1533–1541, The Combustion Institute, Pittsburgh.
Rogg, B. and Peters, N. (1990). The asymptotic structure of weakly strained stoichiometric methane–air flames, *Combust. Flame* **79**, 402–420.
Røkke, N. A., Hustad, J. E., and Sønju, O. K. (1994). A study of partially premixed confined propane flames, *Combust. Flame* **97**, 88–106.
Røkke, N. A., Hustad, J. E., Sønju, O. K., and Williams, F. A. (1992). Scaling of nitric oxide emissions from buoyancy-dominated hydrocarbon turbulent jet diffusion flames, *Twenty-Fourth Symposium (International) on Combustion*, pages 385–393, The Combustion Institute, Pittsburgh.

Ronney, P. D., Haslam, B. D., and Rhys, N. O. (1995). Front propagation rates in randomly stirred media, *Phys. Rev. Lett.* **74**, 3804.

Ronney, P. D. and Yakhot, V. (1992). Flame broadening effects on premixed turbulent flame speed, *Combust. Sci. and Tech.* **86**, 31–43.

Rotta, J. C. (1972). *Turbulente Strömungen*, Teubner-Verlag, Stuttgart.

Ruetsch, G. R., Vervisch, L., and Liñán, A. (1995). Effects of heat release on triple flames, *Phys. Fluids* **7**, 1447–1454.

Rutland, C. J., Ferziger, J. H., and El Tahry, S. H. (1990). Full numerical simulation and modelling of turbulent premixed flames, *Twenty-Third Symposium (International) on Combustion*, pages 621–627, The Combustion Institute, Pittsburgh.

Rutland, C. J and Trouvé, A. (1993). Direct simulations of premixed turbulent flames with non-unity Lewis number, *Combust. Flame* **94**, 41–57.

Sanders, J. P. H., Chen, J.-Y., and Gökalp, I. (1997). Flamelet-based modeling of NO formation in turbulent hydrogen jet diffusion flames, *Combust. Flame* **111**, 1–15.

Sanders, J. P. H. and Lamers, A. P. G. G. (1994). Modeling and calculation of turbulent lifted diffusion flames, *Combust. Flame* **96**, 22–33.

Savaş, Ö. and Gollahalli, S. R. (1986). Stability of lifted laminar round gas-jet flame, *J. Fluid Mech.* **165**, 297–318.

Saxena, V. and Pope, S. P. (1998). Pdf calculations of major and minor species in a turbulent piloted jet flame, *Twenty-Seventh Symposium (International) on Combustion*, pages 1081–1086, The Combustion Institute, Pittsburgh.

Schefer, R. W., Namazian, M., and Kelly, J. T. (1988). Structural characteristics of lifted turbulent-jet flames, *Twenty-Second Symposium (International) on Combustion*, pages 833–842, The Combustion Institute, Pittsburgh.

 (1994a). Stabilization of lifted turbulent-jet flames, *Combust. Flame* **99**, 75–86.

Schefer, R. W., Namazian, M., Filtopoulos, E. E. J., and Kelly, J. T. (1994b). Temporal evolution of turbulence/chemistry interactions in lifted, turbulent-jet flames, *Twenty-Fifth Symposium (International) on Combustion*, pages 1223–1231, The Combustion Institute, Pittsburgh.

Scholefield, D. A. and Garside, J. E. (1949). The structure and stability of diffusion flames, *Third Symposium on Combustion, Flame and Explosion Phenomena*, pages 102–110.

Searby, G. and Quinard, J. (1990). Direct and indirect measurements of Markstein numbers in premixed flames, *Combust. Flame* **82**, 298–311.

Seshadri, K. (1996). Multistep asymptotic analyses of flame structure, *Twenty-Sixth Symposium (International) on Combustion*, pages 831–846, The Combustion Institute, Pittsburgh.

Seshadri, K., Bollig, M., and Peters, N. (1997). Numerical and asymptotic studies of the structure of stoichiometric and lean premixed heptane flames, *Combust. Flame* **108**, 518–536.

Seshadri, K. and Peters, N. (1988). Asymptotic structure and extinction of methane–air diffusion flames. *Combust. Flame* **73**, 23–44.

 (1990). The inner structure of methane–air flames, *Combust. Flame* **81**, 96–118.

Seshadri, K., Peters, N., and Williams, F. A. (1994). Asymptotic analyses of stoichiometric and lean hydrogen–air flames, *Combust. Flame* **96**, 407–427.

Seshadri, K. and Williams, F. A. (1994). Reduced chemical systems and their application in turbulent combustion, in P. A. Libby and F. A. Williams, editors, *Turbulent Reacting Flows*, 153–210, Academic Press, New York.

Sethian, F. A. (1996). Level set methods, *Cambridge Monographs on Applied and Computational Mathematics*, Cambridge University Press, Cambridge, U.K.

Shepherd, I. G. (1996). Flame surface density and burning rate in premixed turbulent flames, *Twenty-Sixth Symposium (International) on Combustion*, pages 373–379, The Combustion Institute, Pittsburgh.

Shy, S. S., Jang, R. H., and Ronney, P. D. (1996). Laboratory simulation of flamelet and distributed models for premixed turbulent combustion using aqueous autocatalytic reactions, *Combust. Sci. and Tech.* **113–114**, 329–350.

Sivashinsky, G. I. (1977a). Diffusional-thermal theory of cellular flames, *Combust. Sci. and Tech.* **15**, 137–146.

(1977b). Nonlinear analysis of hydrodynamic instability in laminar flames – I. Derivation of basic equations, *Acta Astronautica* **4**, 1177–1206.

(1988). Cascade-renormalization theory of turbulent flame speed, *Combust. Sci. and Tech.* **62**, 77–96.

Sivathanu, Y. R. and Faeth, G. M. (1990). Generalized state relationships for scalar properties in non-premixed hydrocarbon/air flames, *Combust. Flame* **82**, 211–230.

Smiljanovski, V., Moser, V., and Klein, R. (1997). A capturing-tracking hybrid scheme for deflagration discontinuities, *Combust. Theory Modelling* **1**, 183–215.

Smith, L. L., Dibble, R. W., Talbot, L., Barlow, R. S., and Carter, C. D. (1995). Laser Raman scattering measurements of differential molecular diffusion in turbulent nonpremixed jet flames of H_2/CO_2 fuel, *Combust. Flame* **100**, 153–160.

Smith, N. S. A., Bilger, R. W., and Chen, J.-Y. (1992). Modelling nonpremixed hydrogen jet flames using a conditional moment closure method, *Twenty-Fourth Symposium (International) on Combustion*, pages 263–269, The Combustion Institute, Pittsburgh.

Smith, T. M. and Menon, S. (1996). Model simulations of freely propagating turbulent premixed flames, *Twenty-Sixth Symposium (International) on Combustion*, pages 299–306, The Combustion Institute, Pittsburgh.

(1997). One-dimensional simulations of freely propagating turbulent premixed flames, *Combust. Sci. and Tech.* **128**, 99–130.

Smooke, M. D. (1991). Reduced kinetic mechanisms and asymptotic approximations of methane-air flames, *Lecture Notes in Physics*, **384**, Springer-Verlag, Berlin.

Smooke, M. D., Ern, A., Tanoff, M. A., Valdati, B. A., Mohammed, R. K., Marran, D. F., and Long, M. B. (1996). Computational and experimental study of NO in a axisymmetric laminar diffusion flame, *Twenty-Sixth Symposium (International) on Combustion*, pages 2161–2170, The Combustion Institute, Pittsburgh.

Smooke, M. D., Puri, I. K., and Seshadri, K. (1986). A comparison between numerical calculations and experimental measurements of the structure of a counterflow diffusion flame burning in diluted air, *Twenty-First Symposium (International) on Combustion*, pages 1783–1792, The Combustion Institute, Pittsburgh.

Sønju, O. K. and Hustad, J. E. (1984). An experimental study of turbulent jet diffusion flames, in *Dynamics of Flames and Reactive Systems*, *Progr. Astronautics and Aeronautics* **95**, 320–339.

Spalding, D. B. (1971). Mixing and chemical reaction in steady confined turbulent flames, *Thirteenth Symposium (International) on Combustion*, pages 649–657, The Combustion Institute, Pittsburgh.

Speziale, C. G. (1998). Turbulence modeling for time-dependent RANS and VLES: A review, *AIAA J.* **36**, 173–184.

Subramaniam, S. and Pope, S. B. (1998). A mixing model for turbulent reactive flows based on Euclidian minimum spanning trees, *Combust. Flame* **115**, 487–514.

Sussman, M., Smereka, P., and Osher, S. (1994). A level set approach for computing solutions to incompressible two-phase flow, *J. Comput. Phys.* **114**, 146–159.

Swaminathan, N. and Bilger, R. W. (1998). Conditional variance equation and its analysis, *Twenty-Seventh Symposium (International) on Combustion*, pages 1191–1198, The Combustion Institute, Pittsburgh.
 (1999). Assessment of combustion submodels for turbulent non-premixed hydrocarbon flames, *Combust. Flame* **116**, 519–545.
Tacke, M. M., Geyer, D., Hassel, E. P., and Janicka, J. (1998). A detailed investigation of the stabilization point of lifted turbulent diffusion flames, *Twenty-Seventh Symposium (International) on Combustion*, pages 1157–1165, The Combustion Institute, Pittsburgh.
Tacke, M. M., Linow, S., Geiss, S., Hassel, E. P., and Janicka, J. (1998). H3 Flame Data Release 1998, Technische Hochschule Darmstadt, Fachgebiet Energie- und Kraftwerkstechnik, http://www.tu-darmstadt.de/fb/mb/ekt/flamebase.html.
Taing, S., Masri, A. R., and Pope, S. B. (1993). Pdf calculations of turbulent nonpremixed flames of H_2/CO_2 using reduced chemical mechanisms, *Combust. Flame* **95**, 133–150.
Takahashi, F. and Goss, L. P. (1992). Near-field turbulent structures and the local extinction of jet diffusion flames, *Twenty-Fourth Symposium (International) on Combustion*, pages 351–359, The Combustion Institute, Pittsburgh.
Takeno, T. and Kotani, Y. (1975). An experimental study on the stability of jet diffusion flames, *Acta Astronautica* **2**, 999–1008.
Terhoeven, P. (1998). Ein numerisches Verfahren zur Berechnung von Flammenfronten bei kleiner Mach-Zahl, Dissertation, RWTH Aachen.
Tien, J. H. and Matalon, M. (1991). On the burning velocity of stretched flames, *Combust. Flame* **84**, 238–248.
Tolpadi, A. K., Correa, S. M., Burrass, D. L., and Mongia, H. C. (1997). Monte Carlo probability density function method for gas turbine combustor flow field predictions, *J. Prop. Power* **13**, 218–225.
Tomlin, A. S., Turányi, T., and Pilling, M. J. (1997). Mathematical tools for the construction, investigation and reduction of combustion mechanisms, in M. J. Pilling, editor, *Low Temperature Combustion and Autoignition*, Comprehensive Chemical Kinetics, **35**, 293–437, Elsevier, Amsterdam.
Trouvé, A. and Poinsot, T. J. (1994). The evolution equation for the flame surface density in turbulent premixed combustion, *J. Fluid Mech.* **278**, 1–31.
Tseng, L.-K., Ismail, M. A., and Faeth, G. M. (1993). Laminar burning velocities and Markstein numbers of hydrocarbons/air flames, *Combust. Flame* **95**, 410–426.
Tsuji, H. and Yamaoka, I. (1971). Structure analysis of counterflow diffusion flames in the forward stagnation region of a porous cylinder, *Thirteenth Symposium (International) on Combustion*, pages 723–731, The Combustion Institute, Pittsburgh.
Turns, S. R. (1995). Understanding NO_x formation in nonpremixed flames: Experiments and modeling, *Prog. Energy Combust. Sci.* **21**, 361–385.
Ulitsky, M. and Collins, L. R. (1997). Relative importance of coherent structures vs. background turbulence in the propagation of a premixed flame, *Combust. Flame* **111**, 257–275.
Vanquickenborne, L. and Van Tiggelen, A. (1966). The stabilization mechanism of lifted diffusion flames, *Combust. Flame* **10**, 59–69.
Vervisch, L., Bidaux, E., Bray, K. N. C., and Kollmann, W. (1995). Surface density function in premixed turbulent combustion modelling, similarities between probability density function and flame surface approach, *Phys. Fluids* **7**, 2496–2503.
Vervisch, L. and Veynante, D. (2000). Turbulent combustion modeling, submitted to *Prog. Energy Combust. Sci.*

Veynante, D., Duclos, J. M., and Piana, J. (1994a). Experimental analysis of flamelet models for premixed turbulent combustion, *Twenty-Fifth Symposium (International) on Combustion*, pages 1249–1256, The Combustion Institute, Pittsburgh.

Veynante, D. and Poinsot, T. J. (1997). Effects of pressure gradients on turbulent premixed flames, *J. Fluid Mech.* **353**, 83–114.

Veynante, D., Trouvé, A., Bray, K. N. C., and Mantel, T. (1997). Gradient and countergradient scalar transport in turbulent premixed flames, *J. Fluid Mech.* **332**, 263–293.

Veynante, D., Vervisch, L., Poinsot, T. J., Liñán, A., and Ruetsch, G. R. (1994b). Triple flame structure and diffusion flame stabilization, in *Proc. of Summer Prog. 1994*, Center for Turbulence Research, Stanford University.

Villermaux, J. and Devillon, J. C. (1972). Représentation de la coalescence et de la redispersion des domaines de ségrégation dans un fluide par un modèle d'interaction phénoménologique, *Proceedings of the Second International Symposium on Chemical Reaction Engineering*, pages 1–13, Elsevier, New York.

Wan, Y. P., Pitsch, H., and Peters, N. (1997). Simulation of autoignition delay and location of fuel sprays under Diesel-engine relevant conditions, SAE Paper 971590, SAE Transaction, *Journal of Engines*, **106**, 1611–1621.

Warnatz, J. (1981). The structure of laminar alkane, alkene, and acetylene flames, *Eighteenth Symposium (International) on Combustion*, pages 369–384, The Combustion Institute, Pittsburgh.

Warnatz, J., Maas, U., and Dibble, R. W. (1996). *Combustion*, Springer-Verlag, Berlin, Heidelberg.

Weller, H. G., Tabor, G., Gosman, A. D., and Fureby, C. (1998). Application of a flame-wrinkling LES combustion model to a turbulent mixing layer, *Twenty-Seventh Symposium (International) on Combustion*, pages 899–907, The Combustion Institute, Pittsburgh.

Wenzel, H. (1997). Turbulent premixed combustion in the laminar flamelet and the thin reaction zones regime, *Annual Research Briefs*, 237–252, Center for Turbulence Research.

(2000). Direkte numerische Simulation der Ausbreitung einer Flammenfront in einem homogenen Turbulenzfeld, Dissertation, RWTH Aachen.

Wenzel, H. and Peters, N. (2000). Direct numerical simulation and modeling of kinematic restoration, dissipation and gas expansion effects of premixed flames in homogeneous turbulence, to appear in *Combust. Sci. and Tech.*

Williams, F. A. (1975). Recent advances in theoretical descriptions of turbulent diffusion flames, in S. N. B. Murthy, editor, *Turbulent Mixing in Nonreactive and Reactive Flows*, pages 189–208, Plenum Press, New York.

(1985a). *Combustion Theory*, Benjamin/Cummins, Menlo Park, CA.

(1985b). Turbulent combustion, in J. Buckmaster, editor, *The Mathematics of Combustion*, pages 97–131, SIAM, Philadelphia.

Wirth, M., Keller, P., and Peters, N. (1993). A flamelet model for premixed turbulent combustion in SI-engines, SAE Paper 932646.

Wirth, M. and Peters, N. (1992). Turbulent premixed combustion: A flamelet formulation and spectral analysis in theory and IC-engine experiments, *Twenty-Fourth Symposium (International) on Combustion*, pages 493–501, The Combustion Institute, Pittsburgh.

Wohl, K., Kapp, N. M., and Gazley, C. (1949). The stability of open flames, *Third Symposium on Combustion, Flame and Explosion Phenomena*, pages 3–21, The Williams & Wilkins Company, Baltimore.

Wouters, H. A., Nooren, P. A., Peeters, T. W. J., and Roekaerts, D. (1998). Effects of micro-mixing in gas-phase turbulent jets, *Int. J. Heat Fluid Flow* **19**, 201–207.

Wu, A. S. and Bray, K. N. C. (1997). A coherent flame model of premixed combustion in a counterflow geometry, *Combust. Flame* **109**, 43–64.

Wünning, J. A. and Wünning, J. G. (1997). Flameless oxidation to reduce thermal NO-formation, *Prog. Energy Combust. Sci.* **23**, 81–94.

Xu, J. and Pope, S. B. (1999). Numerical studies of Pdf/Monte Carlo methods for turbulent reactive flows, *J. Comput. Phys.* **152**, 192–230.

Yakhot, V. (1988). Propagation velocity of premixed turbulent flames, *Combust. Sci. and Tech.* **60**, 191–214.

Yamashita, H., Shimada, M., and Takeno, T. (1996). A numerical study on flame stability at transition point of jet diffusion flames, *Twenty-Sixth Symposium (International) on Combustion*, pages 27–34, The Combustion Institute, Pittsburgh.

Yang, B. and Pope, S. B. (1998). Treating chemistry in combustion with detailed mechanisms – In situ adaptive tabulation in principal directions – Premixed combustion, *Combust. Flame* **112**, 85–112.

Yeung, P. K., Girimaji, S. S., and Pope, S. B. (1990). Straining and scalar dissipation on material surfaces in turbulence: Implications for Flamelets, *Combust. Flame* **79**, 340–365.

Yeung, P. K. and Pope, S. B. (1993). Differential diffusion of passive scalars in isotropic turbulence, *Phys. Fluids* **A5**, 2467–2478.

Yoshida, K. and Takagi, T. (1998). Transient local extinction and reignition behaviour of diffusion flames affected by flame curvature and preferential diffusion, *Twenty-Seventh Symposium (International) on Combustion*, pages 685–692, The Combustion Institute, Pittsburgh.

Zhang, Y., Rogg, B., and Bray, K. N. C. (1995). 2-D simulations of turbulent autoignition with transient flamelet source term closure, *Combust. Sci. and Tech.* **105**, 211–227.

Zimont, V. L. (1979). Theory of turbulent combustion of a homogeneous fuel mixture at high Reynolds numbers, *Combust. Expl. and Shock Waves*, **15**, 305–311.

Zimont, V. L. and Lipatnikov, A. N. (1995). A numerical model of premixed turbulent combustion of gases, *Chem. Phys. Reports* **14**, 993–1025.

Zimont, V. L., Polifke, W., Bettelini, M., and Weisenstein, W. (1998). An efficient computational model for premixed turbulent combustion at high Reynolds number based on a turbulent flame speed closure, *ASME Journal of Engineering for Gas Turbines and Power* **120**, 526–532.

Author Index

A
Abdel-Gayed, R. G., 74, 78, 124, 131, 135, 267
Adalsteinsson, D., 163, 267
Aldredge, R. C., 124, 267
Anand, M. S., 37, 288
Andresen, P., 244, 271
Andrews, G. E., 74, 267
Arlitt, R., 4, 94, 285
Ashurst, W. T., 46, 96, 113, 131, 132, 135, 137, 138, 153, 241, **267**ff., 277, 280
Aung, K. T., 96, 268

B
Baek, S. W., 252, 279
Bai, X. S., 224, 268, 274
Balthasar, M., 222, 224, 268, 274
Baritaud, T. A., 158, 219, 268, 269, 275, 283
Barlow, R. S., 204, 205, 206, 211, 215, 220, 224, 229, 236, **268**ff., 283, 290
Barths, H., 26, 48, 217, 218, 229, 231, 233, 234, **268**ff., 285, 287
Batchelor, G. K., 113, 268
Baulch, D. L., 22, 231, **268**ff.
Baum, M., 28, 278
Bechtold, J. K., 104, 278
Bédat, B., 75, 76, 269
Behrendt, F., 180, 186, 222, 275, 288
Beretta, F., 230, 274
Bergmann, V., 211, 269
Bettelini, M., 162, 293
Bidaux, E., 90, 291
Bilger, R. W., 35, 53, 54, 55, 89, 171, 175, 193, 211, 221, 224, 228, 229, 236, 240, 268, **269**ff., 272, 279, 280, 281, **283**ff., 290, **291**ff.
Birch, A. D., 222, 252, **269**ff.
Bish, E. S., 185, 268
Blint, R. D., 220, 222, 278

Bockhorn, H., 223, 231, 268, **278**ff., 283
Boger, M., 63, 269
Bollig, M., 26, 96, 231, 284, 289
Borghi, R. W., 39, 78, 85, **269**ff., 283
Boudier, P., 158, 269
Boughanem, H., 63, 269
Boukhalfa, A., 90, 96, 274, 276
Bowman, C. T., 22, 269
Bradley, D., 74, 78, 96, 124, 125, 131, 134, 135, 159, 251, 252, **267**ff., **269**ff.
Branley, N., 61, **270**ff.
Bray, K. N. C., xvii, 42, 43, 50, 51, 52, 75, 89, 90, 96, 124, 126, 132, 135, 157, 158, 219, 222, 247, 250, **270**ff., **271**ff., 280, **281**ff., **282**ff., 285, 291, 292, 293
Brehm, N., 48, 229, 234, 268
Breitbach, H., 246, 248, 252, 253, 254, 258, 259, 284
Broadwell, J. E., 51, 242, 243, 271, 283
Brockhinke, A., 244, 271
Brown, D. R., 222, 252, **269**ff.
Brown, M. J., 96, 271
Brown, T. M., 215, 285
Bruneaux, G., 90, 158, 275, 287
Buckmaster, J., 96, 247, 248, **271**ff.
Bui-Pham, M., 26, 271
Burcat, A., 19, 271
Burgess, C. P., 241, 271
Buriko, Yu. Ya., 220, 221, 271
Burke, S. P., 171, 175, 271
Burras, P. L., 229, 291
Buschmann, A., 84, 271
Bushe, W. K., 61, 273

C
Cabot, W. H., 58, 59, 277, 284, 286
Calhoon Jr., W. H., 57, 284
Cambray, P., 96, 274

Author Index

Candel, S., 51, 78, 80, 90, 136, **271**ff., 274, **287**ff.
Cant, R. S., 52, 96, 157, **271**ff.
Carter, C. D., 211, 215, 268, 290
Cavaliere, A., 230, 274
Champion, M., 75, 157, 158, **270**ff.
Chang, W.-C., 228, 229, **272**ff.
Chauveau, C., 33, 90, 274, 276, 277
Cheatham, S., 187, 272
Chelliah, H. K., 186, 272
Chen, H., 41, 272
Chen, J. H., 96, 105, 106, 129, 272, **275**ff., 281, 286
Chen, J. Y., 204, 205, 206, 220, 224, 228, 229, 236, **268**ff., **272**ff., 289, 290
Chen, M., 47, 196, 215, 221, 229, 255, 257, 258, 272, 287
Chen, R.-H., 228, 272, 275
Chen, S., 41, 272
Chen, Y.-C., 85, 86, **272**ff., **282**ff.
Chen, Y.-H., 246, **275**ff.
Cheng, R. K., 74, 75, 76, 90, 110, 111, 158, 269, **272**ff., 274, **281**ff., 287
Chavalier, C., 231, 273
Choi, C. R., 90, 273
Chung, S. H., 187, 251, 255, 273, 280
Clavin P., 93, 95, 96, 102, 120, 149, **273**ff., 285
Clemens, N. T., 240, 273
Cobos, C. J., 22, 231, **268**ff.
Coelho, P. J., 230, 273
Collins, L. R., 113, 140, 274, 291
Colucci, P. J., 61, 273
Cook, A. W., 59, 60, 61, 216, 217, **273**ff., 274
Cook, D. K., 252, 269
Correa, S. M., 37, 223, 229, **273**ff., 291
Corrsin, S., 32, 33, 273
Cox, R. A., 22, 231, **268**ff.
Cremer, M. A., 211, 280
Crespo, A., 219, 282
Cuenot, B., 187, 220, **273**ff.
Curl, R. L., 40, 273

D

Dahm, W. J. A., 185, 242, 243, 269, 271
Damköhler, G., 120, 121, 122, 132, 133, 250, 274
Dandekar, A., 140, 274
Darabiha, N., 51, 220, 271, 274
Daou, J., 248, 274
David, T., 180, 186, 275
Davidson, D. F., 22, 269
de Bruyn Kops, S. M., 60, 274
Dederichs, A. S., 224, 274
de Goey, L. P. H., 96, 149, **274**ff.
de Joannon, M., 230, 274
Dekena, M., 160, 161, 274
Deschamps, B. M., 90, 274

Deshaies, B., 96, 274
DesJardin, P. E., 61, 274
Devillon, J. C., 39, 292
Dibble, R. W., 22, 204, 206, 211, 224, 268, 272, **283**ff., 290, 292
Dinkelacker, F., 84, 271
Dixon-Lewis, G., 180, 186, 275
Dold, J. W., 247, 248, **275**ff., 278
Domingo, P., 238, 246, 275
Donnerhack, S., 200, 225, 226, 227, 228, 241, 242, 252, 258, 275, 286
Dopazo, C., 35, 39, 41, 275
Dowdy, D. R., 96, 275
Drake, M. C., 28, 96, 220, 222, 278, **284**ff.
Driscoll, J. F., 28, 96, 159, 228, 241, 272, **275**ff., 276, 277, **284**ff.
Duclos, J. M., 51, 90, 91, 158, 268, **275**ff., 292

E

Echekki, T., 96, 100, 105, 106, 246, 272, **275**ff.
Effelsberg, E., 180, 186, 198, 215, **275**ff.
Egolfopoulos, F. N., 96, **276**ff.
Eickhoff, H., 239, 254, 276
Elperin, T., 33, 276
El Tahry, S. H., 129, 289
Erard, V., 96, 276
Ern, A., 223, 290
Esposito, E., 51, 274
Evans, D. D., 252, 284
Everest, D. A., 241, 276

F

Faeth, G. M., 96, 97, 178, 268, 281, 290, 291
Fairweather, M., 222, 269, 276
Favier, F., 247, 276
Fedotov, S. P., 219, 276
Feese, J. J., 223, 276
Feikema, D. A., 241, 276
Ferreira, J. C., 215, 223, 228, 276
Ferziger, J. H., 63, 129, 278, 289
Filtopoulos, E. E. J., 244, 289
Forkel, H., 61, 276
Frank, J. H., 211, 268, 279
Frank, P., 22, 89, 231, **268**ff.
Frankel, M. L., 61, 138, 274, 276
Frenklach, M., 22, 231, 269, **276**ff.
Früchtel, G., 215, 223, 286
Fuchs, L., 222, 224, 268
Fukutani, S., 180, 186, 275
Fureby, C., 162, 292
Fusco, A., 158, 268

G

Gagnepain, L., 33, 276
Gao, F., 41, **277**ff.
Garcia, P., 149, 273

Author Index

Gardiner Jr., W. C., 22, 269
Garside, J. E., 240, 289
Gaskell, P. H., 96, 159, 180, 186, 251, 252, **269**ff., 275
Gazley, C., 239, 292
Geiss, S., 223, 291
Germano, M., 58, 59, 60, **277**ff.
Geyer, D., 244, 250
Ghosal, S., 58, 59, 248, **277**ff.
Gibson, C. H., 46, 105, 131, 135, 267, 277
Giovangigli, V., 51, 217, 274, 277
Girimaji, S. S., 41, 49, 62, 215, **277**ff., 293
Givi, P., 61, 273
Glassmann, I., 223, 277
Gökalp, I., 33, 228, 276, 289
Göttgens, J., 69, 70, 151, 202, 203, **277**ff., 279, 286
Golden, D. M., 22, 269
Goldenberg, M., 22, 269
Gollahalli, S. R., 247, 289
Gomez, A., 223, 277
Gonzalez, M., 39, 269
Gore, J. P., 90, 288
Gosman, A. D., 162, 292
Goss, L. P., 96, 240, 275, 291
Gouldin, F. C., 125, 126, 277
Goussis, D. A., 24, 281
Green, D., 248, 275
Gu, X. J., 70, 96, 159, 251, 252, **269**ff., 277
Gülder, Ö. L., 90, 124, 125, 126, 133, 274, **277**ff.
Gulati, A., 159, 223, 273, 277

H
Haq, M. Z., 70, 277
Hamid, M. N., 135, 267
Hammer, J. A., 242, 258, 260, 284
Hanson, R. K., 22, 269
Hargrave, G. K., 222, 252, 269
Hartley, L. J., 248, 275, 278
Hasegawa, T., 230, 279
Haslam, B. D., 124, 289
Hassan, M. I., 96, 268
Hassel, E. P., 215, 223, 236, 244, 250, 268, 286, **291**ff.
Hasselbrink Jr., E. F., 245, **278**ff.
Haworth, D. C., 90, 135, 220, 222, **278**ff., 287
Hawthorne, W. R., 34, 201, 278
Hayman, G., 22, 231, 268
Hélic, J., 252, 278
Herrmann, M., 163, 165, 255, 278
Henriot, S., 158, 269
Hewson, I. C., 231, 278
Heyl, A., 222, 223, 268, **278**ff.
Higuera, F. J., 59, 61, 279
Hilka, M., 28, 278
Hjertager, B. H., 34, 282

Honnery, D., 223, 280
Horch, K., 241, 278
Hottel, H. C., 34, 201, 278
Hůlek, T., 36, 159, **278**ff.
Huang, Z., 104, 278
Huh, K. Y., 90, 273
Hustad, J. E., 203, 226, 227, 242, **288**ff., 290

I
Im, H. G., 63, 96, 129, 272, 278
Ismail, M. A., 96, 291

J
Jaberi, F. A., 61, 273
Jang, R. H., 125, 290
Janicka, J., 40, 49, 61, 194, 215, 216, 223, 225, 236, 241, 244, 250, 268, 276, **278**ff., 286, **291**ff.
Jiménez, J., 59, 61, 279
Jinno, H., 180, 186, 275
Jones, W. P., 11, 41, 61, 159, 194, 195, 222, 224, **270**ff., 276, **279**ff.
Joulin, G., 106, 279
Juneja, A., 194, 279
Just, Th., 22, 231, 268

K
Kakhi, M., 11, 41, 224, **279**ff.
Kalghatgi, G. T., 241, 251, 252, 254, 258, 260, **279**ff.
Kalt, P. A. M., 89, 279
Kaplan, C. R., 252, 279, 284
Kapp, N. M., 239, 292
Karpov, V. P., 96, 124, **279**ff.
Katsuki, M., 230, 279
Kawabata, Y., 124, 126, 280
Kee, R. J., 180, 186, 187, 189, 275, 286
Keller, D., 45, 47, 101, 146, 147, 149, 220, 279, 283
Keller, P., 109, 110, 292
Keller-Sornig, P., 160, 279
Kelly, J. T., 243, 244, 285, **289**ff.
Kennel, C., 69, 279
Kent, J. H., 55, 180, 223, **280**ff., 281
Kerr, J. A., 22, 231, **268**ff.
Kerr, R. M., 131, 267
Kerstein, A. R., 46, 55, 56, 57, 82, 113, 125, 126, 131, 132, 135, 211, 267, 277, **280**ff., 284
Khoklov, A. M., 125, 280
Kim, J. S., 184, 280
Kioni, P. N., 247, 250, 280
Kleeorin, N., 33, 276
Klein, R., 103, 160, 163, 280, 290
Klimenko, A. Yu., 50, 53, 54, 125, **280**ff.
Ko, Y. S., 251, 255, 280
Kobayashi, H. T., 124, 126, **280**ff.

Kohse-Höinghaus, K., 244, 271
Kolbe, W., 40, 278
Kollmann, W., 40, 90, 129, 228, 272, 278, 281, 291
Kortschik, C., 75, 287
Kosály, G., 54, 60, 211, 216, 217, 220, 273, 274, 284, 285
Koszykowski, M., 228, 272
Kostiuk, L. W., 75, 96, 135, 158, 159, 271, **281**ff.
Kotani, Y., 240, 291
Kraichnan, R. H., 41, 272
Kravchenko, A. G., 59, 281
Kronenburg, A., 55, 211, **281**ff.
Kuznetsov, V. R., 43, 44, 220, 221, 271, 281
Kwon, S., 97, 281

L

Lacas, F., 51, 271
Lam, S. H., 24, 281
Lamers, A. P. G. G., 252, 289
Langella, G., 230, 274
Lau, A. K. C., 124, 125, 134, 135, 251, **270**ff.
Law, C. K., 69, 96, 186, 187, 272, 273, **276**ff., 281
Lawes, M., 70, 124, 125, 134, 135, 267, 270, 277
Lawn, C. J., 241, 271
Lentini, D., 46, 193, 196, 223, **281**ff., 283
Lee, J. G., 135, **281**ff.
Lee, L., 59, 286
Lee, S., 58, 59, 284
Lee, T. W., 135, 281
Lenze, B., 239, 254, 276
Lesieur, M., 58, 281
Leuckel, W., 239, 254, 276
Leung, K. M., 22, 281
Libby, P. A., 12, 50, 51, 75, 89, 90, 157, 158, **270**ff., **282**ff.
Liew, S. K., 42, 222, **282**ff.
Liñán, A., 59, 61, 124, 187, 191, 215, 219, 246, 247, 248, 250, 274, 279, 280, **282**ff., 289, 292
Lindstedt, R. P., 22, 36, 158, 159, 222, 276, **278**ff., 281, **282**ff.
Linow, S., 223, 291
Lipatnikov, A. N., 96, 124, 133, 161, **279**ff., 293
Lissianski, V., 22, 269
Littman, M. G., 223, 277
Liu, Y., 59, 287
Long, M. B., 223, 290
Louessard, P., 231, 273
Lucht, R. P., 204, 206, 268
Ludford, G. S. S., 96, 271

Lund, T. S., 59, 63, 278, 284
Lundgren, T. S., 35, 282
Lwakabamba, S. B., 74, 267

M

Maas, U., 22, 24, 42, 269, **282**ff., 292
Mack, A., 48, 229, 234, 268
Maffio, A., 60, 277
Magnussen, B. F., 34, 282
Maistret, E., 51, 271
Majda, A., 100, 101, 282
Mallens, R. M. M., 96, 274
Mansour, M., 75, 85, 86, 110, 111, 163, 249, 250, 251, 255, 257, **272**ff., **282**ff., 287
Mantel, T., 89, 283
Marble, F. E., 51, 283
Mariotti, G., 60, 277
Marquis, A. J., 222, 276
Marracino, B., 196, 283
Marran, D. F., 223, 290
Marti, A. C., 123, 283
Maruta, K., 124, 126, **281**ff.
Masri, A., 204, 205, 224, 225, 236, 240, 268, **283**ff., 291
Mastorakos, E., 159, 219, **283**ff.
Masuya, G., 50, 270
Matalon, M., 93, 95, 96, 104, 149, 187, 247, 248, 271, 272, 278, **283**ff., 291
Matkowsky, B. J., 93, 95, 149, 283
Mauss, F., 47, 69, 70, 220, 222, 223, 224, 231, **268**ff., 274, 277, **283**ff.
McCaffrey, B. J., 252, 284
McLean, I. C., 96, 271
McMurtry, P. A., 56, 211, 280, 284
Meier, W., 211, 269
Melius, C. F., 231, 284
Mell, W. E., 54, 220, 284
Meneveau, C., 59, 135, 153, **284**ff.
Menon, S., 56, 57, 82, 126, **284**ff., **290**ff.
Métais, O., 58, 281
Miake-Lye, R. C., 242, 258, 260, 284
Mikulić, I., 217, 218, 285
Miller, J. A., 180, 186, 231, 275, 284
Mishra, D. P., 96, 284
Mohammed, R. K., 223, 290
Moin, P., 58, 59, 60, 61, 65, 277, 281, **284**ff., **286**ff.
Mongia, H. C., 224, 291
Monin, A. S., 85, 144, 284
Montgomery, C. J., 252, 284
Moser, V., 160, 290
Moss, J. B., 42, 50, 90, 222, **270**ff., **282**ff.
Mueller, C. J., 28, 96, **284**ff.
Müller, C. M., 246, 248, 252, 253, 254, 258, 259, 284
Müller, U. C., 96, 231, 273, 284

Author Index

Mukunda, H. S., 96, 284
Mungal, M. G., 100, 242, 243, 245, 271, 275, **278**ff., 285
Murrells, T., 22, 231, **268**ff.
Musonge, P., 195, 279

N

Najm, H. H., 96, 285
Nakashima, T., 124, 280
Namazian, M., 243, 244, 285, **289**ff.
Nandula, S. P., 215, 285
Neuber, A. A., 215, 223, 286
Nicoli, C., 149, 273
Niioka, T., 124, 280
Nilsen, V., 54, 211, 220, 284, 285
Nooren, P. A., 39, 285, 292
Norris, A. T., 41, 285
North, G. L., 135, 281
Noviello, C., 230, 274
Nye, D. A., 135, 281

O

Oberlack, M., 4, 12, 94, **285**ff.
O'Brien, E. E., 35, 39, 48, 285
Oevermann, M., 214, 223, 285
Oran, E. S., 125, 252, 279, 280, 284
Osher, S., 162, 290
Overholt, M. R., 194, 285

P

Paczko, G., 217, 218, 285
Paul, R. N., 90, 285
Paul, P. H., 240, 273
Paul, P. J., 96, 284
Peeters, T., 39, 236, 268, 292,
Pelce, P., 93, 95, 285
Peters, N., 4, 23, 24, 25, 26, 27, 28, 43, 44, 45, 46, 47, 48, 49, 65, 66, 69, 70, 74, 78, 81, 82, 83, 85, 86, 89, 90, 94, 95, 96, 105, 106, 108, 109, 110, 111, 113, 114, 115, 116, 122, 124, 125, 126, 127, 129, 130, 140, 141, 142, 143, 144, 146, 147, 149, 151, 160, 161, 163, 180, 185, 186, 187, 188, 189, 191, 192, 194, 196, 198, 200, 202, 203, 207, 210, 211, 215, 216, 217, 218, 220, 221, 223, 225, 226, 227, 228, 229, 230, 231, 233, 234, 236, 240, 241, 242, 246, 248, 249, 250, 251, 252, 253, 254, 255, 257, 258, 259, 267, 268, 270, 272, 273, 274, 275, **276**ff., **277**ff., 279, **282**ff., **283**ff., **284**ff., **285**ff., **287**ff., 288, **289**ff., **292**ff.
Pfitzner, M., 48, 229, 234, 268
Pfuderer, D. G., 215, 286
Phillips, H., 247, 286
Piana, J., 91, 292
Pierce, C. D., 59, 60, 61, 65, **286**ff.
Pilling, M. J., 22, 24, 231, 268, 291
Piomelli, U., 58, 59, 277, **286**ff.
Pitsch, H., 24, 26, 47, 61, 62, 196, 210, 211, 214, 215, 217, 218, 221, 223, 229, 231, 233, 234, 268, **287**ff., 292
Pitz, R. W., 215, 285
Pires da Cruz, A., 219, 283
Plessing, T., 75, 110, 111, 163, 230, 249, 250, 251, 255, 257, **287**ff.
Pocheau, A., 125, 287
Poinsot, T. J., 28, 51, 78, 80, 89, 90, 135, 136, 153, 158, 187, 219, 220, 247, 269, **271**ff., **273**ff., 275, 278, 283, 284, **287**ff., 291, 292
Polifke, W., 162, 293
Pope, S. B., 24, 35, 36, 37, 38, 39, 40, 41, 42, 46, 51, 52, 61, 136, 194, 195, 211, 215, 220, 222, **224**ff., 225, 236, **288**ff., 289, 290, 291, **293**ff.
Post, M. E., 96, 275
Prasad, R. O. S., 90, 288
Prasetyo, Y., 159, 279
Prieur, J., 90, 274
Puechberty, D., 96, 276
Puri, I. K., 189, 223, 281, 290

Q

Quinard, J., 96, 289

R

Renz, U., 85, 272
Reuss, D. L., 28, 96, **284**ff.
Réveillon, J., 61, **288**ff.
Rhys, N. O., 124, 289
Riesmeier, E., 211, 223, 287
Riley, J. J., 54, 59, 60, 61, 216, 217, 220, **273**ff., 274, 284
Roberts, W. L., 96, 275
Rodi, W., 12, 288
Roekaerts, D., 39, 292
Rogachevskii, I., 33, 276
Rogers, M. M., 59, 61, 279
Rogg, B., 23, 69, 95, 96, 102, 186, 219, 222, 247, 250, 271, 280, 286, **288**ff., 293
Røkke, N. A., 226, 227, 242, **288**ff.
Ronney, P. D., 124, 125, 267, **289**ff., 290
Rosalik, M. E., 28, 284
Rotta, J. C., 12, 289
Ruetsch, G. R., 247, 250, 289, 292
Rutland, C. J., 129, 135, **289**ff.

S

Sagués, F., 123, 283
Sancho, J. M., 123, 283
Sanders, J. P. H., 228, 252, **289**ff.
Santavicca, D. A., 135, **281**ff.
Savaş, Ö., 247, 289
Saxena, V., 225, 289
Schäfer, T., 84, 271

Schefer, R. W., 243, 244, 285, **289**ff.
Schmitt, F., 222, 268
Schneemann, G. A., 85, 272
Scholefield, D. A., 240, 289
Schumann, T. E. W., 171, 175, 271
Searby, G., 96, 289
Sello, S., 60, 277
Seshadri, K., 26, 27, 70, 151, 186, 188, 189, 191, 271, 272, 277, 287, **289**ff., 290
Sethian, J. A., 91, 100, 101, 163, 267, 282, 289
Shepherd, I. G., 74, 90, 91, 96, 158, 268, 272, 274, 290
Shimada, M., 240, 293
Shy, S. S., 125, 290
Sivashinsky, G. I., 93, 96, 97, 125, 139, **290**ff.
Sivathanu, Y. R., 178, 290
Smallwood, G. J., 90, 274, 278
Smereka, P., 162, 290
Smiljanovski, V., 48, 160, 229, 234, 268, 290
Smith, D. B., 96, 271, 275
Smith, G. P., 22, 57, 269
Smith, L. L., 211, 290
Smith, N. S. A., 228, 229, 268, 290
Smith, T. M., 57, **290**ff.
Smooke, M. D., 25, 27, 28, 135, 180, 186, 189, 217, 223, 267, 275, 277, **290**ff.
Snelling, D. R., 90, 274
Sønju, O. K., 203, 226, 227, 242, **288**ff., 290
Spalding, D. B., 12, 33, 288, 290
Speziale, C. G., 58, 290
Squires, K., 58, 59, 284
Steiner, H., 61, 62, 287
Stricker, W., 211, 269
Subramaniam, S., 41, 290
Sussman, M., 162, 290
Sutkus, D. J., 96, 275
Swaminathan, N., 54, 55, 221, **291**ff.

T
Tabor, G., 162, 292
Tacke, M. M., 61, 223, 244, 250, **291**ff.
Taing, S., 224, 291
Takagi, T., 211, 293
Takahshi, F., 240, 291
Takeno, T., 240, 291, 293
Talbot, L., 211, 290
Tamura, T., 124, 281
Tanoff, M. A., 223, 290
Taylor, A. M. K. P., 159, 283
Taylor, S. C., 96, 271, 275
ten Thije Boonkkamp, J. H. M., 96, 149, **274**ff.
Terhoeven, P., 105, 106, 163, 249, 250, 251, 255, 257, 286, 287, 291
Tien, J. H., 95, 291
Tolpadi, A. K., 229, 291
Tomlin, A. S., 24, 291

Trinité, M., 96, 276
Troe, J., 22, 231, 268
Trouvé, A., 51, 63, 89, 90, 135, 252, 269, 274, 278, 287, 289, 291, 292
Tseng, L.-K., 96, 97, 281, 291
Tsuji, H., 186, 188, 291
Turányi, T., 24, 291
Turns, S. R., 225, 227, 223, 276, 291

U
Ulitsky, M., 113, 291
Uryvsky, A. F., 220, 221, 271

V
Vaezi, V., 124, 267
Valdati, B. A., 223, 290
Vanquickenborne, L., 239, 254, 291
van Tiggelen, A., 239, 254, 291
Váos, E. M., 158, 159, **282**ff.
Vervisch, L., 60, 61, 90, 238, 246, 247, 248, 250, 275, 276, 277, **288**ff., 289, **291**ff., 292
Veynante, D., 28, 51, 60, 63, 78, 80, 89, 90, 91, 247, 269, 271, 275, 278, **288**ff., **291**ff.
Villermaux, J., 39, 292
Volkov, D. V., 220, 221, 271

W
Wagner, H. Gg., 180, 280
Walker, R. W., 22, 231, **268**ff.
Wan, Y. P., 233, 234, 287, 292
Wang, H., 231, 276
Warnatz, J., 22, 180, 186, 222, 231, **268**ff., 273, 275, 276, 288, **292**ff.
Weber, R., 248, 271
Weddell, D. S., 34, 201, 278
Weisenstein, W., 162, 293
Weller, H. G., 162, 292
Wenzel, H., 89, 130, 131, 132, 144, 285, 286, **292**ff.
Wheeler, J. C., 125, 280
Whitelaw, J. H., 159, 283
Williams, F. A., 12, 18, 19, 21, 24, 25, 26, 27, 42, 69, 70, 89, 90, 92, 95, 96, 113, 123, 124, 132, 144, 184, 192, 226, 227, 240, 241, 268, 271, 273, 280, 282, **286**ff., 288, **289**ff., **292**ff.
Wirth, M., 109, 110, **292**ff.
Wohl, K., 239, 292
Wolanski, P., 96, 279
Wolff, D., 211, 269
Wolfrum, J., 84, 271
Woolley, R., 70, 277
Wouters, H. A., 39, 292
Wruck, N., 85, 272
Wu, A. S., 75, 90, 293
Wünning, J. A., 230, 293

Author Index

Wünning, J. G., 230, 287, 293
Wyckoff, P. S., 96, 285

X
Xu, J., 38, 293

Y
Yaglom, A. M., 85, 144, 284
Yakhot, V., 125, 289, 293
Yamaoka, I., 186, 188, 291
Yamashita, H., 240, 293
Yang, B., 42, 293

Yeung, P. K., 211, 215, **293**ff.
Yoon, Y., 228, 275
Yoshida, K., 211, 293

Z
Zaitsev, S. A., 220, 221, 271
Zimont, V. L., 84, 124, 133, 161, 162, 279, **293**ff.
Zhang, Y., 219, 293
Zhou, Y., 62, 277
Zhu, D. L., 96, 276
Zolver, M., 158, 275

Subject Index

A
A priori testing, 60, 61

B
Balance equation, **18**ff., 21, 69, 175
Binary flux approximation, 18, 211
BML model, **50**ff., 63, 64, 67, **87**ff., 146, 157
Bunsen burner, **71**ff., 85, 98
Buoyancy, xiii, xvi, 11, 125, 171, 179, 198, 202, 226
Burning velocity diagram, 124, 133, 135
Burke-Schumann solution, **176**ff.

C
Candle flame, 171, **179**ff.
Cascade hypothesis, **1**ff., 59, 125, 144, 168
Chapman-Rubesin parameter, 201
Chemical equilibrium, 4, 24, 66, 87, 171, 178, 205, 221
Chemical source term, 18, **21**ff., 36, 37, 43, 45, 48, 49, 50, 54, 60, 64, 90, 104, 147, 157, 168, 175, 208
Coal dust, 170
Coherent Flame Model (CFM), 51, 63, 64, 67, **87**ff., 130, 158
Coherent structure, 113, 241
Conditional Moment Closure (CMC), 47, **53**ff., 63, 64, 212, 228
Conditional scalar dissipation rate, 47, 54, 214, **216**ff., 221, 232, 241, 255, 256
Conditional velocity, 54, 115, 255, 257
Conserved scalar, 21, 61, 92, 171, 175, 236
Conserved Scalar Equilibrium Model (CSEM), **48**ff., 60, 61, 63, 64, 196, 198
Continuity equation, 10, 69, 101, 199
Counterflow flames, 75, 102, **180**ff.

Counter-gradient diffusion, **50**ff., 63, **89**ff., 137, 159
Crossing length scale, 51, 90, 126
Cross-over temperature, xv, 27, 66, 180

D
Damköhler number (scaling), 3, 133, 255, 259
Darrieus-Landau instability, 119, **137**ff.
Diesel engine, xiv, 47, 170, 172, 217, **229**ff., 231, 234, 238, 247
Diffusional-thermal instability, **96**ff., 139
Diffusion flux, 18, 20
Dynamic Modeling, **58**ff.

E
Eddy cascade hypothesis, **1**ff., 12, 14, 168
Eddy viscosity, 11, 30, 200
Eulerian Particle Flamelet Model, 48, 65, **229**ff.
Element mass fraction, **175**ff., 211
Elementary reaction mechanism, **22**ff., 69, 250
Emission (index), xiii, 2, 158, 172, 223, **226**ff.
Energy equation, 12, 21, 101, 256
Energy spectrum, **17**ff.
Equilibrium (chemical) solution, 48, 176
Explosions, 97, 144
Extinction, xv, 3, 22, 46, 55, 85, 96, 153, 159, **186**ff., 205, 211, 217, 218, 221, 223, 230, 240, **241**ff.

F
Fire, 171
Flame brush thickness, 73, 111, **118**ff., 131, 141, 144, 160, 165, 256
Flame curvature, 76, 91, 95, 99, 151, 211, 248
Flame length, 2, 170, 198, **201**ff., 226

Author Index

Wünning, J. G., 230, 287, 293
Wyckoff, P. S., 96, 285

X
Xu, J., 38, 293

Y
Yaglom, A. M., 85, 144, 284
Yakhot, V., 125, 289, 293
Yamaoka, I., 186, 188, 291
Yamashita, H., 240, 293
Yang, B., 42, 293

Yeung, P. K., 211, 215, **293**ff.
Yoon, Y., 228, 275
Yoshida, K., 211, 293

Z
Zaitsev, S. A., 220, 221, 271
Zimont, V. L., 84, 124, 133, 161, 162, 279, **293**ff.
Zhang, Y., 219, 293
Zhou, Y., 62, 277
Zhu, D. L., 96, 276
Zolver, M., 158, 275

Subject Index

A
A priori testing, 60, 61

B
Balance equation, **18**ff., 21, 69, 175
Binary flux approximation, 18, 211
BML model, **50**ff., 63, 64, 67, **87**ff., 146, 157
Bunsen burner, **71**ff., 85, 98
Buoyancy, xiii, xvi, 11, 125, 171, 179, 198, 202, 226
Burning velocity diagram, 124, 133, 135
Burke-Schumann solution, **176**ff.

C
Candle flame, 171, **179**ff.
Cascade hypothesis, **1**ff., 59, 125, 144, 168
Chapman-Rubesin parameter, 201
Chemical equilibrium, 4, 24, 66, 87, 171, 178, 205, 221
Chemical source term, 18, **21**ff., 36, 37, 43, 45, 48, 49, 50, 54, 60, 64, 90, 104, 147, 157, 168, 175, 208
Coal dust, 170
Coherent Flame Model (CFM), 51, 63, 64, 67, **87**ff., 130, 158
Coherent structure, 113, 241
Conditional Moment Closure (CMC), 47, **53**ff., 63, 64, 212, 228
Conditional scalar dissipation rate, 47, 54, 214, **216**ff., 221, 232, 241, 255, 256
Conditional velocity, 54, 115, 255, 257
Conserved scalar, 21, 61, 92, 171, 175, 236
Conserved Scalar Equilibrium Model (CSEM), **48**ff., 60, 61, 63, 64, 196, 198
Continuity equation, 10, 69, 101, 199
Counterflow flames, 75, 102, **180**ff.

Counter-gradient diffusion, **50**ff., 63, **89**ff., 137, 159
Crossing length scale, 51, 90, 126
Cross-over temperature, xv, 27, 66, 180

D
Damköhler number (scaling), 3, 133, 255, 259
Darrieus-Landau instability, 119, **137**ff.
Diesel engine, xiv, 47, 170, 172, 217, **229**ff., 231, 234, 238, 247
Diffusional-thermal instability, **96**ff., 139
Diffusion flux, 18, 20
Dynamic Modeling, **58**ff.

E
Eddy cascade hypothesis, **1**ff., 12, 14, 168
Eddy viscosity, 11, 30, 200
Eulerian Particle Flamelet Model, 48, 65, **229**ff.
Element mass fraction, **175**ff., 211
Elementary reaction mechanism, **22**ff., 69, 250
Emission (index), xiii, 2, 158, 172, 223, **226**ff.
Energy equation, 12, 21, 101, 256
Energy spectrum, **17**ff.
Equilibrium (chemical) solution, 48, 176
Explosions, 97, 144
Extinction, xv, 3, 22, 46, 55, 85, 96, 153, 159, **186**ff., 205, 211, 217, 218, 221, 223, 230, 240, **241**ff.

F
Fire, 171
Flame brush thickness, 73, 111, **118**ff., 131, 141, 144, 160, 165, 256
Flame curvature, 76, 91, 95, 99, 151, 211, 248
Flame length, 2, 170, 198, **201**ff., 226

302

Subject Index

Flamelet equations, **146**ff., 207
Flamelet model, **42**ff., 64, 87, **222**ff., **229**ff., 235, 251
Flamelet regimes, **78**ff., **190**ff.
Flame stretch, 51, **95**ff., 134, 135, 159, 180, 252
Flame surface area ratio, **127**ff.
Flame surface density, **51**ff., 63, 90, 136, 158
Flame thickness, 28, 69, 70, **78**ff., 92, 119, 123, 125, 146
Flame tip, 73, 99, 166
Fractal dimension, 125
Fuel consumption (layer), **27**ff., 43, 44, 171, 190
Furnace, 20, 170, 171

G
Gas expansion effects, 97, **137**ff.
Gas turbine, 65, 66, 162, 170, **229**ff., 254
Gibson (length) scale, **81**ff., 114, 125, **126**ff.
Gradient transport approximation, **30**ff., 48, 50, 58, 89, 116, 154, 158, 165, 194, 213, 230

H
Heat capacity, 19, 70, 95, 217

I
Ignition, xiv, 3, 5, 22, 23, 43, 46, 55, 67, 158, 187, **217**ff., 230, **232**ff., 238, 247
Inertial range invariance, **1**ff., 13, 46, 121, 209
Inertial range scaling, 56, 65
Inner cut-off, **125**ff.
Inner layer, **27**ff., 43, 80, 82, 85, 104, 151, 152, 153, 156, 179, 188
Integral length scale, 12, **15**ff., 81, 85, 119, 122, **132**ff.
Integral time scale, 40, 51, 84, 113, 119, **132**ff.
Intrinsic Low-Dimensional manifold, 24, 42

J
Jet (diffusion) flame, **198**ff., 222, 238
Joint probability density function (pdf), **7**ff., 9, 35, 46, 54, 88, 136, 153, 156, 159, 212

K
Karlovitz number, **78**ff., 135, 167, 193, 219, 245
k–ε model, 10, **11**ff., 224
Kinematic balance, 52, 72, **91**ff., 250, 258
Kinematic restoration, **113**ff., 116, 122, 129, 131, 141, 143, 144, 167, 168
Kolmogorov scales, **15**ff., 28, 78, 126, 193

L
Lagrangian Flamelet Model, **47**ff., 55, 61
Lagrangian particle, 24, **37**ff., 38, 41, 64, 65

Laminar burning velocity, **69**ff., 73, 77, 95, 121, 122, 132
Large Eddy Simulation (LES), 18, 43, **57**ff., 162
Large scale instabilities, 17, 57, 65, 144
Large scale turbulence, 68, 120, **121**ff., 133, 244
Layered mixture, 53, 248, 249, 261
Lean-burn gas turbine, 66
Level set approach, 52, 68, **91**ff., 146, 160, 161, 162, 168
Lewis number, **19**ff., 95, 96, 135, 155, 211, 214
Lift-off height, 2, **238**ff., 255, 258
Linear Eddy Model, **56**ff., 63, 64

M
Mach number, 19, 69, 100, 196
Markstein diffusivity, **97**ff., 114, 135
Markstein length, **95**ff., 102, 108, 114, 139
Mass flux, **120**ff., 149, 155, 173
Mixing layer, 47, **184**ff., 213, 214, 247
Mixing length scale, **83**ff., 153
Mixture fraction, 43, 45, 53, 55, 59, **172**ff., 194, 200, 207, 212, 253
Modeling constant, 12, 34, 116, 130, **133**ff., 157
Momentum equation, 9, **10**ff., 69, 182, 183, 184, 202
Monte-Carlo method, **37**ff., 230
Multiple (flame) crossing, 93, 112, **128**ff.

N
Navier–Stokes equation, **10**ff., 53, 57, 102, 137, 162, 231
NO_x formation, 2, 221, 223, **225**ff., 230

O
Obukhov–Corrsin scale, **82**ff., 85, 114, 126
Outer cut-off, 125
Oxidation layer, **27**ff., 28, 153, 156

P
Pdf Transport Equation Model, **35**ff., 63, 64, 159, **224**ff., **229**ff., 235
Poisson equation, 101
Potential equation, 101
Presumed (shape) pdf approach, 48, 109, **156**ff., 166, **196**ff., 255
Probability density function (pdf), **5**ff., 34, **35**ff., 46, 48, 54, **109**ff.
Progress variable, **50**ff., **87**ff., 158, 159

Q
Quasi-Steady-State Assumption, **23**ff.

R

Radiation, 69, 195, 196, 227, 229, 236
Random stochastic variable, 6ff.
Reactive scalar, **21**ff., 29, 46
Renormalization group, 125
Reynolds Averaged Navier–Stokes Equations (RANS), **10**ff., 18, 43, 53, 57, 60, 63, 65
Reynolds number independence, 3, 39, 65, 263
Reynolds stress model, **11**ff., 12, 159, 162, 224, 252

S

Sample space variable, **6**ff., 36, 48
Scalar dissipation rate, **31**ff., 45, 47, 48, 54, 60, 114, 123, 129, 130, 131, 155, 167, 182, 184, 185, 188, 194, 195, 212, 214, 216, 217, 231, 234, 240, 247
Scale separation, **4**ff., 27, 29, 32, 40, 50, 53, 55, 65, 87, 168, 193, 264
Scale similarity assumption, **59**ff.
Shear layer, 13, 57, 166, 185, 186, 197, 240, 241
Sivashinsky's integral, **139**ff.
Smagorinsky model, **58**ff., 61
Small scale turbulence, 68, 120, 122, **123**ff., 133, 241
Spark-ignition engine, **66**ff., 67, 90, 109, 153, 160, 161, 237, 261
Spectrum function, **142**ff.
Spray, xiii, 170, 234, 235, 237
Strain rate, **95**ff., 103, 106, 149, 182

Stretch factor, 51, 90, 134, 135, 160
Subgrid variance, **59**ff.
Supernovae, 125
Surface related averages, **152**, 155

T

Taylor length scale, **16**ff., **126**ff., 146
Temperature equation, **21**ff., 104, 177
Thermal diffusivity, **19**ff., 21, 104, 107, 176, 210
Turbulent burning velocity, 69, 73, 117, 118, **119**ff., 132, 135, **137**ff., 161, 168, 253, 255, 261
Turbulent kinetic energy, **11**ff., 12, 15, 17, 53, 57, 158, 162, 163, 196
Two-point correlation, 8, 12, **13**ff., 16, 113, **141**ff.

U

Universal gas constant, 20, 29

V

Viscous dissipation, **12**ff., 35, 129
Viscous stress tensor, **10**ff., 14

W

Weak swirl burner, **75**ff., 110

Z

Zeldovich number, 248